JMP® Essentials

An Illustrated Guide for New Users

Third Edition

Curt Hinrichs
Chuck Boiler
Susan Walsh

§sas®

sas.com/books

The correct bibliographic citation for this manual is as follows: Hinrichs, Curt, Chuck Boiler, and Susan Walsh 2020. *JMP® Essentials: An Illustrated Guide for New Users, Third Edition.* Cary, NC: SAS Institute Inc.

JMP® Essentials: An Illustrated Guide for New Users, Third Edition

Copyright © 2020, SAS Institute Inc., Cary, NC, USA

ISBN 978-1-64295-650-4 (Hard cover)
ISBN 978-1-64295-389-3 (paperback)
ISBN 978-1-64295-390-9 (Web PDF)
ISBN 978-1-64295-391-6 (epub)
ISBN 978-1-64295-392-3 (kindle)

SAS Institute Inc., SAS Campus Drive, Cary, NC 27513-2414

March 2020

Contents

About the Book

JMP Essentials was written for the new user of JMP software who needs to get the best results right away. If you have data and problems to solve, JMP helps you make sense of them to gain understanding and arrive at a good decision. It is the goal of this book to help you with this task.

We often find new users of JMP simply trying to complete a specific task with their data. Perhaps you just need to generate a certain graph for a Microsoft PowerPoint presentation or to quickly see how patterns in your data will lead you to an important discovery. If these scenarios sound close to home, you have come to the right place. This book is task-oriented and will help you access your data, identify that graph or statistic you need, and quickly and easily create and share it with others.

JMP software is built around the workflow of the problem solver. One of its outstanding features is that it consistently provides the correct graphs and statistics for the data that you are working with (something we refer to as the JMP smart interface). JMP leads you down the path to the best result, provided your data is properly classified and you have an idea of what questions you are trying to answer.

JMP is easy to use. In most cases, generating a graph or result will take you seconds, maybe minutes, to complete but never hours or days. Much of this efficiency is due not only to the small number of steps required but to the ability to navigate intuitively toward the right solution quickly, rather than through repetitive and time-consuming trial-and-error. This book provides you with the essential knowledge to get to your solution even faster.

Though JMP contains state-of-the-art visualization and advanced statistical and data mining tools, we also believe it is the right tool for the user who might be less confident in his or her analytic abilities. While there is a time and place for every feature in JMP, we have tried to include only those topics that a typical new user—a manager, engineer or analyst, for example— would need. If it has been a while since you have studied statistics or dealt with its terminology, do not worry. We will present the key ideas and fill in the gaps as needed.

Audience

While this third edition of *JMP Essentials* is designed for new users, we have also focused on the most fundamental and commonly used features of JMP so that it may also serve as a valuable reference to the occasional user too. This focus is what we like to think of as the 20% of JMP that is used 80% of the time. Throughout the book, we have provided the necessary instruction to generate results quickly. Each key step in this process is illustrated with screenshots to help you see the result and develop your confidence in using JMP. We do not assume any formal background in statistics. Instead, we emphasize the intuition of concepts over statistical theory.

If you require deeper statistical understanding, we recommend the documentation included in JMP and some excellent textbooks in the Bibliography.

Most new users of JMP have one or more of the following distinct needs:

1. They need to get their data into an appropriate and structured format so it may be effectively analyzed or visualized. (See Chapter 2).

2. They have a good idea of what graph they need and recognize it by its illustration or name and simply want to create it. (See Chapter 3.)

3. They really do not know what the data will say, and need help with exploring or with summarizing it. (See Chapter 4.)

4. They need to interpret or answer some specific questions about the data they have. (See Chapters 5 and 6.)

We also believe that the complete book is well suited as a reference guide to the following groups of users:

- **Spreadsheet users** who are looking for a convenient way to produce nice visualizations of their data or to supplement a spreadsheet's statistical capabilities. JMP reads and writes data from a variety of programs including Microsoft Excel. This book provides a quick and easy way to make your spreadsheet data come alive and enables you to fully and interactively explore that data.

- **Students enrolled in introductory statistics courses** who may need JMP instruction. JMP is the ideal tool for students because its navigation reinforces the basic assumptions taught in an introductory course. This book provides an overview of the JMP tools needed in most first-year courses.

- **SAS users** who want to take advantage of JMP's data visualization tools. We have provided Appendix A to illustrate the features of how JMP integrates with SAS.

Approach and Features

We have found that the best way to learn JMP is by using it and getting value out of it quickly. Our goal is to present the materials in this book in the most user-oriented approach possible. So, we have made every effort to organize the presentation around the new user's common needs and questions and the most direct and concise means to answer them. We also recognize that the most basic use of data is in generating graphs of data rather than performing more complex statistical analyses. The following features are included for this purpose:

- We present the material with a show-and-tell approach. In most cases, we show you what the results look like alongside the conditions and steps required to produce them. We think this approach is especially useful for JMP users who have a good idea of what they want from JMP and just need the steps to create it.

- When appropriate, we provide an example-driven context for each JMP platform that explains its use, value, and general application to problems. We have tried to distill these contexts down to typical or easy-to-understand cases.

- We organize the contents into easily manageable chunks of information. While the entire book is designed to cover a fairly complete overview of the basics, each chapter represents one family of tasks (such as importing data, creating graphs, and sharing graphs).

- We hope you will keep this book near your computer. Within each chapter, we have designed each section to be self-contained enabling you to quickly find and execute the steps required to complete your task.

- No matter what your professional background, this book assumes only that you have a basic working knowledge of Microsoft Windows. Virtually all of the information in this book applies to using JMP on the Macintosh operating system, but only the Windows version of JMP is used in the examples.

Software Used to Develop This Content

This third edition of JMP Essentials was developed with JMP 15. Users of JMP Pro, JMP Student Edition or earlier releases of JMP will find nearly all of the instructions in this book suitable for their needs.

Organization

This book is designed like a cookbook. Find what you need and follow the steps. We have organized the contents of this book to reflect both the process of analyzing data (getting data, analyzing it, and sharing the results) and the progression from the very basic features in JMP to more specialized ones. We hope this organization offers the most value to the reader. Much of our judgment in this regard comes from our experience working one-on-one with new JMP users.

- Chapter 1 covers the preliminary material you will need for the rest of the book. The chapter identifies the conventions we use and introduces you to JMP menus, windows, and preferences.

- Chapter 2 covers the first step in any analysis: getting your data into JMP. With the exception of some material in Chapter 2, other chapters are self-contained, and you can read them in any order.

- Chapters 3 through 6 cover graphing and analysis:

  ○ Chapter 3 is for the user who knows what graph he or she wants.

  ○ Chapter 4 is for the user who does not know what the data says and needs to explore it to find an appropriate graph or summary. This chapter introduces maps.

  ○ Chapters 5 and 6 are for the user who needs to solve a problem and answer questions using analytics and graphs.

- Chapter 7 covers topics related to sharing your graphs or results in a presentation, document, or through a browser.

- Chapter 8 covers additional resources that are available within JMP, online, and from outside resources, such as training, books, and user groups.

Figure 1 Chapter Organization

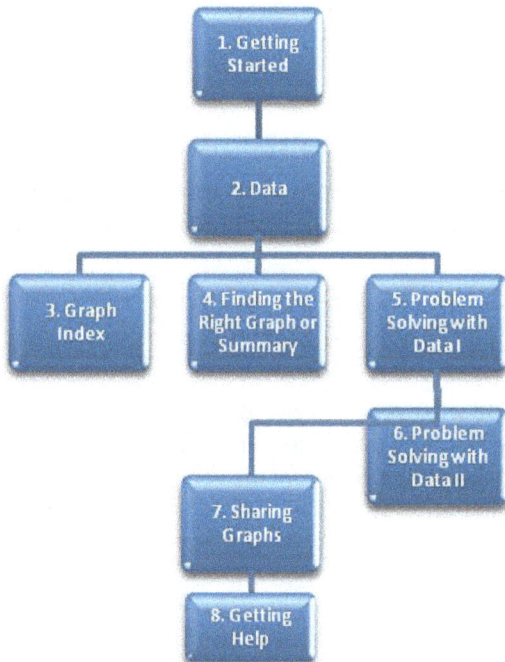

New to the Third Edition

It is gratifying to hear that previous editions have been useful to many new users of JMP. Our goal in writing this third edition is two-fold; first, to improve upon the core features and to that end, we have made many refinements, corrections and additions throughout the book based on feedback from users, colleagues, and reviewers. Secondly, in the five years since the second edition was published, JMP has added several important new features that we believe are

Essential to the new user and we thought should be included. While there are many changes throughout the book, here is our Top Ten:

- Getting data out of database is a common task. **Query Builder** provides an easy, point and click interface to extract needed data. It also automatically creates SQL code that you can reuse on new data.

- An example of **Join by matching columns** is included. This is imperative when combining data tables where data in the tables to be joined is not in the same row order or does not have a one-to-one match.

- New graphs including **Packed Bars**, **Parallel Plots,** and **Dot plots**. JMP has been an innovator in visualization for many years. This continues with Packed Bar charts that are useful to summarize categorical variables with many (hundreds) of levels. Parallel plots are a useful way to express multivariate data. Finally, dot plots have been added. These plots commonly used in introductory statistics courses are simple but useful alternatives to histograms.

- Using **Text Explorer to create a Word Cloud** to visualize un-structured text that might arise as a result of customer comments, product reviews, or warranty claims for example.

- A new section, **Comparing Two Continuous Columns**, presents an introduction to simple linear regression analysis, which provides an equation for a line fit to the data. This equation provides a slope that defines the magnitude of the increase (or decrease) in a response column as the predictor column increases.

- Expansion of the discussion of the **Fit Model** platform and Prediction Profiler to include both continuous and categorical inputs.

- Revised information about the JMP **Dashboard Builder**. Combining multiple graphs and even filters into a single window, or dashboard, has been greatly improved and illustrated in this section.

- **JMP Public or JMP Live** for use in publishing reports. This edition introduces these new products and ways to share interactive graphics and results with those who may not have JMP.

- Recording **animated graphs for exporting to PowerPoint** for presentation. This is an alternative to flash objects that appeared in the previous edition but will no longer be supported by Adobe.

- A more detailed look at the **JMP website** and the myriad of tools available for obtaining more information about the software from the JMP team and other JMP users like yourself.

We Want to Hear from You

SAS Press books are written *by* SAS Users *for* SAS Users. We welcome your participation in their development and your feedback on SAS Press books that you are using. Please visit sas.com/books to do the following:

- Sign up to review a book
- Recommend a topic
- Request information about how to become a SAS Press author
- Provide feedback on a book

Do you have questions about a SAS Press book that you are reading? Contact the author through saspress@sas.com or https://support.sas.com/author_feedback.

SAS has many resources to help you find answers and expand your knowledge. If you need additional help, see our list of resources: sas.com/books.

Learn more about these authors by visiting their author pages, where you can download free book excerpts, access example code and data, read the latest reviews, get updates, and more:
http://support.sas.com/hinrichs
http://support.sas.com/boiler
http://support.sas.com/walsh

About the Authors

CURT HINRICHS joined SAS in 2006 to develop and launch the JMP Academic Program that provides faculty, researchers, and students with easy access to JMP software and learning resources. Curt and his team are charged with developing JMP users among the next generation of data analytics professionals. They work directly with faculty, authors, and administrators and partner with leading academic societies, publishers, and service providers to support the effective use of JMP software in the classroom. Prior to joining SAS, Curt worked as an editor and publisher for Thomson Learning, and its Mathematics and Statistics Publishing Group. He holds a degree in economics from San Diego State University.

CHUCK BOILER is a US Systems Engineer Manager for JMP, a business unit of SAS. Since joining SAS, Chuck has held management roles with JMP and helped develop solutions for conducting many types of analysis, including design of experiments for the semiconductor industry, quality control for pharmaceutical manufacturing, and marketing applications for survey analysis using JMP software. He now works with field engineering staff and customers to help them solve problems and discover hidden opportunities in the data. Prior to joining SAS, Chuck worked as technical services manager and software quality assurance manager for Abacus Concepts. A member of the American Society of Quality, Chuck received a bachelor's degree in education from the University of Oregon and has done graduate work in ancient philosophy at the Graduate Theological Union in Berkeley, California. He is a graduate of Gallup University's Great Manager Program.

SUE WALSH worked for 10 years as a SAS Technical Support Statistician supporting JMP. Prior to working in Technical Support, she worked at SAS as a Statistical Training Specialist, teaching both SAS and JMP to users, and as an Analytic Consultant, supporting the use of SAS in colleges and universities across the country. Sue has over 10 years experience teaching mathematics and statistics at community colleges. She was the first woman commissioned from the Air Force ROTC program at Manhattan College. She retired with a total of 23 years of service in the US Air Force and Reserve. Sue holds a master of business administration from Rensselaer Polytechnic Institute in Troy, New York and a master of science in applied statistics from Wright State University in Dayton, OH.

Learn more about these authors by visiting their author pages, where you can download free book excerpts, access example code and data, read the latest reviews, get updates, and more:

http://support.sas.com/hinrichs
http://support.sas.com/boiler
http://support.sas.com/walsh

Acknowledgments

We would first like to thank John Sall for creating a great product in JMP. Who knew that statistics could be so fun? If you are new to JMP, we hope it inspires you as it has us. Thanks also to Jon Weisz, Dave Richardson, Todd Hoffman, and Diana Levey for supporting the idea for this third edition.

We have been very fortunate to work with the outstanding professionals at SAS Press. In particular, we would like to thank Catherine Connolly who has been our principal source of advice and support throughout the development process of this third edition. To Suzanne Morgen who helped us get to the essentials of what needed to be said. Thanks also to Denise Jones and Robert Harris for making it all work and look good, and to Sian Roberts for her encouragement and bringing this work to market.

The manuscript for this book was substantially improved by the insightful suggestions of our reviewers. The book contains many new features and refinements due to their input. Our reviewers include:

Chris Albright, Indiana University
Mark Bailey, SAS Institute
Kristen Bradford, SAS Institute
Peter Bruce, Statistics.com
Rob Carver, Brandeis University
Laura Higgins, SAS Institute
Ruth Hummel, SAS Institute
Bob Lamphier, SAS Institute
Sheila Loring, SAS Institute
Paul Marovich, SAS Institute
Gail Massari, SAS Institute
Tonya Mauldin, SAS Institute
Don McCormack, SAS Institute

Di Michelson, SAS Institute
Kemal Oflus, SAS Institute
Chris Olsen, Grinnell College
Jeff Perkinson, SAS Institute
Lori Rothenberg, North Carolina State University
Heath Rushing, SAS Institute
Mia Stephens, SAS Institute
Scott Wise, SAS Institute
Annie Dudley Zangi, SAS Institute
Richard Zink, SAS Institute

We also wish to thank Sam Savage of Stanford University for testing the first edition in his class, and David Shultz and Mary Loveless for using the book with customer training.

This book began with a desire to help the new JMP user and evolved into a labor of love. But without the love and incredible support of our families, friends, and colleagues this project would have never materialized. Thank you!

Curt Hinrichs
Chuck Boiler
Sue Walsh

San Francisco, CA and Raleigh, NC January 2020

Chapter 1: Getting Started

JMP was developed to help people with questions about their data get the answers that they need through the use of graphs and numerical results. For most people, memories of statistics can be a very unpleasant, if not forgotten, part of their education. If you see yourself as a new, occasional, or even reluctant user of data analysis, we want you to know that we have written this book for you.

It is important to note that throughout the historical development of statistics as a scientific discipline, people had real problems that they needed to solve and developed statistical techniques to help solve them. Statistics can be thought of as sophisticated common sense, and JMP takes a practical, commonsense approach to solving data-driven problems.

JMP was designed around the workflow of analyzing data rather than as a collection of tools only a statistician can understand. When you think about your data analysis problem, try to formulate the questions that might help you address it. For example, do you need to describe the variation in selling prices of homes in a city or understand the relationship of customer satisfaction with service waiting times? With this mindset, you will find the menus and navigation in JMP to be very compatible with the questions that you are trying to answer.

Displaying graphs (or pictures) of data is one of JMP's strengths. For most people, an effective graph can convey more information more quickly than a table of numbers or statistics. In any JMP analysis, graphs are presented first, and then the appropriate numerical results follow. This is by design. JMP also provides a **Graph** menu that contains additional visualization tools that are independent of numerical results, at least initially. The goal of this chapter is to introduce you to JMP and its basic navigation. We cover the menus and windows and introduce you to the conventions used throughout the book.

1.1 Using JMP Essentials

All but one chapter in this book (Chapter 3, "Index of Graphs") is laid out in a consistent manner to help you generate results quickly. The format of the book has been designed to be used alongside your computer where JMP is installed. After an introduction to the concept, we have designed each section to be self-contained. That is, with few exceptions, the steps required to produce a result begin and end without having to flip through several pages.

We provide numbered steps that generate the result illustrated in the figure that follows. (See Figure 1.1.)

Figure 1.1 Book Layout

Importing a Text File

If you are importing a text file, another handy wizard is included in the **Data with Preview** file option. Like the Excel Import Wizard, this wizard allows you to view your data and specify how you want it to appear before importing it into a JMP data table. It also provides options to convert your text file if it is delimited by commas, tabs, or spaces:

1. Select **File ▶ Open**.

    The **Big Class.txt** file, illustrated here, can be found by selecting **C: ▶ Program Files ▶ SAS ▶ JMP ▶ 15 ▶ Samples ▶ Import Data ▶ Big Class.txt**.

2. Select **Text Files (*.txt, *.csv, *.dat, *.tsv)** in the **Files of Type** drop-down menu. Select the file.

3. Select **Data (Using Preview)** (see Figure 2.8).

    Figure 2.8 Data using Preview

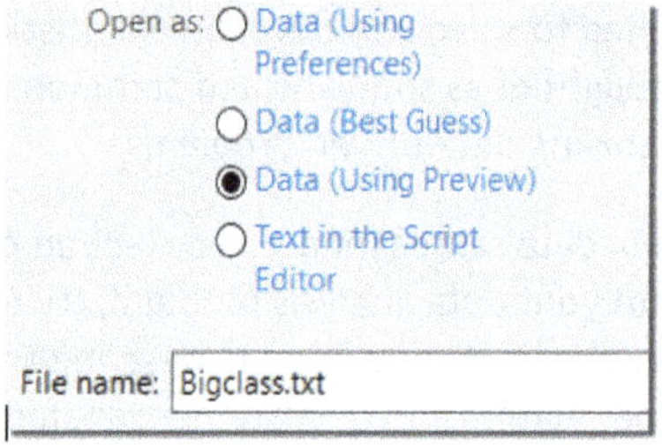

    Open as: ○ Data (Using Preferences)

    ○ Data (Best Guess)

    ● Data (Using Preview)

    ○ Text in the Script Editor

    File name: Bigclass.txt

4. Select **Open**.

Conventions

We are confident that, having made it this far, you know the basic terminology associated with operating a computer, including click, right-click, double-click, drag, select, copy, and paste. We use these terms and they appear in numbered steps. (See Figure 1.2.) When there is a single or self-evident step, these instructions are included in the body of the text. Each step or action appears in bold type.

Figure 1.2 Selection Path Example

---

1. Select **File ▶ Open**.
2. The **Big Class.xls** file, which is illustrated here, can be found by selecting **C: ▶ Program Files ▶ SAS ▶ JMP ▶ 15 ▶ Samples ▶ Import Data ▶ Big Class.xls**.
3. From the **Files of Type** drop-down menu, select **Excel Files**.
4. Select the file that you want, then select **Open** which will launch the Excel Import Wizard dialog with a view of your data. If it looks correctly structured, select **Import**.

In writing this book, we have adopted the same conventions contained in JMP documentation to ease your transition to using the documentation.

Menu items such as Graph are associated with a JMP command such as Graph Builder. We use a right pointed arrow symbol (▶) to indicate the next step in an operation. Thus, **Graph ▶ Graph Builder** indicates that you should select the **Graph Builder** command (or platform) from the **Graph** menu. (See Figure 1.3.)

Figure 1.3 Menu Conventions

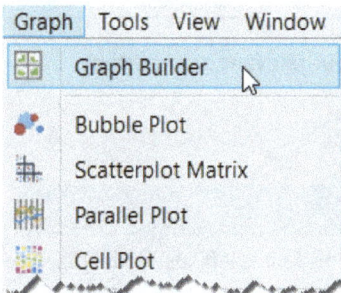

Book Features

Most chapters feature one or more examples to illustrate the procedures within that chapter. (See Figure 1.4.) All of the examples have corresponding data tables that are included in JMP's built-in Sample Data directory (**Help ▶ Sample Data Library**).

Figure 1.4 Data Table Description

Example 2.1 Big Class Families

We will be using the Big Class Families.jmp data file to illustrate the steps in this section. This data set consists of 40 middle-school students and their image, name, height, weight, gender, age and other miscellaneous information. You can access this data set in the Sample Data folder that is installed with JMP: **File ▶ Open ▶ C: ▶ Program Files ▶ SAS ▶ JMP ▶ 15 ▶ Samples ▶ Data ▶ Big Class Families.jmp**. Alternatively, you can select **Help ▶ Sample Data Library ▶ Big Class Families.jmp**.

Important definitions are in bold for easy reference. (See Figure 1.5.)

Figure 1.5 Definitions

Data

refers to any values placed within a cell of a JMP data grid. Examples include numeric and/or text descriptions: 3.6, $2500, Female, Somewhat Likely, or 7/7/19.

Data type

refers to the nature of the data. The data type is usually either numeric (numbers) or character (often words and letters but sometimes also numbers). Other special purpose data types include expression (used for images and matrices) and row state.

Modeling type

refers to how the data within a column should be used in an analysis or a graph. JMP uses three distinct and primary modeling types: continuous, nominal, and ordinal. (JMP also includes three additional special purpose modeling types: Multiple response, Unstructured text and None. Since these are less common, we will only summarize them in the note at the end of this section).

We include notes, tips, and cautions where appropriate to point out relevant or important information. (See Figure 1.6.)

Figure 1.6 Note and Tip Box

Note

Numeric data is right-justified in the data table, whereas character data is left-justified. This can be useful to check whether data contains errors.

The appendices offer reference material including Appendix A, an introduction to using JMP and SAS together; Appendix B, a glossary of terms used in this book; and Appendix C (see Figure 1.7), a JMP 15 Quick Guide that provides essential menu steps to perform a specific analysis (if you know what you want).

Figure 1.7 JMP Quick Guide

Task	Menu Selection
Adding Labels	*Click on column heading;* **Cols > Label/Unlabel**
ANOVA, One Way	**Analyze > Fit Y by X; ▼ > Means/Anova**
ANOVA, two or more factors	**Analyze > Fit Model**
Bar Chart	**Graph > Graph Builder**

1.2 Launching JMP

Let's begin by launching JMP. To launch JMP from the Microsoft Windows Start menu:

1. Select the **Start** menu.
2. Scroll to JMP 15.
3. Select **JMP 15 ▶ JMP 15** (see Figure 1.8).

Note: Windows 10 users will begin with the Start Screen.

Figure 1.8 Opening JMP in Windows

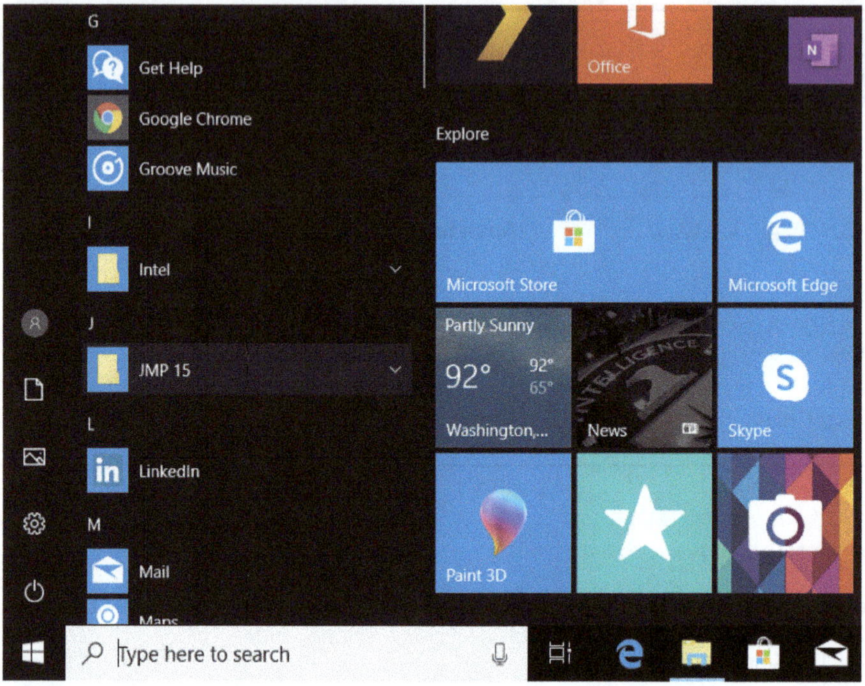

> **Note**
>
> JMP is offered in two versions: JMP and JMP Pro. JMP Pro contains more advanced predictive modeling tools that are beyond the scope of this book. Thus, you will find the steps that we cover in this book are identical in both versions.

Macintosh users can click the JMP icon (see Figure 1.9) to launch JMP from the application dock. If the icon does not appear on the dock, select **Finder ▶ Applications ▶ JMP 15**.

Figure 1.9 Accessing JMP on a Mac

After JMP has launched, you might notice that two windows have also opened: Tip of the Day and JMP Home Window.

Tip of the Day

The Tip of the Day window is the first thing you see because it addresses the most common questions that new users ask such as, "How do I do X?" Well, the X in these common questions is represented and answered in more than 60 different Tip of the Day windows. You can scroll through them by clicking **Next Tip** at the bottom of the window (Figure 1.10). Some of the Tip boxes contain important and basic navigational hints, while others only apply to more advanced features in JMP.

Figure 1.10 Tip of the Day

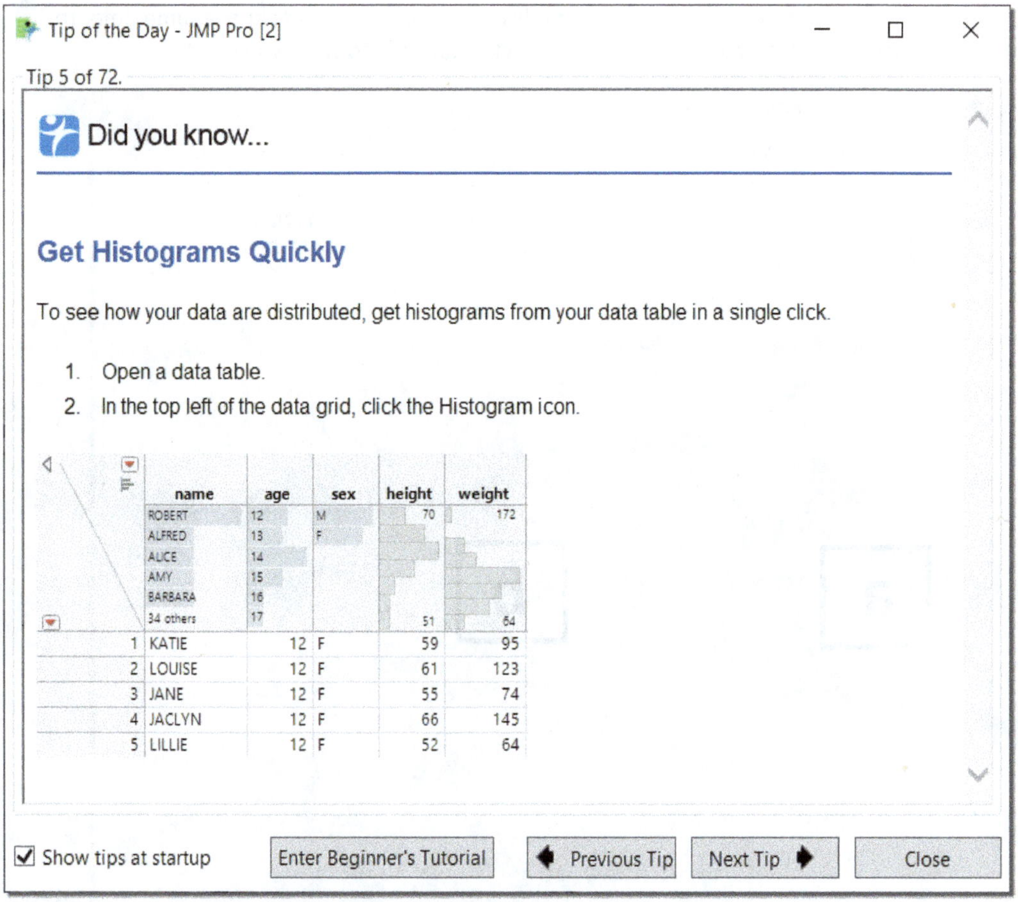

Note the **Enter Beginner's Tutorial** button. This tutorial walks you through a basic analysis of data, from opening data tables to creating graphs and results. JMP contains several other tutorials that are directed toward more specific types of problems and are found in the Help menu.

> **Note**
>
> If you do not want to see the Tip of the Day window every time you launch JMP, you can simply uncheck the **Show tips at startup** box in the lower left corner of the window.

The JMP Home Window

When you launch JMP, the Home Window appears (Figure 1.11). The Home Window organizes and helps you navigate data tables, documentation, and open files and any results that you have generated. If you tend to have several data tables and analyses running at the same time, the Home Window provides a convenient way to quickly navigate to what you want.

Figure 1.11 The JMP Home Window

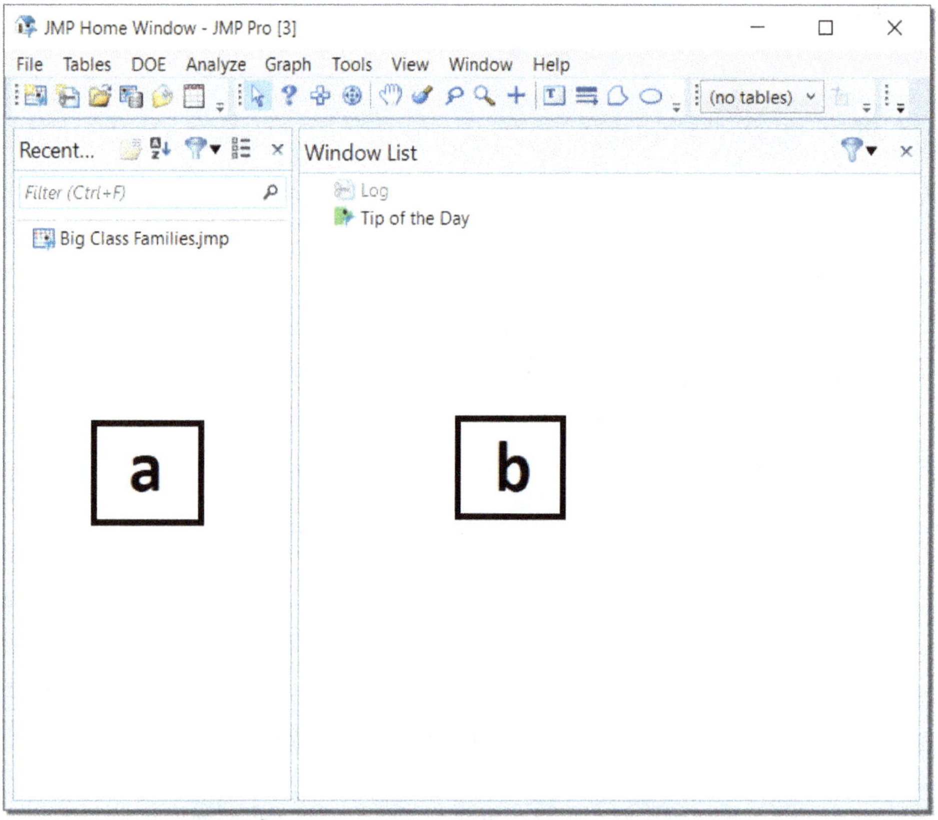

By default, the Home Window is divided into two panels, which are:

    a. The left panel contains recent files that you have accessed, listed from the most recently opened. If you are opening JMP for the first time, this panel should be blank.

    b. The right panel titled "Window List" contains a list of open data tables files and their associated results. In JMP, you can have any number of data tables and results open, but only one active data file can be analyzed at any one time. You can double click on any item in this panel to activate it and bring it to the forefront.

While the Home Window enables you to navigate directly to a file or result, each data table and results window also provides shortcuts back to the Home Window. At the lower right of each window, select the icon that looks like a house to return to the Home Window (Figure 1.12).

Figure 1.12 Shortcut Back to the Home Window

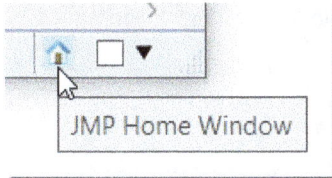

A results window has a second icon, which is also the same icon used to denote ".jmp" formatted files called JMP Data Tables. Click on a Data Table icon and you will be taken to the corresponding data table for that results window (Figure 1.13). Note that if you are looking at a data table, you will not see this second icon because you are already in the data table window.

Figure 1.13 Shortcut Back to the Data Table

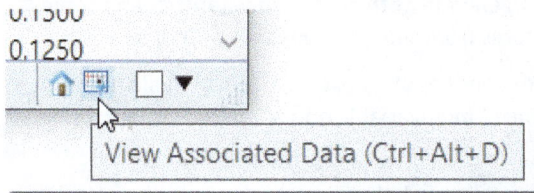

The check box with the down arrow button next to it enables you to combine multiple results windows or graphs into a single window or "dashboard." We will discuss creating dashboards in Chapter 7.

1.3 JMP Menus

At the top of the Home Window, you will see a series of menus (File, Edit, Tables, and so on). These are the menus that we use to illustrate the concepts in this book. They are also the same menus that we refer to as JMP's native menus because they have been present in JMP since its first release.

These menus serve to open or import data, to edit or structure it, and to create graphs and analyses of your data. They are also a valuable source for assistance through the Help menu, which is discussed later. The menus are logically sequenced from left to right (Figure 1.14).

Figure 1.14 JMP Native Menus

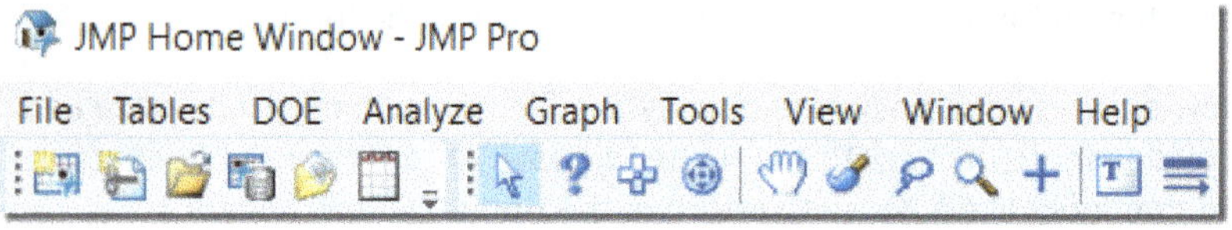

- **File** is where you go to open or import data and to save, print, or exit JMP. It is also where you can customize the appearance or settings within JMP through **Preferences** (explained in Section 1.5).

- **Edit** will appear when needed and provides the usual cut, clear, copy, paste, and select functions, as well as undo, redo, and special JMP functions.

- **Tables** provides the tools to manage, summarize, and structure your data. (See Section 2.6.)

- **DOE** contains the Design of Experiments tools. These tools are used to design experiments that are then used to collect data to eventually analyze. In this book, we assume you already have data in some form and thus will not cover DOE. For more information, see **Help ▶ JMP Documentation Library ▶ DOE Guide.**

- **Analyze** contains the analysis tools that generate both graphs and statistics and serves as the home for all of JMP's statistical tools from simple to advanced (Chapters 5 and 6).

- **Graph** contains graph tools that are independent of statistics (at least initially). Graphs in this menu include basic charts to advanced multivariable and animated visualization tools and maps (Chapters 3 and 4).

- **Tools** enables you to transform your cursor into a help tool, a brushing tool, a selection or scrolling tool, and much more (Section 7.2).

- **View** provides options to control which windows, menus and toolbars are visible including the JMP Starter (Section 8.3).

- **Window** helps you manage windows within JMP.

- **Help** provides resources for learning and using JMP. Let's start with an introduction to the Help menu.

> **Note**
>
> Additional menu items including "Add-ins" and "SAS" may appear if and when you have tools of these types installed.

The Help Menu

The Help menu (see Figure 1.15) provides access to learning resources that you can use as you expand your knowledge of JMP features, learn about statistics, and learn how to interpret results. These resources include searchable indexes, guided tutorials, tips of the day, and links to printable books including *Using JMP*. Data tables used in this book and in all JMP documentation are included in the Sample Data directory. Chapter 8 covers the features of the Help menu in greater detail.

Figure 1.15 The Help Menu

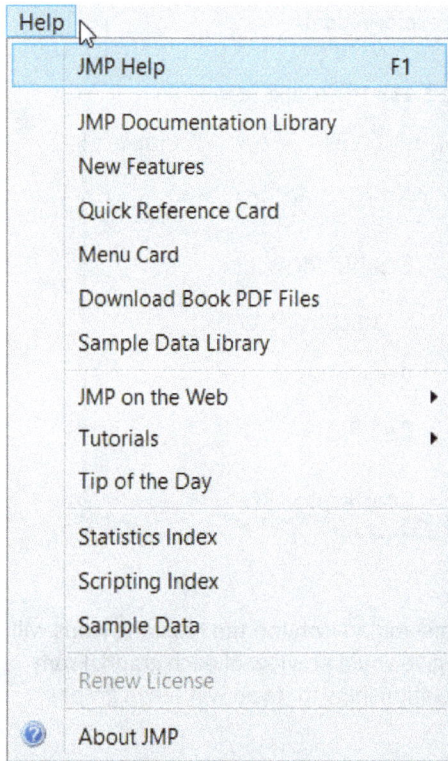

Help
JMP Help   F1
JMP Documentation Library
New Features
Quick Reference Card
Menu Card
Download Book PDF Files
Sample Data Library
JMP on the Web ▶
Tutorials ▶
Tip of the Day
Statistics Index
Scripting Index
Sample Data
Renew License
About JMP

JMP also features context-specific help, meaning that when you use the JMP Help Tool in any graph or statistical result, you are directed to the right spot in the online documentation to assist you in understanding the result. For more information about the JMP Help Tool, see section 8.1. In statistical results, JMP provides Hover Help that reveals context-specific interpretation of statistical results. See Chapter 5 for more information.

Interpretation can be straightforward for descriptive graphs or basic summary statistics, but as you dig deeper into an analysis or use more advanced methods, it is vitally important that you understand the meaning of the results, particularly when they are shared or presented. The documentation under **Help ▶ JMP Documentation Library** includes over 6,000 pages of reference material in fourteen books that address the needs of professional statisticians and analysts. If you encounter results that you do not understand, however, we strongly recommend that you seek assistance from experienced data analysts.

The Analyze and Graph Menus

Because most graphs or statistical results begin with the Analyze and Graph menus, let's explore the structure within these two menus a little bit more.

Click the **Analyze** menu at the top of the window. Glance at the choices on the menu. Top-down, the platforms are organized from the basic to more advanced tools. Next, click the **Graph** menu at the top of the window. Glance at the graph choices. The menus in JMP – specifically the Analyze and Graph menus (see Figures 1.16a and 1.16b) – are designed to provide both a description and visual cues for analyzing, graphing, and exploring data.

Figure 1.16a The Analyze Menu

Analyze	Graph Tools View Wind
	Distribution
	Fit Y by X
	Tabulate
	Text Explorer
	Fit Model
	Predictive Modeling ▶
	Spreadsheet M

Figure 1.16b The Graph Menu

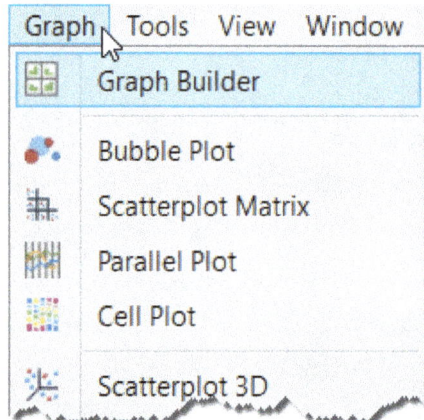

Graph	Tools View Window
	Graph Builder
	Bubble Plot
	Scatterplot Matrix
	Parallel Plot
	Cell Plot
	Scatterplot 3D

Note that each entry under these menus has both a name and an icon (on the Mac, the icons will not appear). The icons next to the Graph menu options give you a preview of each graph. From the Analyze menu, the icons depict the description or relationships that you will see in graphs and statistical results (Figure 1.17).

Figure 1.17 Visual Cues Provided for Basic Analysis

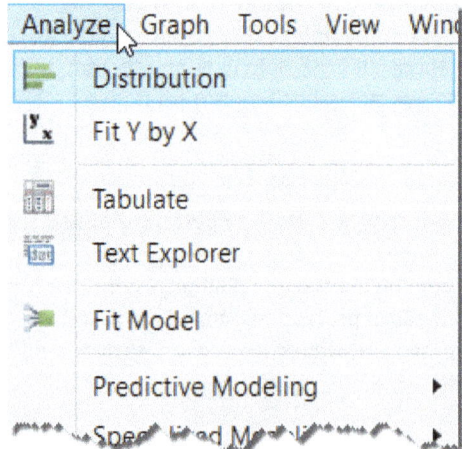

Analyze	Graph Tools View Wind
	Distribution
	Fit Y by X
	Tabulate
	Text Explorer
	Fit Model
	Predictive Modeling ▶
	Spreadsheet M

The **Analyze** menu items produce both graphs and statistical results, while the **Graph** menu items produce only graphs, at least initially.

Framework of the Analyze Menu

There is a problem-solving framework to the Analyze menu that we will discuss in detail in Chapter 5. As mentioned in the introduction, your exploratory objective will translate to these menu items. This structure streamlines the analysis process; in order to select the correct menu item, you only need to count how many columns you are interested in and know whether you are trying to describe, compare, or understand their relationship. (See Figure 1.18.)

Figure 1.18 Framework of the Analyze Menu

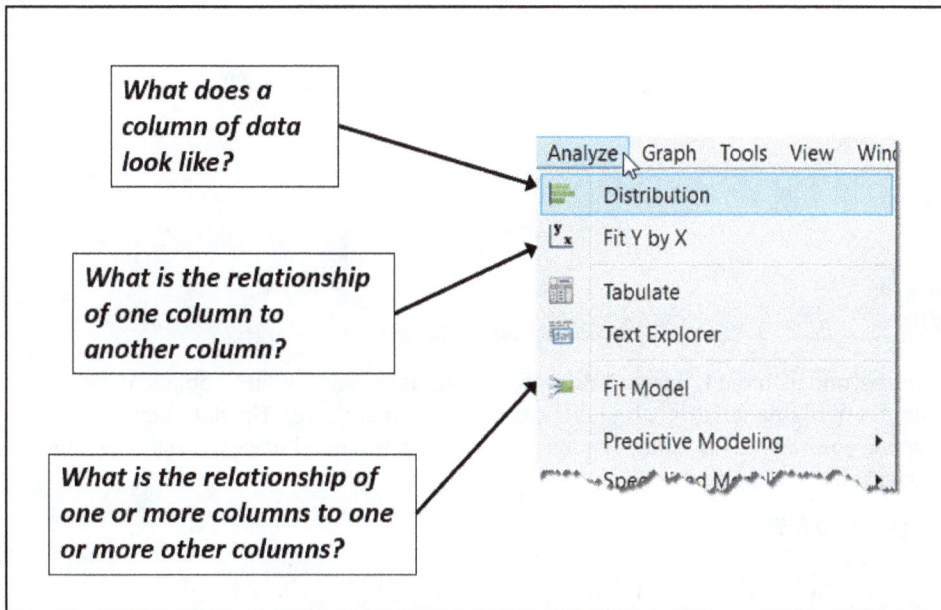

This framework cues you to the correct analysis choice on the menu without exposing you to many statistical terms until you need them. Make no mistake; you still get the statistics when you want them, but you do not have to know all the statistical terms or assumptions in order to access them.

JMP's Analyze menu contains terms such as Distribution and Fit Y by X that might be unfamiliar, but the ideas behind them are very straightforward. We describe them in simple terms as needed throughout the book. Many items under the Analyze and Graph menus are referred to as platforms or commands through this book. For example, Distribution and Fit Y by X are referred to as platforms.

1.4 Elements of Using JMP

Before we launch JMP for the first time, let's look at the four common elements of a JMP analysis. All JMP analyses contain these elements, and they follow a consistent process.

1. The first is the **JMP Home Window**, where you begin a JMP session (Figure 1.19). This is your mission control center. As described earlier in this chapter, from here you can open or create a data table or easily navigate between data tables, results, and help.

    Figure 1.19 The JMP Home Window

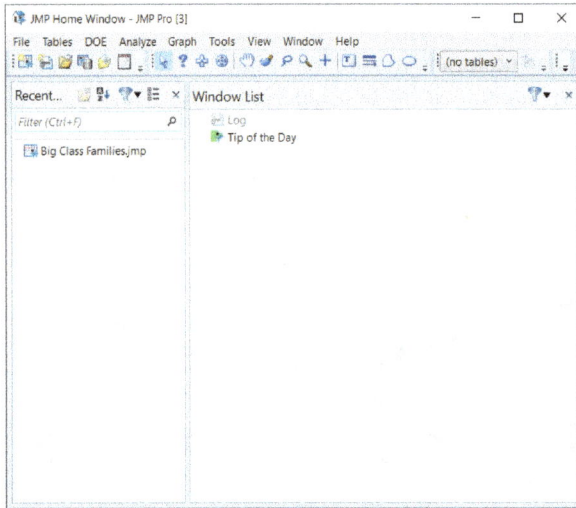

2. The second element is a **Data Table** where your data reside, which you might have imported or opened through the Home Window (Figure 1.20). The data table is also where you will usually initiate an analysis or graph described next. We will cover the Data Table in Chapter 2.

    Figure 1.20 A JMP Data Table

3. Once you have a data table open in JMP, you will want to select a task through the JMP menus. These tasks (or commands as we call them in JMP) generate a **Launch Window** to execute your desired command (Figure 1.21). You will notice that the columns or variables from your data table are pre-populated in the launch window. Chapters 3 through 6 will explore these tasks and their results.

Figure 1.21 A JMP Launch Window

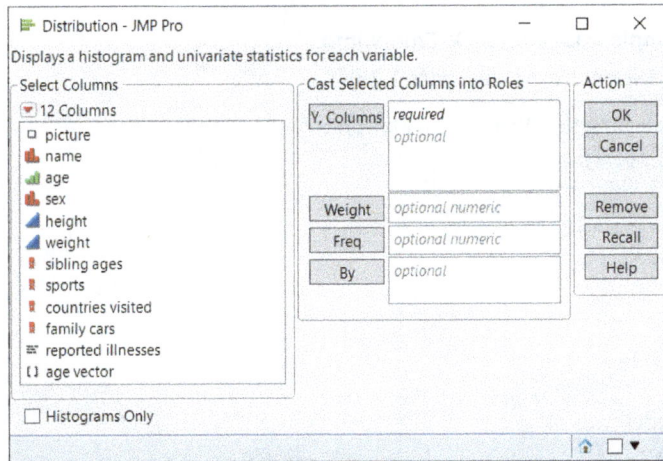

4. The result of any executed command is called the **Report Window,** which contains the graphs and statistics that you have asked JMP to glean from your data (Figure 1.22). We will be seeing Report Windows throughout this book as we illustrate JMP's features, but Chapter 7 will focus on how to share these graphs and reports with others.

Figure 1.22 A Report Window

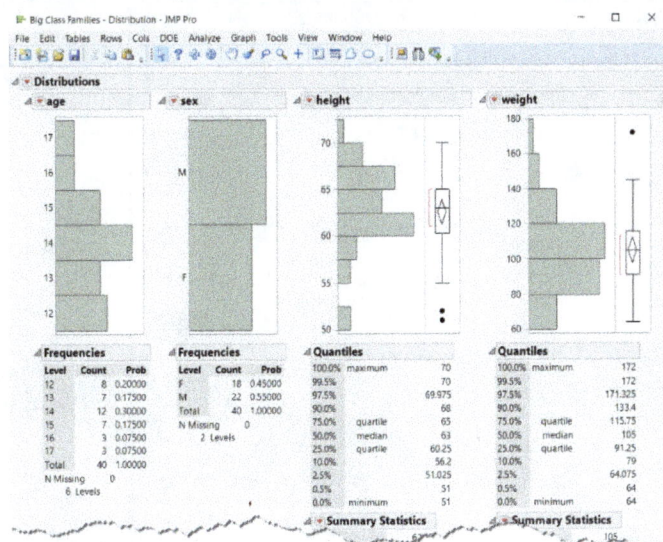

1.5 JMP Launch Dialog Windows

Throughout this book, each set of instructions used to create a graph or an analysis is prompted by a launch window that follows a consistent format and execution. To launch a window, however, you must first open a data table.

For purposes of illustration, we will open the Equity.jmp data table:

1. Select **Help ▶ Sample Data Library ▶ Equity.jmp**.
2. Select **Analyze ▶ Distribution** (see Figure 1.23).

Figure 1.23 Selecting the Distribution Platform

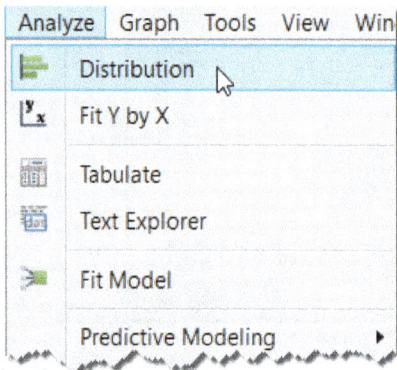

3. This generates the Distribution window with the columns (variables) from the Equity.jmp data table populated in the Select Columns list (see Figure 1.24).

Figure 1.24 The Distribution Launch Window

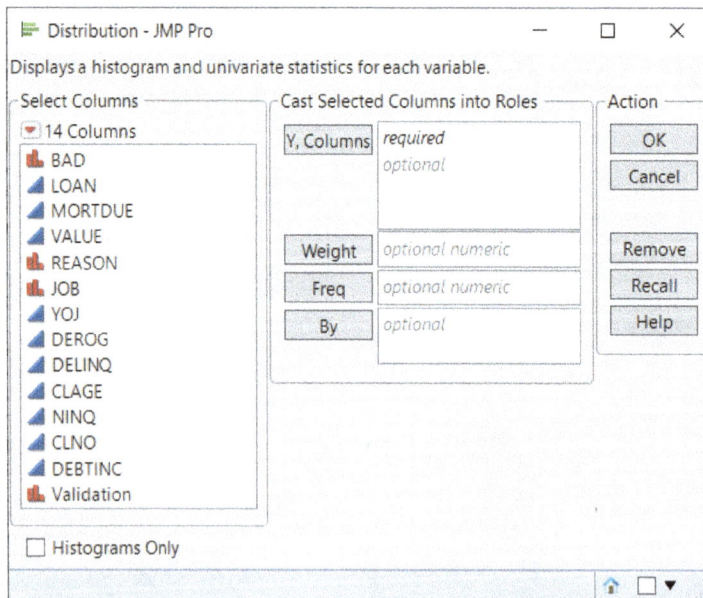

Most JMP launch windows consist of three main elements, organized from left to right (see Figure 1.25):

Figure 1.25 Launch Window Basics

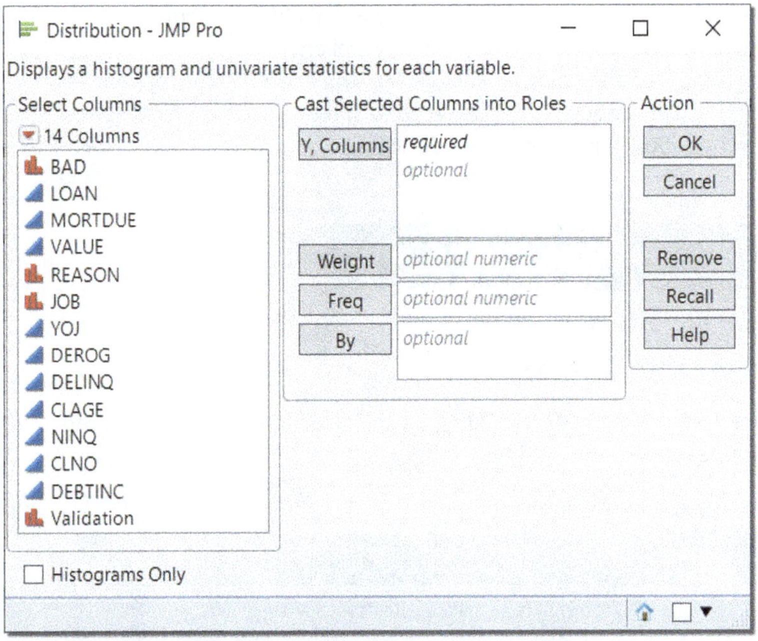

- **Available columns** (or variables) of data to analyze from your data table. These appear on the left under **Select Columns**.
- **Roles** that you want to place (or cast) on the column(s). In this area, you see buttons and empty areas under **Cast Selected Columns into Roles**. Within these empty areas, you are given a hint in italics about which columns are required and which are optional to run the analysis.
- **Action buttons** to execute commands.

To use this Distribution window or almost any other in JMP, click on a column and select the role (or click and drag the column into that role's empty space). Once you are satisfied with your selections, select **OK**.

Almost every analysis and graph window in JMP appears in this way. Now that you have learned this format, you are ready to handle just about any command window in JMP.

> **Note**
>
> The Y, Columns role refers to what column you want to place on the vertical, or y, axis. In other windows, such as Fit Y by X, you also have an X role to select that corresponds to the horizontal, or x, axis. The Weight, Freq, and By roles are more specialized, but can streamline your analyses often without the requirement of reshaping your data (For more information, see **Help ▶ JMP Documentation Library ▶ Using JMP ▶ Get Started ▶ Launch Windows**).

1.6 The Excel Add-In (Optional)

We find that many new users of JMP are often Microsoft Excel users too. JMP can easily import Excel data, which we will describe in greater detail in Chapter 2, but one feature that Excel power users might appreciate is the JMP add-in for Excel. The Excel add-in is a convenient Windows-only way to launch JMP platforms from within the Excel environment. If Excel is installed on your Windows computer and you then install JMP, the add-in should appear as a new tab along the top of your Excel window. (See Figure 1.26.) If it does not, go to "Add-ins" within Excel and select the check box next to the JMP add-in item.

Figure 1.26 The JMP Add-In Tab in Excel

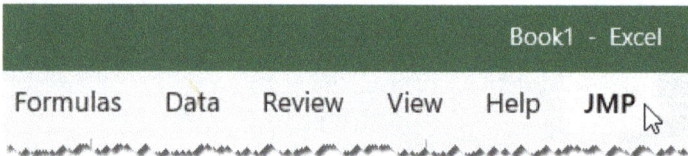

Selecting the JMP tab will reveal a JMP ribbon providing a good selection (but not all) of the commonly used JMP platforms (see Figure 1.27).

Figure 1.27 The JMP Ribbon in Excel

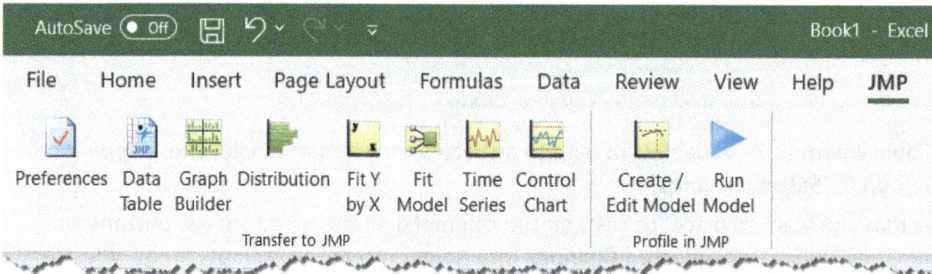

Because the JMP environment offers dynamic and visual exploration of your data, each JMP platform option will launch JMP, convert your Excel worksheet into a JMP data table, and set up the corresponding launch window within the JMP environment. Let's briefly summarize their functions.

1. **Preferences** helps bring your data to JMP in the right format. Here, you can specify the number of header rows in your Excel worksheet and whether to bring over hidden rows or columns.

2. **Data Table** automatically converts your Excel worksheet into a JMP Data Table. Note that it will use the preferences that you have set. If your data does not transfer correctly, change your preferences accordingly or use the Excel Import Wizard discussed in Chapter 2.

3. **Graph Builder** is an easy-to-use data visualization platform. Selecting this option will convert your worksheet into a JMP data table, launch the Graph Builder platform, and populate the dialog box with your variables or columns so that you are ready to visualize your data.

4. **Distribution**, **Fit Y by X**, **Fit Model**, **Time Series,** and **Control Chart** will convert your worksheet into a JMP data table and launch the corresponding platform with your variables ready to be assigned into roles.

5. **Create/Edit Model** and **Run Model** enable you to visualize your spreadsheet models using JMP's profiler. If you are interested in performing "what-if" analysis on your spreadsheet models, the profiler enables you to do so visually. This is a great tool for presenting models because you can interact with the model and immediately visualize the effect of change. It also contains Monte Carlo simulation to explore how uncertainty will affect your model and fine-tune it to achieve desired results. (See Figure 1.28.)

Figure 1.28 The Excel Profiler

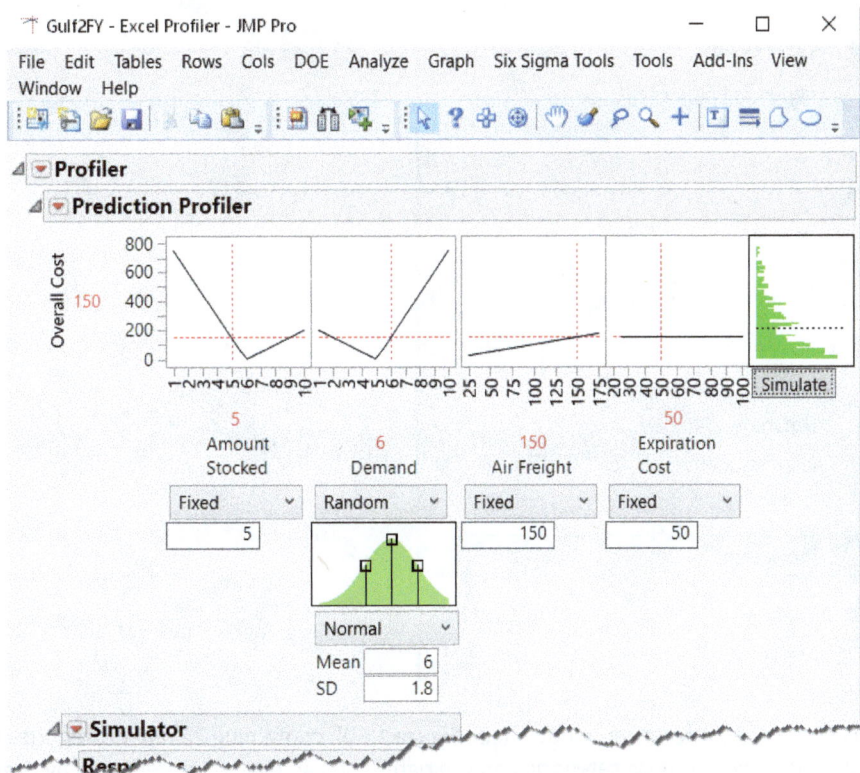

1.7 JMP Preferences

JMP's **Preferences** determine the way JMP appears or behaves on your machine. JMP has been carefully crafted to support the workflow of the data analyst. Its defaults have been selected to reflect common use, which we use in this book. However, JMP also provides options to customize the software to corporate standards or individual tastes. In this section, we will explore how one can customize the look, feel, and options that appear in JMP. Preferences (**File ▶ Preferences** on Windows; **JMP ▶ Preferences** on Mac) are the primary means of setting or changing the defaults in JMP that you will see each time you operate the software – think global settings here. Virtually any function in JMP can be set as a default, including specific tests within any platform, the look of graphs, color schemes, font sizes and styles, and how JMP works with other products such as SAS.

To view the preferences, choose **File ▶ Preferences** or **JMP ▶ Preferences** on a Mac. (See Figure 1.29.)

Figure 1.29 Accessing Preferences from the File Menu

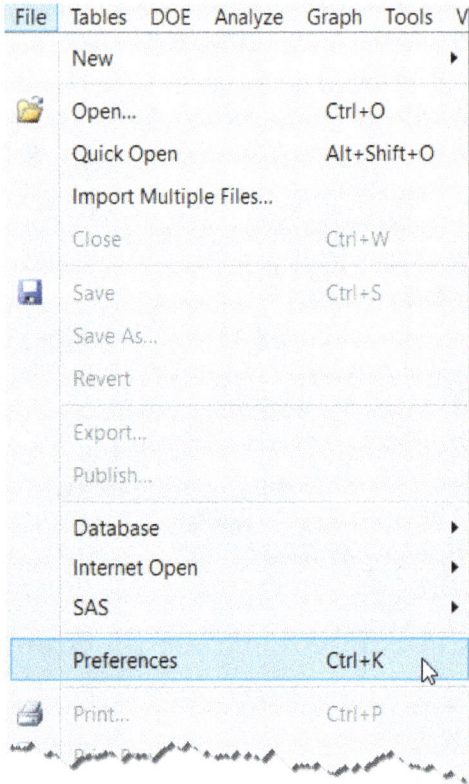

This opens the Preferences window (see Figure 1.30), containing 22 main categories on the left and options within those categories on the right. You can change preferences by checking or unchecking the boxes within the categories on the right or by selecting items from drop-down menus. Changing preferences can affect such things as the graph or result format, the font, the location of a file, and much more, each and every time you use those features in JMP. If you are unsure about making a change to the preferences, we recommend that you wait until you have a need to do so.

Figure 1.30 The Preferences Dialog Window

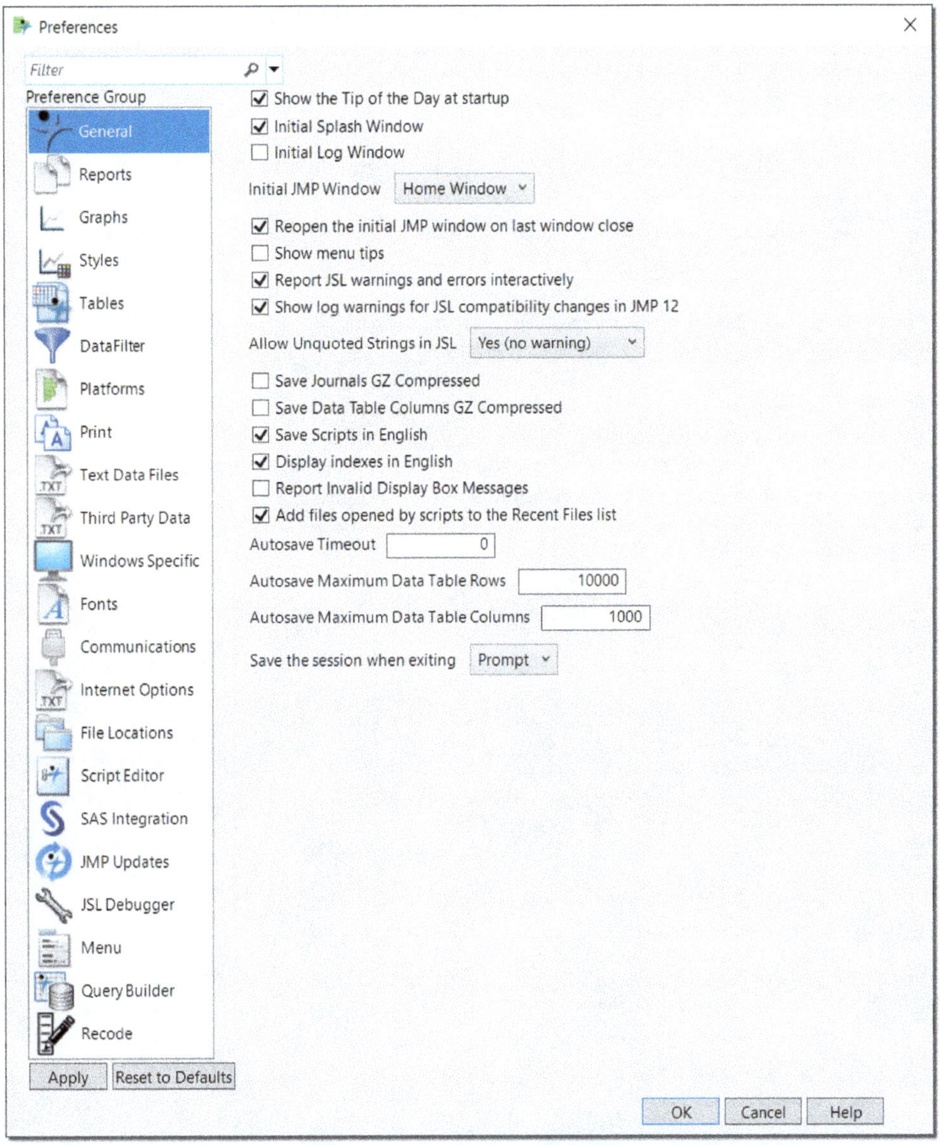

> **Note**
>
> If you need to make a change within a single graph or result, note that JMP also provides many of these formatting options within the graphs themselves.

Let's see how this works. New users on Windows often prefer to turn off the menu auto-hide option (which by design, provides a little more window real estate for graphics and statistics power users), making it a little easier to find the menu options described in this book.

Below we have an illustration of the menu hidden and unhidden (Figure 1.31). Notice the File, Edit, and other menus appear when they are not hidden. When the menus are hidden, you see an ellipsis where the menus would be. Holding your cursor over the ellipsis displays the menus.

Figure 1.31 Illustration of Menu Hidden and Unhidden

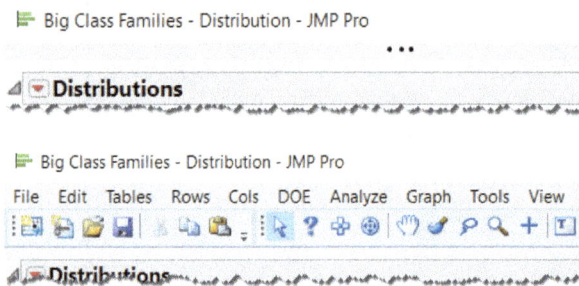

To change this auto-hide default to always show the menus, select **File ▶ Preferences ▶ Windows Specific ▶ Autohide menus and toolbar ▶ Never** (Figure 1.32).

Figure 1.32 Removing Menu Auto-Hide

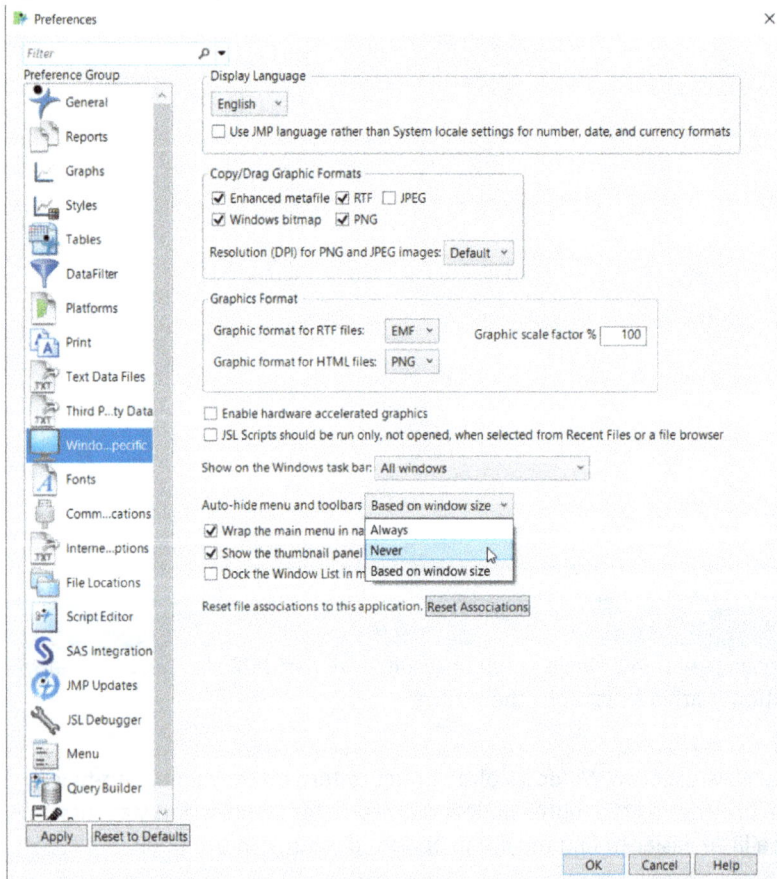

If you want to change the default marker size, style, or color themes used in graphs, select **File ▶ Preferences ▶ Graphs**. Included is a handy preview to see how your selections will appear (Figure 1.33).

Figure 1.33 Graph Preferences

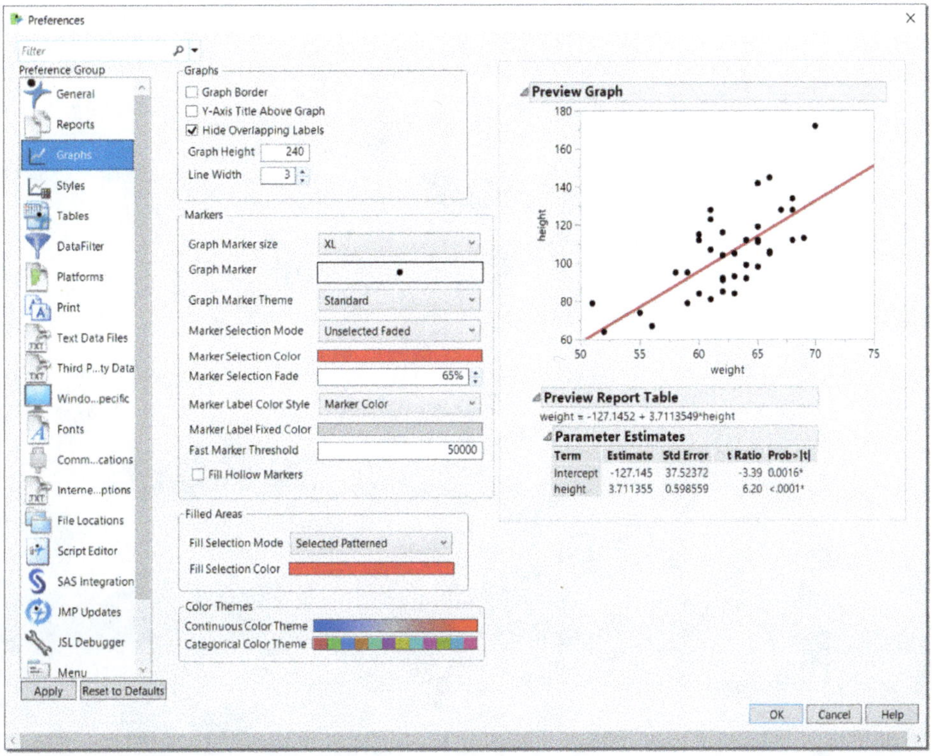

1.8 Summary

JMP was developed to help business professionals, scientists, or engineers get answers to the questions and problems that they encounter. The navigation and menus within JMP provide a natural extension of your problem-solving and a direct means to explore your data and generate the results that you need. This book uncovers the structure of JMP's menus and provides easy steps for producing results. The standardized format of the windows in JMP prompts you through most analysis and graphing. Results can be customized using global detailed preferences.

Chapter 2: Data

The first step in creating a graph or analysis is to get your data into JMP. With JMP, you can easily import data from many different sources such as Microsoft Excel or ODBC-compliant databases, or you can enter your data directly into a JMP data table. Because most readers already have data in one form or another, this section focuses on getting that data into JMP from another file format. Sometimes data is not in the best condition when you import it. Later in this chapter, we discuss what you can do to format data or deal with missing data. JMP also supports unstructured text and shape files (that can be used to create maps). We will describe the special requirements of using these file types.

As mentioned in the previous chapter, we use Windows as our default operating system to illustrate JMP and its native menus as shown in the below image. JMP instructions for Windows and Macintosh are basically the same, though some operating system differences are noted when they occur.

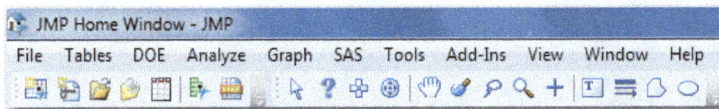

Example 2.1 Big Class Families

We will be using the Big Class Families.jmp data file to illustrate the steps in this section. This data set consists of 40 middle-school students and their image, name, height, weight, gender, age, and other miscellaneous information. You can access this data set in the Sample Data folder that is installed with JMP: **File ▶ Open ▶ C: ▶ Program Files ▶ SAS ▶ JMP ▶ 15 ▶ Samples ▶ Data ▶ Big Class Families.jmp**. Alternatively, you can select **Help ▶ Sample Data Library ▶ Big Class Families.jmp**.

2.1 Getting Data into JMP

Getting your data into JMP is a familiar process. Like many other desktop applications, you can simply select **File ▶ Open** to import your data into JMP. JMP can handle many different data formats. Table 2.1 shows the default formats JMP recognizes. Other previously installed applications could contain proprietary formats that might also appear as import options. You can import files with these formats as well.

In this section, we show you how to open JMP data tables and how to import Microsoft Excel spreadsheets and text files in JMP. Each of these file formats follows the same basic procedure, but each has special options that enable you to import exactly what you want. JMP interfaces with databases using Open DataBase Connectivity standard (ODBC). Through the Query Builder dialog box, you can easily set up queries of your data and automatically create SQL code that can be saved and repurposed. And you don't need to learn SQL to do this. We illustrate only the essential connectivity here; more information about querying your data is available in the JMP documentation (**Help ▶ Search ▶ SQL**). At the end of this section, we show you how to create a new data table in JMP.

Table 2.1 Default File Formats Supported by JMP

File Type	File Extension
JMP Files	.jmp, .jsl, .jrn, .jrp, .jmpprj, .jmpmenu, .jmpaddin, .jmpapp, .jmpquery
Excel Files	.xls, .xlsx, .xlsm
Text Files	.txt, .csv, .dat, .tsv
SAS Data Sets	.sas7bdat, .sas7bxat, .xpt, .stx
SAS Program files	.sas
R Code	.r, .R
MATLAB Code	.m, .M
SPSS Data Files	.sav
Minitab Worksheet Files	.mtp
Shapefiles	.shp
HTML	.htm, .html
XML Data files	.xml
JSON	.json

File Type	File Extension
SQLite 3.0 or Higher	.sqlite, .db, .sqlite3, .db3
Teradata Database	.trd
xBase Data Files	.dbf
Triple-S Files	.sss, .xml
Flow Cytometry v2 or 3	.fcs
Hierarchical Data Format v5	.h5

Note

JMP can be used to initialize or pull data from other third-party applications: Python, ODBC, R, MATLAB, and so on when such integrations are implemented in JMP. The third-party application must match in architecture. The term architecture is sometimes referred to as "bitness" (32-bit or 64-bit). One can verify the architecture of JMP via **Help ▶ About JMP** (Windows).

Opening a JMP File

Let's start by opening a JMP data table. At the top left of the JMP window is the **File** menu:

1. Select **File ▶ Open** (see Figure 2.1). A dialog box opens. (On a Mac, select **File ▶ Open** and locate your file in the appropriate folder.)

   Figure 2.1 Opening a File

   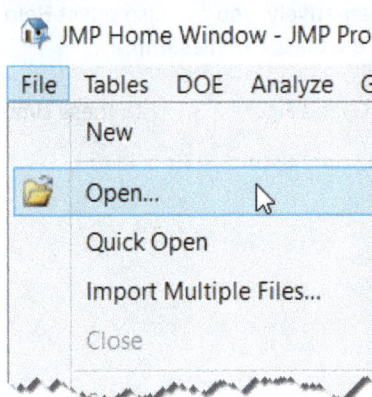

   We will use the Big Class Families data table described earlier. Click the **Big Class Families.jmp** file and select **Open**. (See Figure 2.2.)

Figure 2.2 Open File Dialog Box

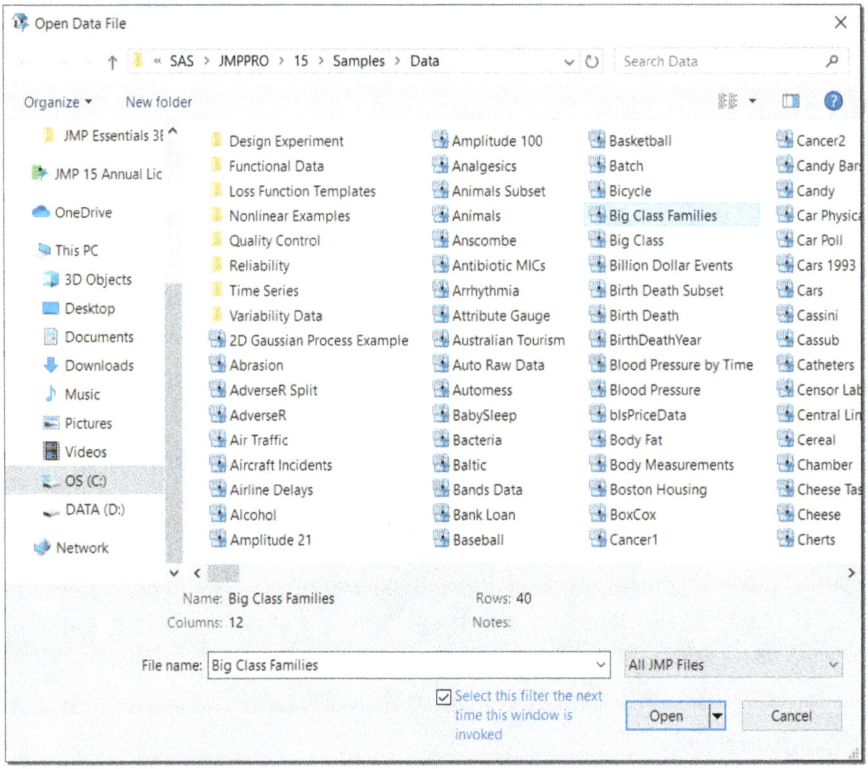

To locate this file for the first time, select **File ▶ Open ▶ C: ▶ Program Files ▶ SAS ▶ JMP ▶ 15 ▶ Samples ▶ Data ▶ Big Class Families.jmp**. Alternatively, you can also select **Help ▶ Sample Data ▶ Open the Sample Data Directory ▶ Big Class Families.jmp**.

These steps open the JMP data table Big Class Families. (See Figure 2.3.) With these simple steps, you are now ready to analyze or visualize this data.

Figure 2.3 The JMP Data Table

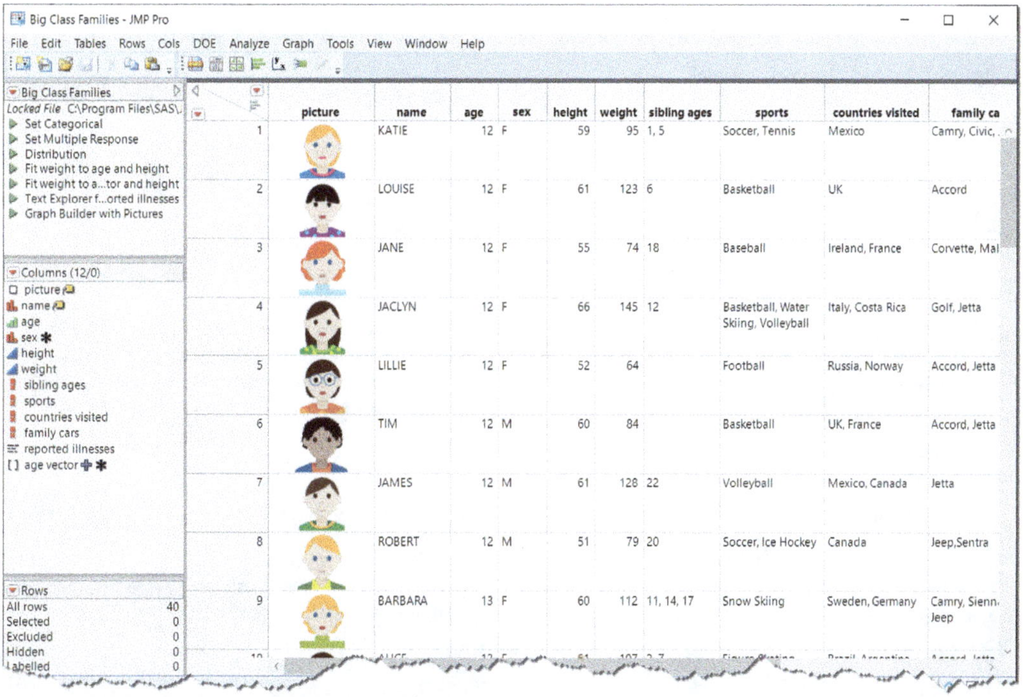

This spreadsheet-like table is referred to as the *JMP data table,* which is JMP's common data format regardless of where the data comes from. Section 2.2 discusses the components of the data table.

> **Note**
>
> In the Big Class Families example, each row of the table contains information about a single student in the class and is, therefore, one observation. Each column of the table is one piece of information (or variable) collected on each student. This structured format of the data table is required for most graphs and analyses in JMP. The importing examples that follow assume that your data already exists in this format. Section 2.4 introduces some tools to use if your data does not conform to this structure.

Importing Data into JMP

Importing data into JMP from another file format is similar to opening a JMP file. Within the **File ▶ Open** pop-up window, the **Files of Type** drop-down menu indicates **All JMP Files** as the default.

If you are importing another file type, simply click on the down arrow and select the correct type. You can also select **All Files** from the drop-down menu. (See Figure 2.4.) Select the file that you want, and then click **Open**. On the Mac, select **File ▶ Open** and available files will be highlighted. On the Mac, files that JMP cannot open will be dimmed.

Figure 2.4 Selecting All File Types

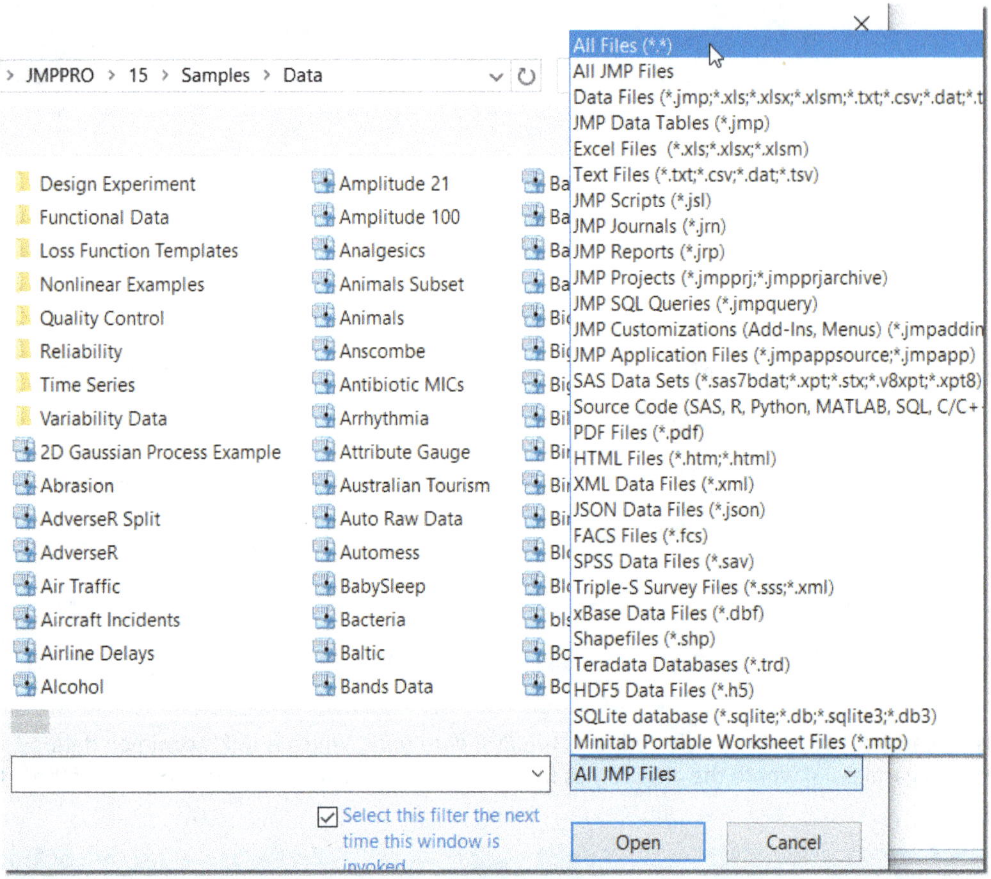

Note

If you know the format of your data, first select the correct format from the Files of Type drop-down menu. You will see the available files of that type within the folder. Once you have located the right file, select the file and click **Open**.

Importing an Excel File

Importing an Excel file is easy if your variables are in columns and your cases or observations are in rows. Ideally, any variable names should appear in the row directly above the first row of data, as shown in Figure 2.5. The import process automatically opens and converts the data into a JMP data table and uses your variable names as column headings:

Figure 2.5 Importing an Excel File

1. Select **File ▶ Open**.

   The **Big Class.xls** file, which is illustrated here, can be found by selecting **C: ▶ Program Files ▶ SAS ▶ JMP ▶ 15 ▶ Samples ▶ Import Data ▶ Big Class.xls**.

2. From the **Files of Type** drop-down menu, select **Excel Files**.

3. Select the file that you want, then select **Open** to launch the Excel Import Wizard dialog box with a view of your data. If it looks correctly structured, select **Import**.

Shortcut

If you have an Excel worksheet or workbook on your desktop, you can simply drag the file over the JMP shortcut icon on your desktop to launch JMP and the Excel import wizard.

The Excel Import Wizard

While the previous example was simple and straightforward, a common characteristic of Excel worksheets is that data does not always conform to the essential column/row structure that is required by JMP. For example, you might have multiple nested headers where one row might represent year and the next row contains months within that year. The Excel Import Wizard has made importing this worksheet and maintaining the month-within-year structure much easier.

This wizard also provides options to specify which rows should be headers for columns, to specify hidden or merged columns, and to replicate these settings or merge data from multiple worksheets within a workbook.

To illustrate this feature, let's try another example with one of these characteristics. We will use the Team Results.xlsx worksheet from the Import Data folder. This worksheet has headers/column names that appear to be in the second row. (See Figure 2.6.) The Excel Import Wizard will help JMP decide how to import this data.

Figure 2.6 The Excel Import Wizard

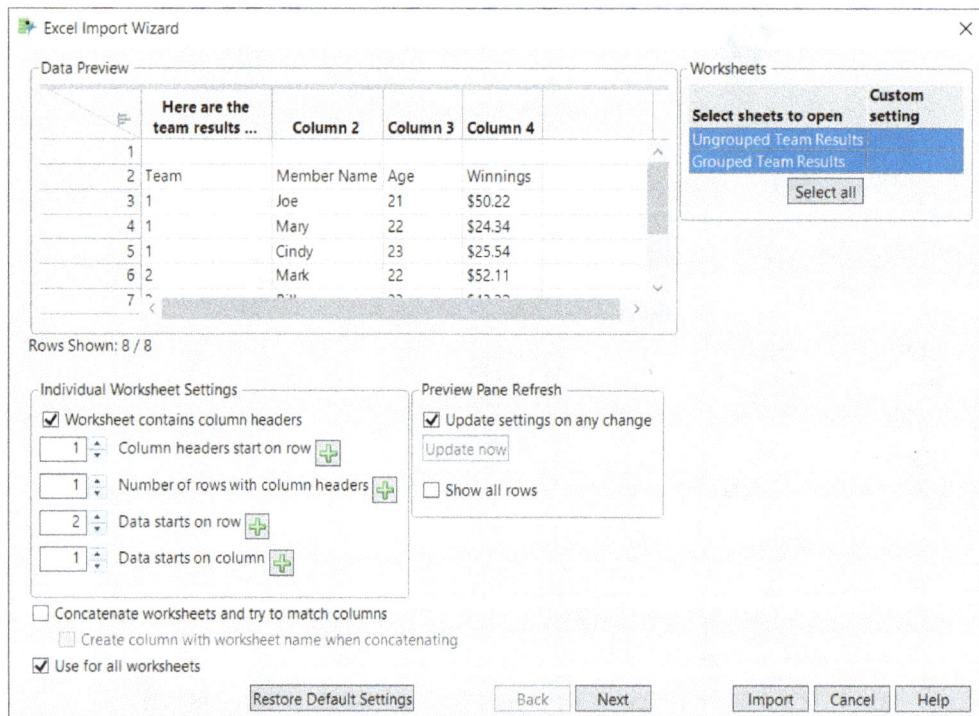

1. Select **File ▶ Open**, then select an Excel workbook. In this example, we are using **Team Results.xlsx** from the **C: ▶ Program Files ▶ SAS ▶ JMP ▶ 15 ▶ Samples ▶ Import Data** folder.
2. Select **Team Results.xls ▶ Open** to launch the Excel Import Wizard with an initial display of your data in the window (Figure 2.6).

3. As you can see, the column headers begin in row 3 (when you include the note that appears as a header in the preview) and the first set of observations in row 4. To get this into the right format, adjust the **Individual Worksheet Settings** as we have done in Figure 2.7.

Figure 2.7 Adjusting the Worksheet Settings

Individual Worksheet Settings

☑ Worksheet contains column headers

3	Column headers start on row
1	Number of rows with column headers
4	Data starts on row
1	Data starts on column

Data Preview

	Team	Member Name	Age	Winnings
1	1	Joe	21	$50.22
2	1	Mary	22	$24.34
3	1	Cindy	23	$25.54
4	2	Mark	22	$52.11
5	2	Bill	23	$43.32
6	2	Jennifer	24	$11.23

Rows Shown: 6 / 6

4. Once you have adjusted these settings, you should see your data take its proper shape in the Data Preview panel. Once you are satisfied with the adjustments you have made, select **Import**.

Importing a Text File

If you are importing a text file, another handy wizard is included in the **Data with Preview** file option. Like the Excel Import Wizard, this wizard enables you to view your data and specify how you want it to appear before importing it into a JMP data table. It also provides options to convert your text file if it is delimited by commas, tabs, or spaces:

1. Select **File ▶ Open.**

   The **Big Class_L.txt** file, illustrated here, can be found by selecting **C: ▶ Program Files ▶ SAS ▶ JMP ▶ 15 ▶ Samples ▶ Import Data ▶ Big Class_L.txt.**

2. Select **Text Files (*.txt, *.csv, *.dat, *.tsv)** in the **Files of Type** drop-down menu. Select the file.

3. Select **Data (Using Preview)** (Figure 2.8).

   Figure 2.8 Data Using Preview

   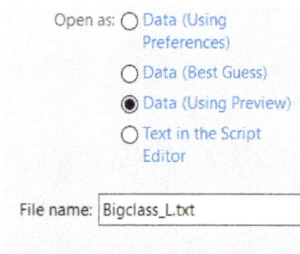

4. Select **Open.**

5. Choose the settings that you may need to structure your data for import. (See Figure 2.9.) Click **Nex**t and then **Import.**

   Figure 2.9 Text Data Preview

   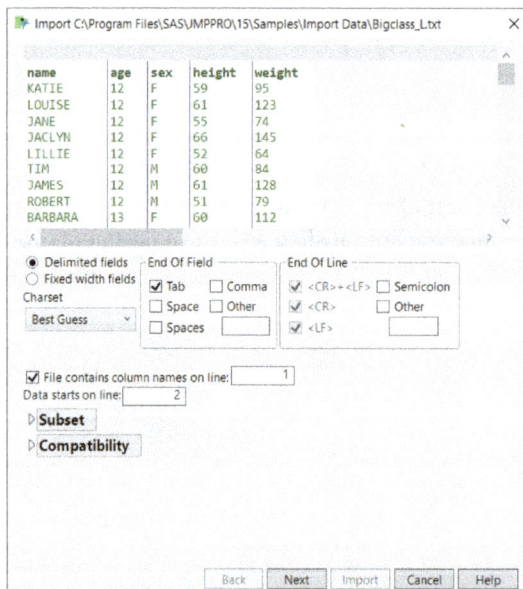

Importing a Database File

Options to import data extracted from a database are available through ODBC within JMP. To access this data, first connect to the database (the data source should already be defined) and then specify the table of interest. You can also create custom queries of your data using Query Builder (explained in the next section) or the **Advanced** button. If you need more help defining your data source, select **Help ▶ JMP Documentation Library▶ Using JMP ▶ Ch. 3 Import Data from a Database.**

1. Select **File ▶ Database ▶ Open Table** (Figure 2.10).

   Figure 2.10 Database Open Menu

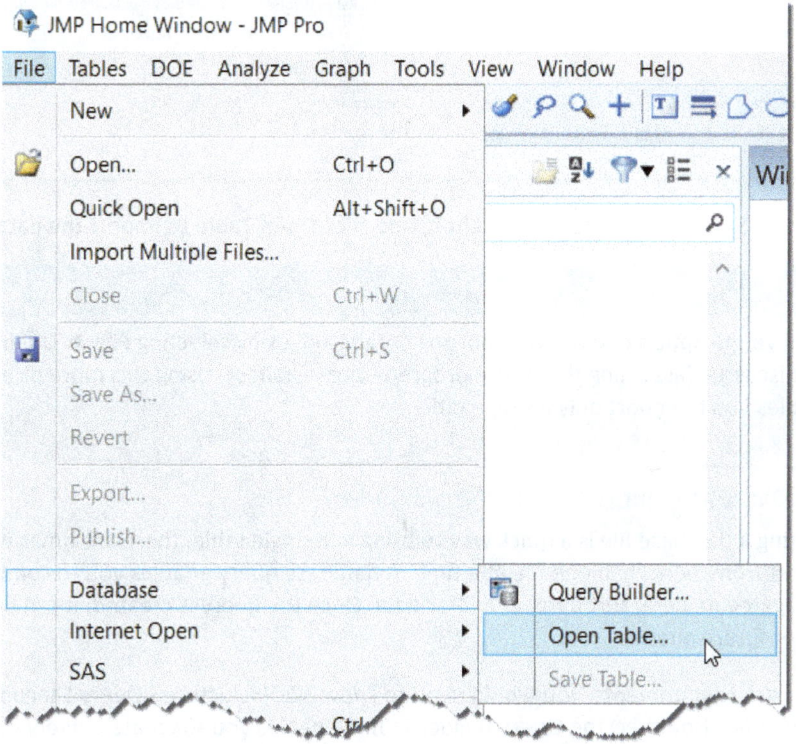

2. The Database Open Table window appears. (See Figure 2.11.) It prompts you to connect to your database and either open a data table or specify a query. Clicking the **Connect** button launches the Select Data Source window to locate and connect to your database.

Figure 2.11 Database Open Dialog Box

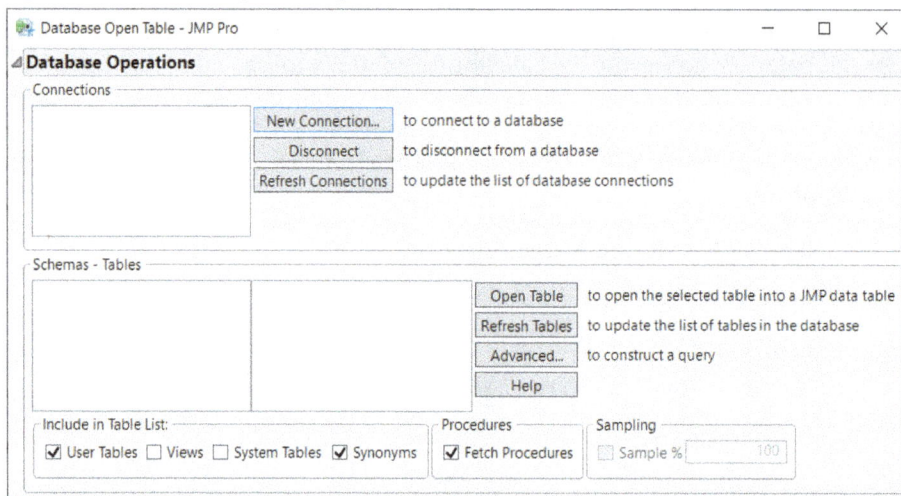

3. Locate the table of interest, highlight it, and click **Open Table** to import the data.

Note: You have the option of directly importing database files by selecting **File ▶ Open** as previously discussed (assuming that these programs are installed). Using this more direct option enables you to import only a single table.

Creating a Database Query Using Query Builder

While importing a database file is a quick way to bring in a single table, there are times when you want to import from more than one file at a time. A database query enables you to combine data tables, aggregate data, sort data, and filter data. Once the query is created, it can be saved and reused for future queries.

In other software environments, you would need to know SQL (Structured Query Language) to create such queries. However, the Query Builder in JMP enables you to create a query in an easy, point-and-click environment that creates the SQL code for you. As with importing a single database file, you must first connect to the database. The data source should already be defined. Once the connection is established, you can choose the data tables of interest and specify the parts of the data tables that you want to extract into your resulting JMP table.

1. Select **File ▶ Database ▶ Query Builder**.
2. Choose **New Connection** and connect to your database (this example is for illustration purposes only since you will be unable to connect to this database).

For this example, you want to create a data table with information about historical SAT scores for states in the South and Southwestern regions. You would like to include the following information in your data table:

- From the data table SATByYear: the average SAT Math and SAT Verbal scores for each state.

- From the data table SATStateInfo: the Student/Faculty ratio for each state.

- From the data table StateToRegion: the Region each state is assigned to.

1. Select the **SATByYear** data table and assign it as the **Primary** table for the query.
2. Select **SATStateInfo** and **StateToRegion** data tables and assign them as the **Secondary** tables for the query. (See Figure 2.12.) Note that the column State appears in each data table and is designated as a "key" variable. Key variables are automatically used as the matching column to join data table in the query.

Figure 2.12 Database Query

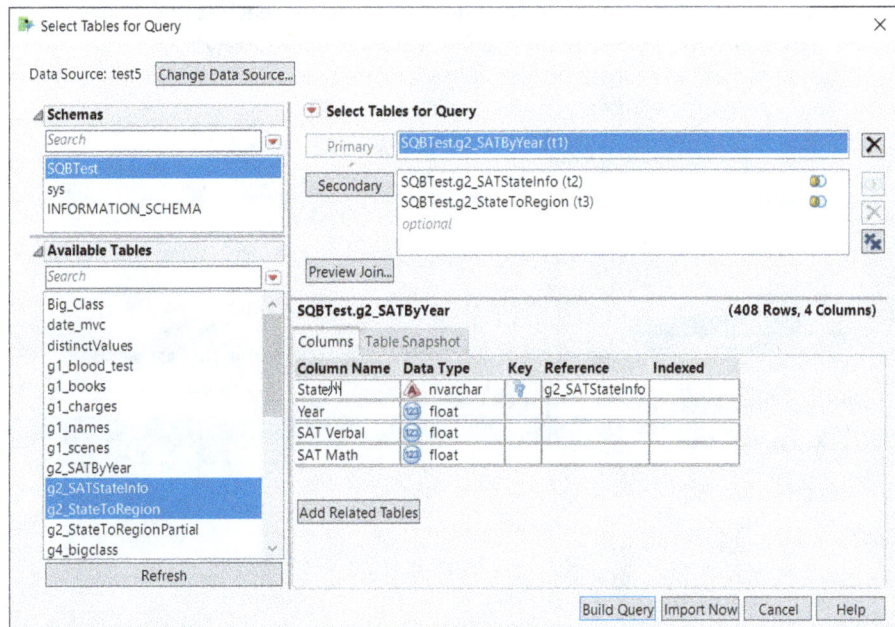

3. Select Build Query.
4. In the Tables panel, select the SATByYear data table.
5. Choose the **State**, **SAT Verbal**, and **SAT Math** from the Available Columns panel and click the **Add** button. (See Figure 2.13.) (Because these are small data tables, you might want to turn on the **Update preview automatically** option. This option can slow performance down if the data tables are very large.)

Figure 2.13 Database Query Building

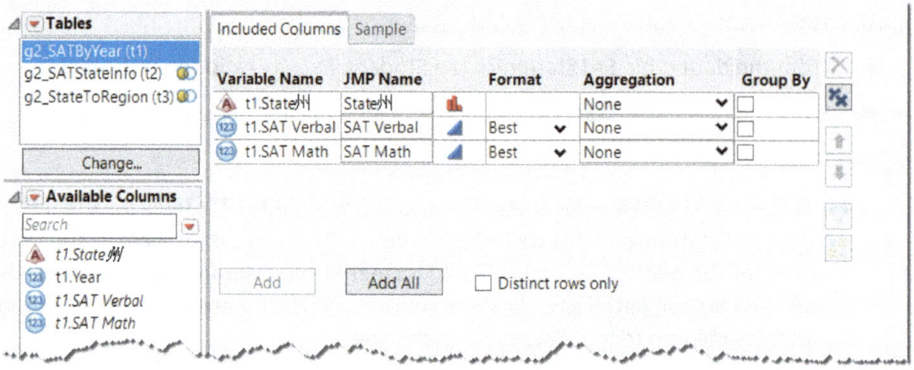

6. Since you are interested in the average SAT scores for each state, change the **Aggregation** for the SAT Math and SAT Verbal columns to **Average**. Note that Group By was automatically turned on for State. Examine the sample shown in the lower panel. You have one row for each state. (See Figure 2.14.)

Figure 2.14 Aggregating Data

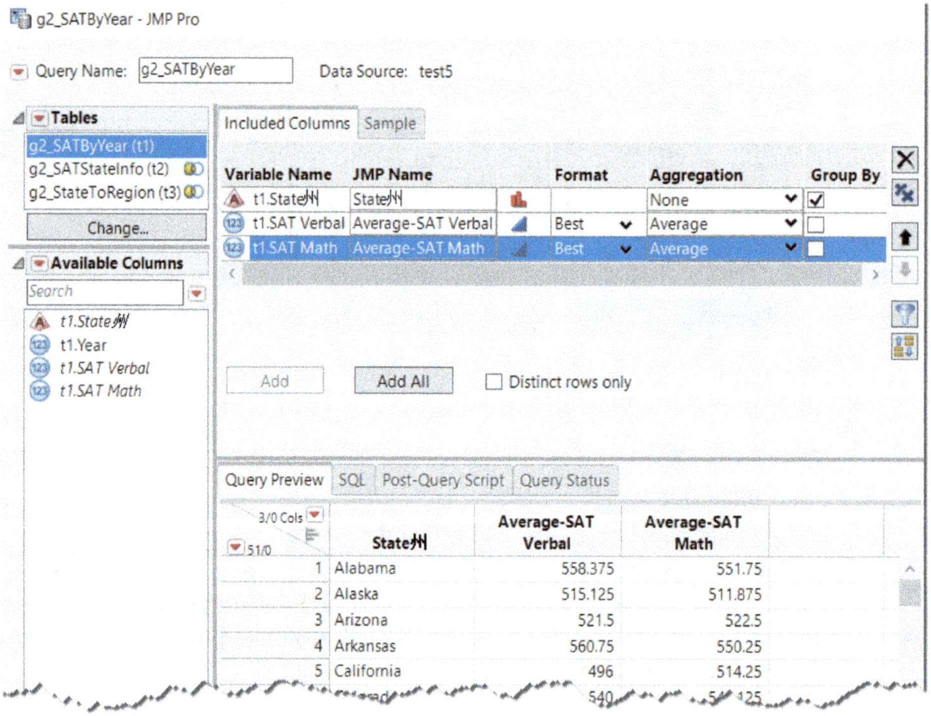

7. Select the **SATStateInfo** data table in the Tables panel. Add the **Student/Faculty Ratio** column to the included columns.

8. Select the **StateToRegion** data table in the Tables panel. Add the **Region** column to the included columns. (See Figure 2.15.)

Figure 2.15 Defining Included Columns

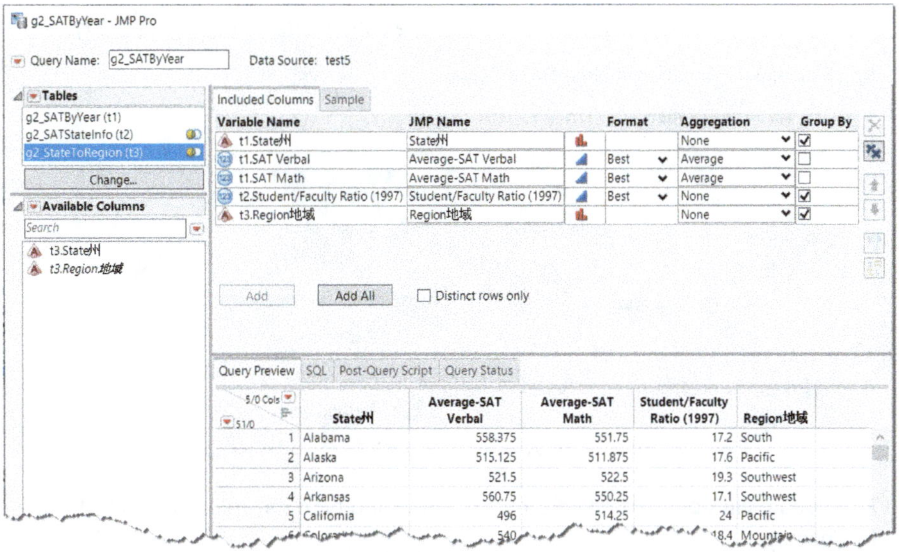

9. To include only the states in the South and Southwestern regions, select **Region** in the Included Columns list and drag it to the **Filter** pane on the right side of the window.
10. Control-click to select both South and Southwest in the Filters list.
11. Click **Save** to save the query as a .jmpquery file for use later.
12. Click the **Run Query** button to create the new data table. (See Figure 2.16.)

Figure 2.16 Resulting Data Table

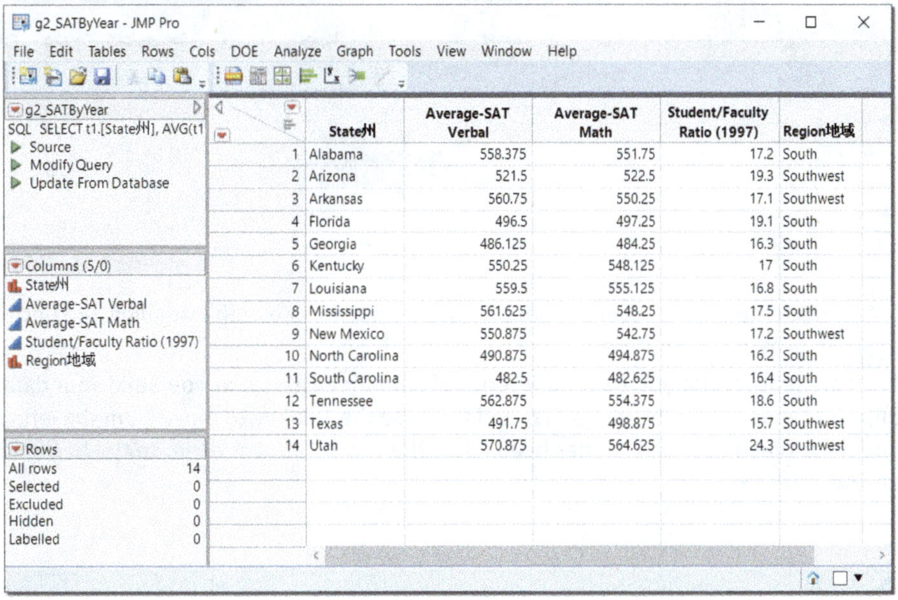

	State州	Average-SAT Verbal	Average-SAT Math	Student/Faculty Ratio (1997)	Region地域
1	Alabama	558.375	551.75	17.2	South
2	Arizona	521.5	522.5	19.3	Southwest
3	Arkansas	560.75	550.25	17.1	Southwest
4	Florida	496.5	497.25	19.1	South
5	Georgia	486.125	484.25	16.3	South
6	Kentucky	550.25	548.125	17	South
7	Louisiana	559.5	555.125	16.8	South
8	Mississippi	561.625	548.25	17.5	South
9	New Mexico	550.875	542.75	17.2	Southwest
10	North Carolina	490.875	494.875	16.2	South
11	South Carolina	482.5	482.625	16.4	South
12	Tennessee	562.875	554.375	18.6	South
13	Texas	491.75	498.875	15.7	Southwest
14	Utah	570.875	564.625	24.3	Southwest

Creating a JMP Data Table from Scratch

1. Select **File ▶ New ▶ Data Table** to create a new data table (Figure 2.17):

   Figure 2.17 Creating a New Data Table

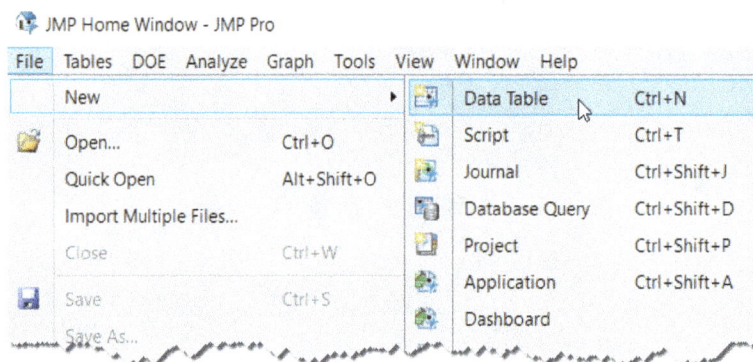

2. Double-click on the first column's heading and type the column name (the variable name).

3. Press **Enter** and type the data into the first cell directly below the heading. Press **Enter** again, type the data, and repeat as needed. Rows within JMP are consecutively numbered as observations or cases. (See Figure 2.18.)

   Figure 2.18 Add a New Column

4. To create another column, double-click on the next column's heading and enter the data as you did before.

If it is more practical for you to enter a series of data for each row as you build your data table, set up all of your column headings first and then use the Tab key to move from the left columns to the right. When each column has been filled, the Tab key moves down to the beginning of the next row.

> **Note**
>
> JMP will recognize the type of data you are entering and assign a data type to the column, either numeric or character. It also assigns an icon next to the columns (or variables) in the box on the left. These icons are the modeling type and are discussed in Section 2.3.

2.2 The JMP Data Table

The JMP Data Table looks very much like any spreadsheet. (See Figure 2.19.) In JMP, column headings indicate variables (what you have measured or counted), and rows indicate individual cases or observations. JMP requires your data to be structured in this way. If it is not, JMP can help you reformat your data. (See Section 2.4.)

Figure 2.19 The JMP Data Table

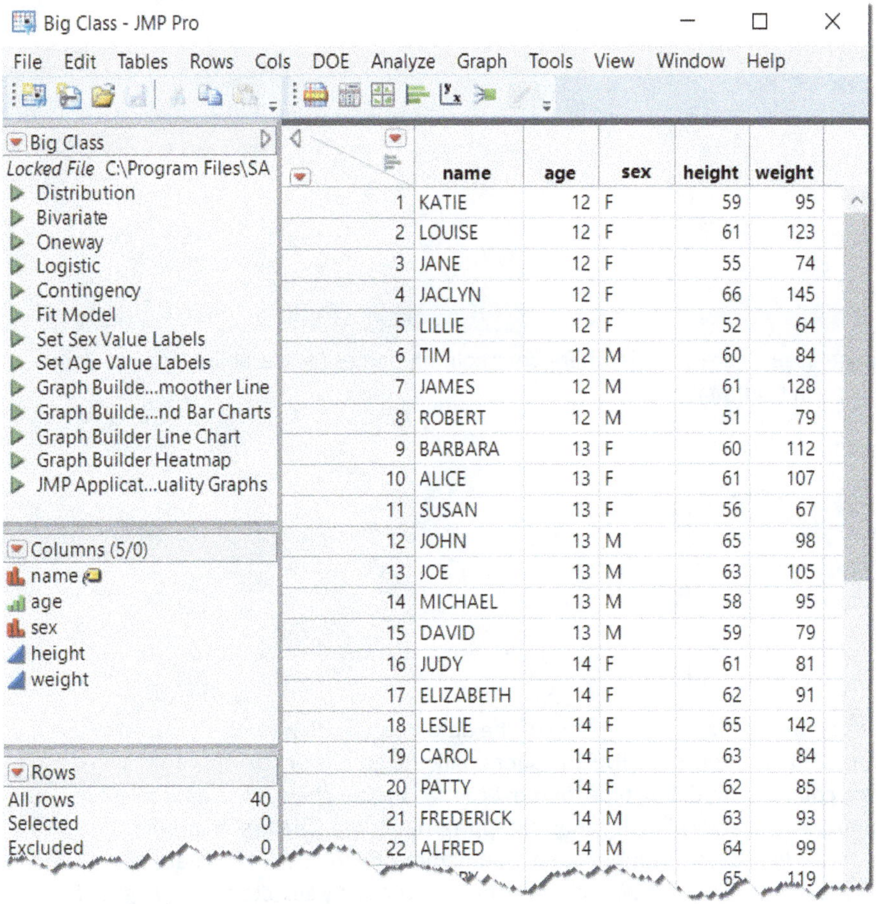

Data Table refers to the spreadsheet-like grid where your data resides.

The data grid can contain any number of columns (your variables) or rows (observations or cases). In this sense, we refer to data within the JMP data table as structured data.

In addition to the data grid, notice the three panels to the left of the data table. These panels contain information about your data (metadata). They provide vital information about your data as well as options to streamline and save your analyses.

The first and upper-most panel contains the name of the data table. (See Figure 2.20.) This panel stores references, notes, and/or scripts. Scripts enable you to save, automate, and customize analyses. If you perform a regular analysis or scheduled task, you will want to learn more about JMP scripts. (See the JMP Scripting Guide at **Help ▶ JMP Documentation Library ▶ JMP Scripting Guide**.)

Figure 2.20 The Table Panel

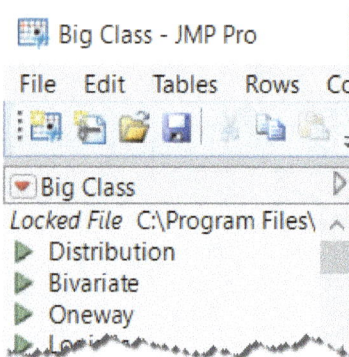

The Columns panel (see Figure 2.21) is where your column names (or variables) appear. Each column has an icon in front of it.

Figure 2.21 The Columns Panel

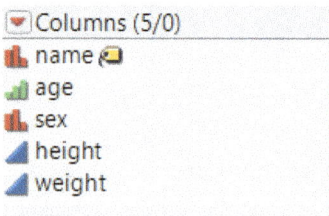

These icons correspond to the modeling type of the data in each column. As discussed in the next section, this is vitally important. JMP produces only the graphs or statistics that are appropriate for a column's modeling type. In most cases, you can change the modeling type by simply clicking on the icon and selecting another appropriate type. The first number in parentheses represents the total number of column names or variables in the data table. The second number represents any of those columns that are currently selected or highlighted.

The bottom panel is the Rows panel. (See Figure 2.22.) The Rows panel indicates how many rows (observations) are in your data table. This panel also indicates the number of selected, hidden, or excluded rows, if any.

Figure 2.22 The Rows Panel

♥ Rows	
All rows	40
Selected	0
Excluded	0
Hidden	0
Labelled	0

When rows are hidden, the observations are not included in graphs. When rows are excluded, they are not included in analyses. This row state is effective when you want to see or analyze a subset of your data. You can also both hide and exclude specific rows, which effectively removes the row(s) from your analyses and graphs, but not from your data table. Section 2.5 provides more information about row states including hiding and excluding rows.

Note

Multiple data tables can be open at any time, but only one active data table can be analyzed at a time. If you have multiple data tables open within JMP and you want to switch to another open data table, go to the **Home Window** (see Figure 2.23) and select the desired data table under **Windows List**.

Figure 2.23 The Home Window

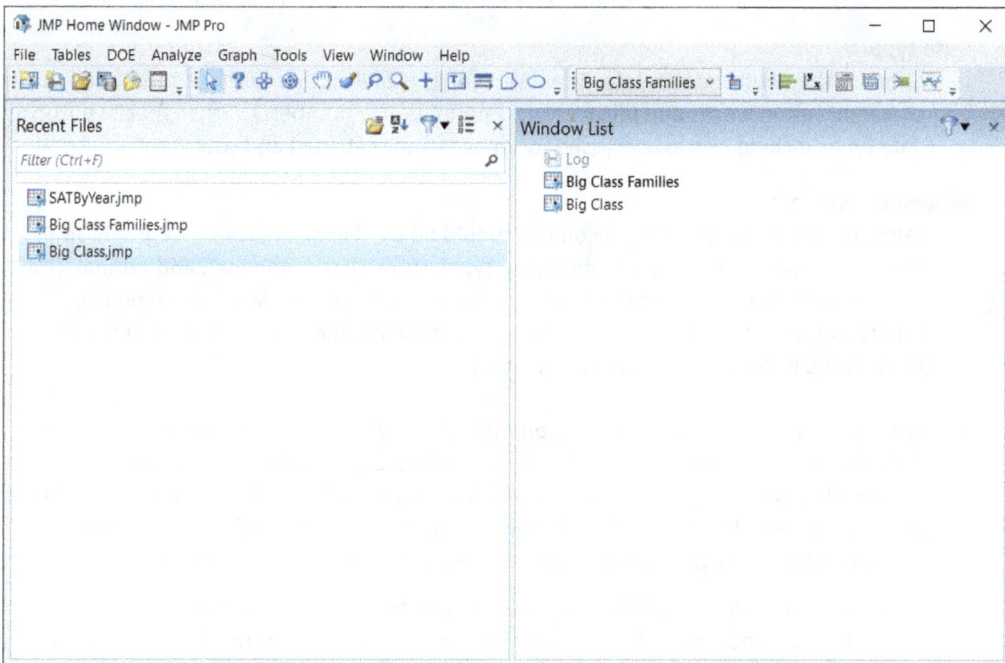

A special type of data table is shape files, which are used to create thematic maps. These data tables consist of two tables including a "Name" and corresponding "Boundary" table. These are stored in the Maps folder: **C: Program Files ▶ SAS ▶ JMP ▶ 15 ▶ Maps**. Section 2.7 covers some of the basics about shape files.

> **Note**
>
> There is no practical limit on the size of the data table that you can analyze. However, because JMP runs in your computer's local memory, the amount of RAM that you have determines the upper size limit of your data table. Your computer should be equipped with at least twice as much memory as the size of the data table. Thus, if you have just 8 GB of RAM, you can analyze a 4GB data table or about a 10-variable data set with 4 million rows! More details about JMP system requirements can be found at www.jmp.com.

2.3 Data and Modeling Types

One of JMP's great features is the ability to produce graphs and statistics that make sense for the data that you are analyzing. This feature assumes that your data is correctly classified in the data table. So, what do we mean by data type and modeling type? Let's define a few terms.

Data
> refers to any values placed within a cell of a JMP data grid. Examples include numeric and/or text descriptions: 3.6, $2500, Female, Somewhat Likely, or 7/7/19.

Data type
> refers to the nature of the data. The data type is usually either numeric (numbers) or character (often words and letters but sometimes also numbers). Other special purpose data types include expression (used for images and matrices) and row state.

Modeling type
> refers to how the data within a column should be used in an analysis or a graph. JMP uses three distinct and primary modeling types: continuous, nominal, and ordinal. (JMP also includes three additional special purpose modeling types: Multiple response, Unstructured text, and None. Since these are less common, we will only summarize them in the note at the end of this section.)

- *Continuous* data (also referred to as *quantitative*, *ratio*, or *interval scale data*) takes a numeric form and is often thought of as some type of measurement. For example, home selling prices, income earned, costs per square foot, and dates are all examples of continuous data. As a rule of thumb, continuous data can be used in calculations. For example, calculating the average cost per square foot would be meaningful.

- *Nominal* data is categorical data (also referred to as *qualitative*, *discrete*, *count*, or *attribute data*) and can take on either a character or numeric form. Nominal data fits into categories or groups such as car type, gender, department, and sales territory and also includes indicator variables such as yes/no or 0/1. In nominal data, it is helpful to count the frequency of the occurrence of values, but otherwise, nominal data is not used in calculations. For example, calculating the average car type would not be meaningful.

- *Ordinal* data is categorical data that has an inherent order or hierarchy. For instance, Likert scales (such as levels of satisfaction) in a survey and grade levels in school (freshman, sophomore, junior, senior) are examples of ordinal data. That is, they represent categories that have some sequence or order that should be retained in any analysis. Ordinal data is less common than continuous and nominal data, but there are a few analyses designed specifically for it. In most JMP analyses, nominal and ordinal data are treated the same way.

Table 2.2 Data Types Appropriate for Modeling Types

Modeling Type	Data Type	
	Numeric	**Character**
Continuous	Yes	No
Nominal	Yes	Yes
Ordinal	Yes	Yes

Note

Numeric data is right-justified in the data table, whereas character data is left-justified. This can be useful to check whether data contains errors.

In our example, **Big Class** contains five variables (or columns) representing each of these modeling types. (See Figure 2.24.) Let's briefly explain why they are classified by their data and modeling types:

- **Name** is nominal because it is a character data type and the student's name is arbitrary.
- **Age** is ordinal because the values are rounded down and we want to retain the six ordered age groups (12 to 17) in our analysis.
- **Sex** is nominal because its data type is character (M or F) and it has no order.
- **Height** and **weight** are continuous because they are both numeric and represent a measurement.

Note

Age could also be considered continuous because the values are numeric, but this would treat age differently and yield different results.

Figure 2.24 Understanding Modeling Types in the Data Table

	name	age	sex	height	weight
1	KATIE	12	F	59	95
2	LOUISE	12	F	61	123
3	JANE	12	F	55	74
4	JACLYN	12	F	66	145
5	LILLIE	12	F	52	64
6	TIM	12	M	60	84
7	JAMES	12	M	61	128
8	ROBERT	12	M	51	79
9	BARBARA	13	F	60	112
10	ALICE	13	F	61	107
11	SUSAN	13	F	56	67
12	JOHN	13	M	65	98
13	JOE	13	M	63	105
14	MICHAEL	13	M	58	95
15	DAVID	13	M	59	79
16	JUDY	14	F	61	81
17	ELIZABETH	14	F	62	91
18	LESLIE	14	F	65	142
19	CAROL	14	F	63	84
20	PATTY	14	F	62	85
21	FREDERICK	14	M	63	93
22	ALFRED	14	M	64	99

Big Class - JMP Pro

File   Edit   Tables   Rows   Cols   DOE   Analyze   Graph   Tools   View   Window   Help

Big Class
Locked File   C:\Program Files\SA
▶ Distribution
▶ Bivariate
▶ Oneway
▶ Logistic
▶ Contingency
▶ Fit Model
▶ Set Sex Value Labels
▶ Set Age Value Labels
▶ Graph Builde...moother Line
▶ Graph Builde...nd Bar Charts
▶ Graph Builder Line Chart
▶ Graph Builder Heatmap
▶ JMP Applicat...uality Graphs

Columns (5/0)
name
age
sex
height
weight

Rows
All rows   40
Selected   0
Excluded   0

Note

Let's briefly review some of the more specialized data and modeling types and what they are used for.

- **Row State** is a *data type* that enables you to store and manage information about a row of data. (See Section 2.5.)

- **Expression** is a *data type* that enables you to store images or matrices in a column.

- **Multiple Response** is a *modeling type* that is commonly used in surveys where one may be asked for more than one answer.

- **Unstructured Text** is a *modeling type* that is used for documents such as customer reviews, wine tasting notes, or an entire book. This modeling type is used in text mining with JMP's Text Explorer.

- **None** is a *modeling type* that tells JMP not to use that column in an analysis. This might be used for a column that represents an identifier for each row of data, for example, a student ID number, or patient number.

For more information, select **Help ▶ JMP Documentation Library ▶ Using JMP ▶ Ch. 5, About Modeling Types**.

Changing the Modeling Type

When you import data, the JMP default selects and assigns one of two modeling types based on whether the data is numeric or character. Numeric data becomes continuous and character data becomes nominal. Sometimes you might want to change the default modeling type of your data to generate results that are more meaningful.

For example, if we imported the **Big Class** data from Excel, age as numeric data would be imported as a continuous column. We might want to change that to ordinal. Changing the modeling type is simple in JMP. Click the column's corresponding icon in the Columns panel in the data table and select the correct type. (See Figure 2.25.)

Figure 2.25 Changing the Modeling Type

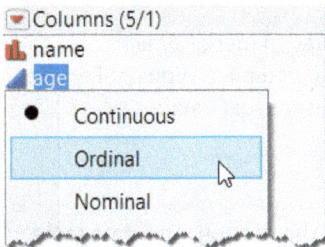

If the **Continuous** option is grayed out, your data type is classified as character. To change the data type, double-click on the column heading and change the data type to numeric. (See Figure 2.26.) In this window, you can also change the modeling type along with a host of other formatting options, which are described in the next section.

Figure 2.26 Changing the Data Type

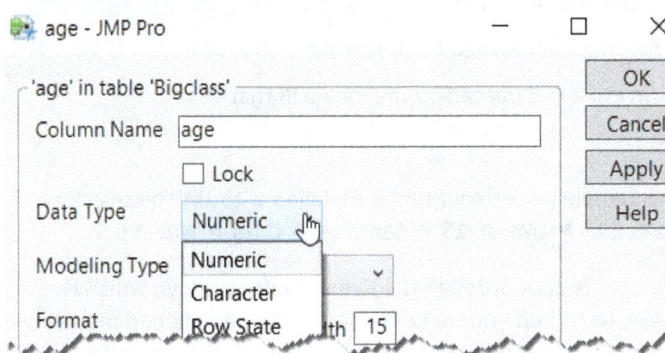

For more information, select **Help ▶ Books ▶ Using JMP ▶ Chapter 5, Set Column Properties ▶ About Data and Modeling Types**.

2.4 Cleaning and Formatting Data

Sometimes data is not in the best shape or in the right form when it is imported. Fortunately, JMP has extensive column formatting abilities. This section focuses on the most common features, including:

- Cleaning up your data format, such as decimal places, dates, times, and currency. We will use the Column Info window to accomplish these tasks.

- Introducing the Formula Editor, which enables you to create new columns from old ones, add IF statements, and transform data using basic or more advanced functions. We will introduce a basic example in this section. For more information, select **Help ▶ JMP Documentation Library.**

- Learning to use the RECODE command, which is a handy way to merge similar categorical responses into a single category. For example, if you have Woman, Female, and Girl as responses, you can merge these into a single response: Female.

Example 2.2 Movie Rentals

We will use the **Movies.jmp** data table to illustrate the concepts in this section. This data table consists of the 277 top-grossing movies released between 1937 and 2003. The columns are:

- **Movie:** name of movie
- **Type:** genre/category of movie (for example, comedy, family)
- **Rating:** US movie rating system (for example, general audience [G], adult [R])
- **Year:** year of movie release (for example, 1937)
- **Domestic $:** US domestic revenue in $ earned by the movie in that year
- **Worldwide $:** Worldwide revenue in $ earned by the movie in that year
- **Director:** director of movie

You can access this data table in the Sample Data folder that is installed with JMP by selecting **File ▶ Open ▶ C: ▶ Program Files ▶ SAS ▶ JMP ▶ 15 ▶ Samples ▶ Data ▶ Movies.**

Getting your data into a standard format is done through the Column Info window, which is accessed from the Cols menu. Options to format your data are driven by the data and modeling types specified for that column of data. You can change these types, if necessary, to meet the requirements of your analysis. Recall that changing these types affects the graphs or statistics that you can generate from that column. (See the previous section.) Let's begin by opening the **Movies.jmp** data table:

1. Open the **Movies.jmp** data table.
2. Select the **Domestic $ column**, and then select **Cols ▶ Column Info.**
3. Because Domestic$ is a numeric value, you see the **Format** drop-down menu (see Figure 2.27), which leads to several options. It is also our starting point for the next items we

will discuss. Note that if you select a Character column, the **Format** menu does not appear in the Column Info window.

Figure 2.27 Column Info Format Menu – Continuous Variables

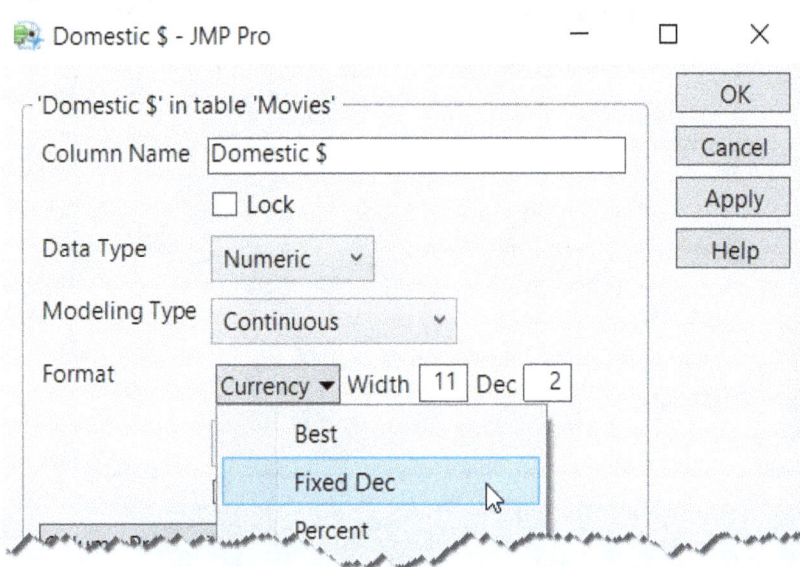

You can also either double-click on the column name as mentioned in the previous section, or right-click on the column header and select Column Info from the menu.

Formatting Decimal Places

To change the number of decimal places displayed in a column of data, do the following:

1. Click on the column of interest. In our example, it is **Domestic$**.
2. Select **Cols ▶ Column Info.** JMP will make a best guess on the format of the data; in our example, **Currency** was correctly specified. (See Figure 2.27.) You can easily change this format by selecting another format from the menu.
3. To the right of the **Format** menu are two boxes, Width and Dec. **Width** refers to the number of characters that can be in the column, and **Dec** refers to the number of decimals right of the point. In our example, type "0" in the **Dec** box, then select **Apply** and **OK**. (See Figure 2.28.)

Figure 2.28 Formatting Decimal Places

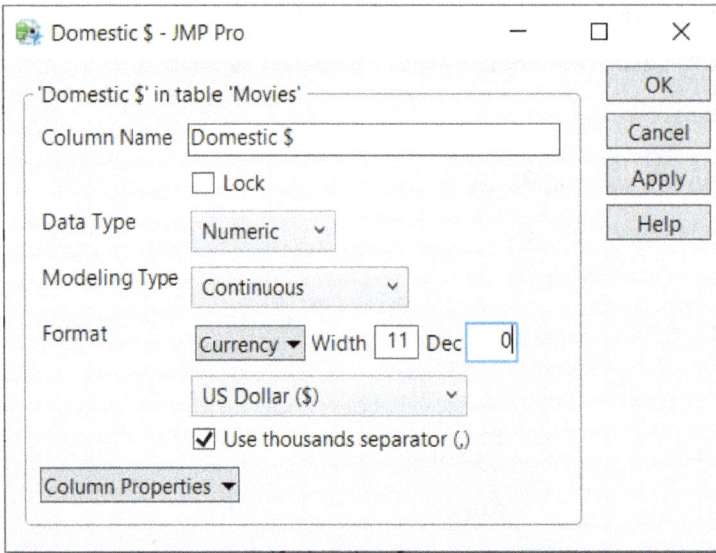

Note
Formatting decimal places affects what is viewed in the data table and in output but does not affect the precision of the data when used in calculations.

Formatting Dates, Times, and Durations

Dates are numeric values in JMP, which allows them to be transformed into other date formats and calculated for duration or elapsed time. If you are importing data that contains dates, ensure that the data type is numeric.

The Column Info (**Cols ▶ Column Info**) window provides several date format options. (See Figure 2.29) When a date is selected from the **Format** menu, a secondary drop-down menu for the display format appears, along with a similar drop-down menu for the input format of your imported data. The format of your imported data needs to match one of JMP's input format options, which can then be transformed into any format among the display format options.

Figure 2.29 Column Info Window for a Date Column

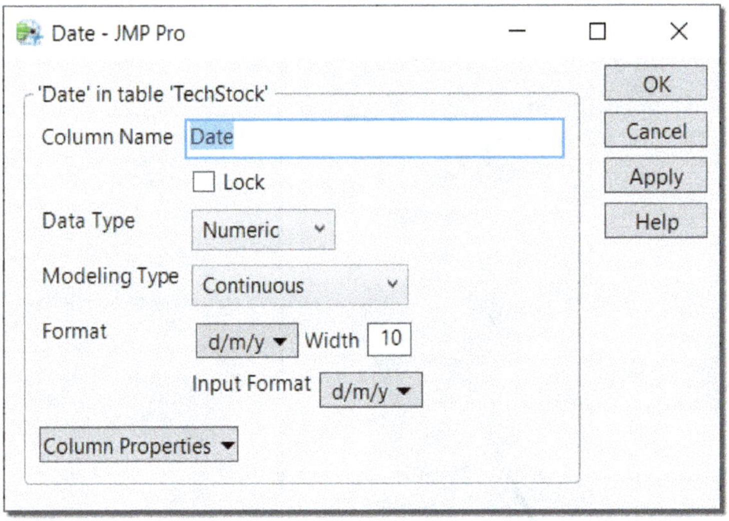

Let's walk through a new example, TechStock, to illustrate this concept.

Example 2.3 TechStock

We will use the TechStock data table to illustrate dates in this section. This data set contains the stock price of the NASDAQ 100 (QQQ) at the high, low, and close for each trading day during the period 11/27/2000 to 2/26/2001. You can access this data set in the Sample Data folder that is installed with JMP: **File ▶ Open ▶ C: ▶ Program Files ▶ SAS ▶ JMP ▶ 15 ▶ Samples ▶Data ▶ Techstock.jmp.**

1. Open the **TechStock.jmp** data table.
2. Click the **Date** column name**.**
3. Select **Cols ▶ Column Info**. Open the **Format** drop-down menu and select **Date**, which displays how the dates will appear in the data table.
4. It is currently displayed as d/m/y (see Figure 2.26), as indicated by the check mark. Change the format to **Monddyyyy.** Click **Apply** or **OK.** The date is now displayed as abbreviated month, day, and year in the Date column. (See lower part of Figure 2.30.)

Figure 2.30 Changing the Display Format of Dates

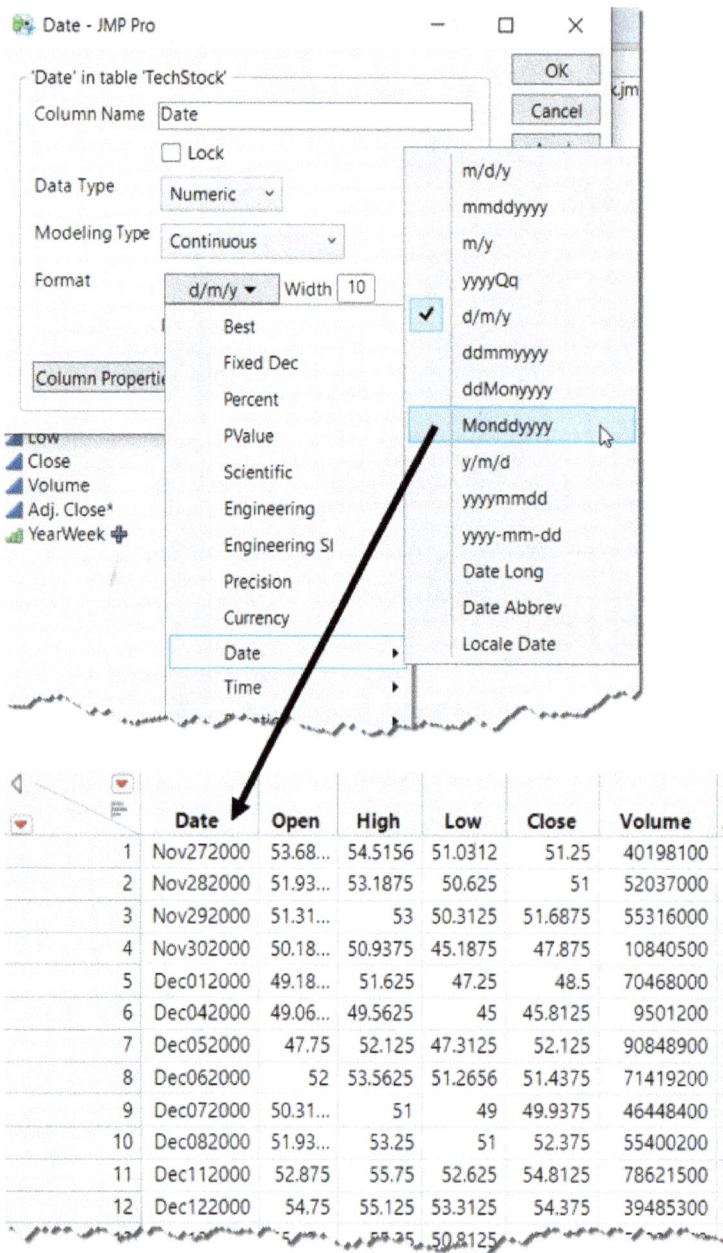

You can also format time and duration from this window provided you have these types of columns.

Column Properties Menu

Column Properties, another useful tool in the Column Info window (see Figure 2.31), enables you to add formulas, check ranges of values for auditing, and assign customized ordering to the data, among other tasks.

Figure 2.31 Column Properties

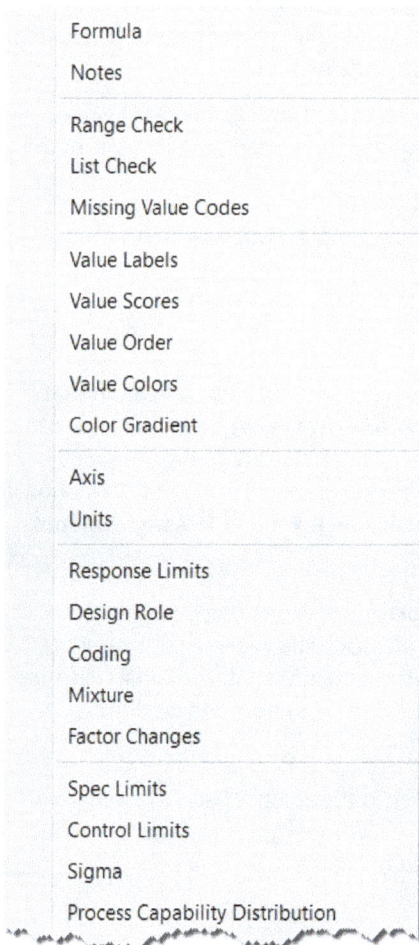

| Formula |
| Notes |
| Range Check |
| List Check |
| Missing Value Codes |
| Value Labels |
| Value Scores |
| Value Order |
| Value Colors |
| Color Gradient |
| Axis |
| Units |
| Response Limits |
| Design Role |
| Coding |
| Mixture |
| Factor Changes |
| Spec Limits |
| Control Limits |
| Sigma |
| Process Capability Distribution |

Some of the more commonly used column properties are:

- **Missing Value Codes:** used to identify values placed in columns that represent unknown (or missing) data. By default, JMP uses a blank cell for missing character data and a • for missing cells in numeric columns. In cases where the original data used a different missing identifier (such as 999 or unknown), these values can be entered in the missing value codes property so that JMP recognizes the values as missing.

- **Value Labels:** used when values are entered into the data table rather than the actual label. For example, an analyst might code Low, Medium, High as 1, 2, 3 for ease of data entry. However, in analysis results, the labels are preferred. Using this property will enable you to enter the data as a code, but provide the labels in the appropriate display.

- **Value Order:** used to specify the order in which column values should be sorted. For example, using the order Poor, Good, Better, Best rather than the alphabetical ordering of these data values. An example is provided later in this section.

- **Axis:** allows the specification of axis properties (such as formatting, minimum value, maximum value, increment, tick marks, and grid marks) to be used for all graphs of the column.

These functions are described in detail in the *Using JMP* book (**Help ▶ JMP Documentation Library ▶ Using JMP ▶ Ch. 5 ▶ Set Column Properties**).

Formula Editor

JMP's formula editor is handy and flexible. Use it when you need to create a new column that contains values that are calculated or derived from existing columns in your data table. You can also transform your data, add conditional statements, and much more. Due to the advanced nature of these features, we will cover only the most basic features here. For more information, see *Using JMP* at **Help ▶ JMP Documentation Library ▶ Using JMP ▶ Ch. 5 ▶ Assign Column Properties**.

One of the common operations performed with the Formula Editor is creating a new column of data that contains a calculation from existing columns. To illustrate this feature, let's return to our **Movies.jmp** data table. For example, suppose we want to obtain the international revenues from these movies by subtracting the domestic revenues (Domestic $) from the worldwide revenues (Worldwide $).

1. First, we need to create a new column. Double-click in the column head to the right of our last populated column (Director). (See Figure 2.32.) Type **"International $"** in the heading, and press **Enter**. Click in the column head to highlight.

Figure 2.32 Creating a New Column

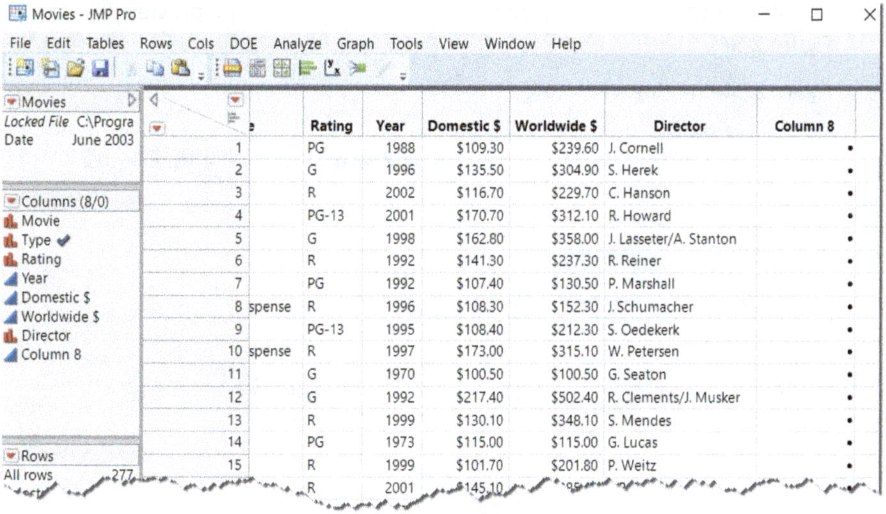

2. With the new column selected, select **Cols ▶ Column Info**. The Formula window appears in the **Columns Properties** menu. (See Figure 2.33.) Select **Formula** and **Edit Formula**. You see a list of columns or variables on the left side of the window.

Figure 2.33 Opening the Formula Editor

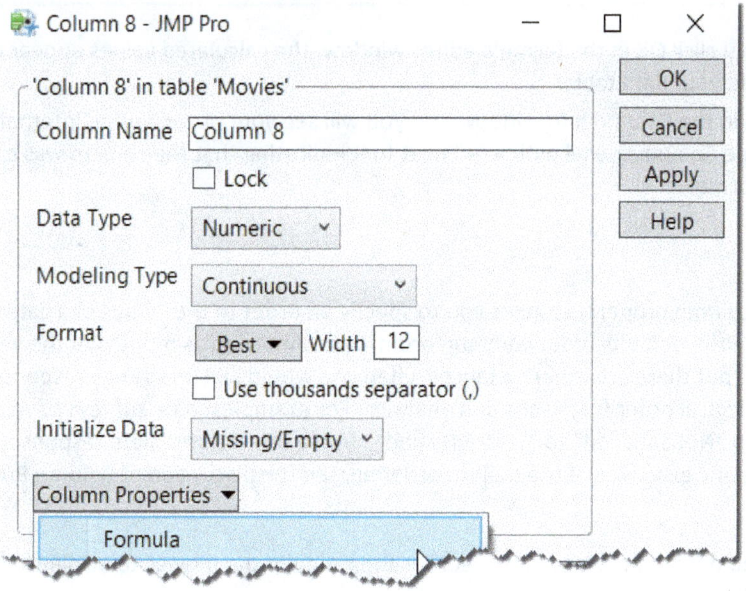

3. Click the **Worldwide $** column, then the "-" symbol in the upper palette, and then the **Domestic $** column. You see your formula take shape in the preview window. (See Figure 2.34).

Figure 2.34 Creating a Formula

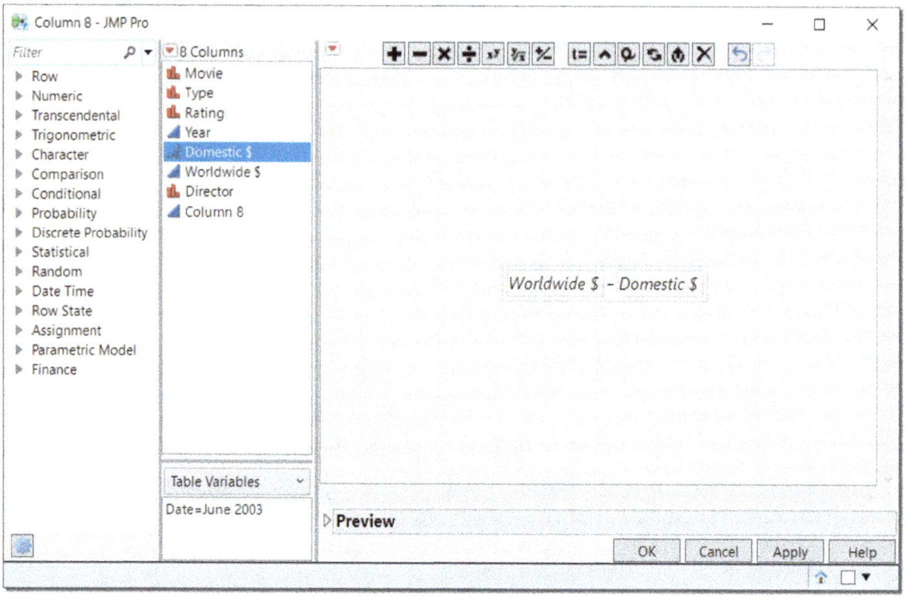

4. When you click **OK** in the formula editor window, the calculated values appear in a new column in your data table.

5. Click **OK** in the Column Info window and you will see your new column, International $, within the Columns panel with a "+" next to it indicating that the column was created with a formula.

Value Order

The value order column property enables you to specify an order to the values of a categorical column. JMP's built-in defaults order common ordinal columns such as months of the year or days of the week, but there are other instances when you would like to arrange responses (values) in some logical order for graphs and analyses. For example, some surveys have a range of responses from "Not Satisfied" to "Very Satisfied," with a few intermediate responses in between. For specific examples of the built-in ordering, see the discussion of Column Properties in *Using JMP*.

In the **Movies.jmp** data table, we want to reorder the rating of the movies to display in this order: R, PG-13, PG, and G.

1. Click the **Rating** column. Select **Cols ▶ Column Info**, and then select **Column Properties**.

2. Select **Value Ordering**. (See Figure 2.35.) A new window appears with the available responses from the Rating column. Select a response and move it up or down or reverse the order, whichever is appropriate. Select **Reverse** for this example. (See Figure 2.36.)

Figure 2.35 Value Order-Column Properties Menu

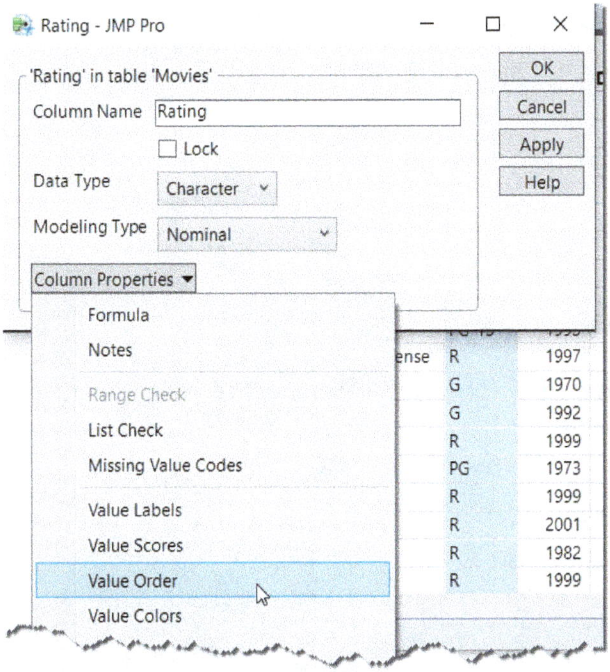

Figure 2.36 Specifying the Order of Categorical Variables

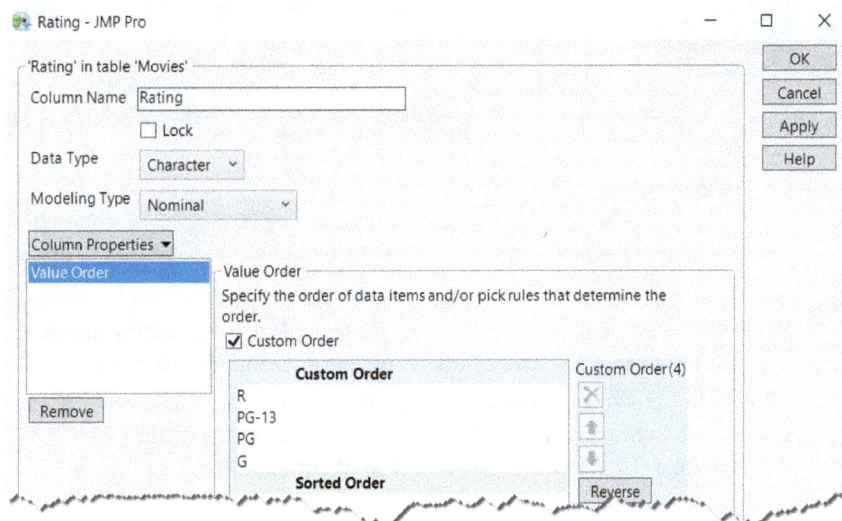

3. When you are satisfied with the order, click **Apply** and **OK**.

    Now let's see the results of this exercise by using the distribution platform, which is discussed in Chapters 3 and 5. Here's a preview:

4. Go to **Analyze ▶ Distribution** (Figure 2.37).

Figure 2.37 Launching the Distribution Platform

5. In the Distribution window, select **Rating**, **Type**, and **Year**, and click **Y, Columns**. Click **OK**. (See Figure 2.38.)

Figure 2.38 The Distribution Launch Dialog Box

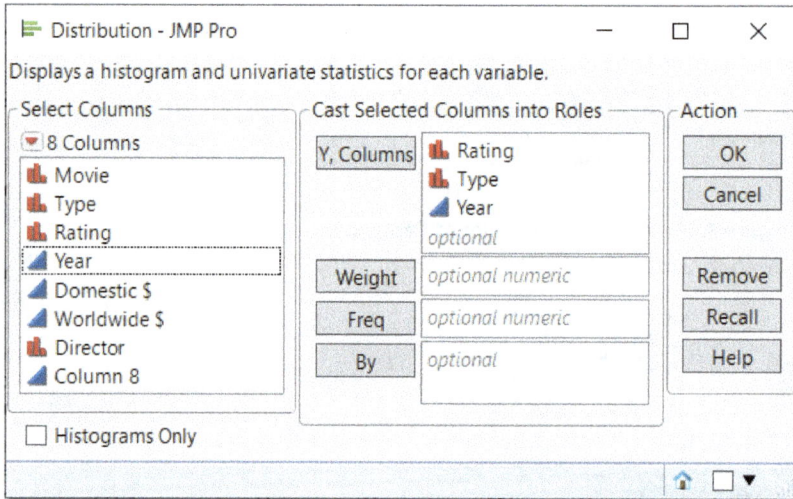

6. Three bar charts appear side by side. Although "R" was the first response listed in the Value Order window, it appears at the bottom of the Distribution graph. If we did not make this change, the order of these responses would be reversed.

7. Click on the green **G** bar under Rating. (See Figure 2.39.) The G responses are now highlighted in the G bar as well as those same responses reflected by Type and Year. This dynamic visual feature is available in all JMP graphs.

Figure 2.39 The Distribution Results Window

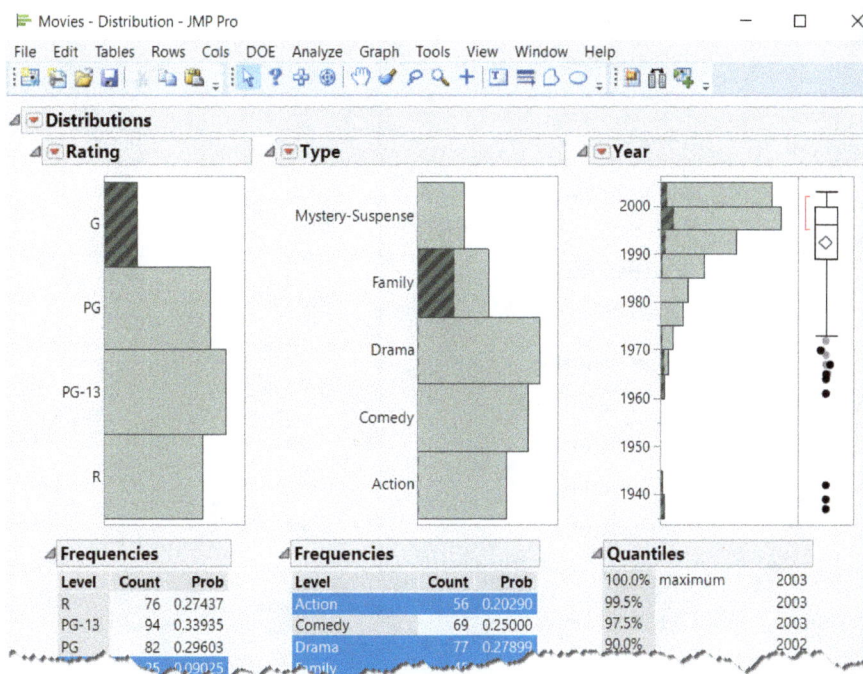

Recode

The Recode command is useful when you have a column of data containing values that you would like to rename or consolidate. For example, if you have data labeled Apple Music, Pandora, and Spotify, you might want to consolidate these into one response: Streaming Music. Recoding assigns a specified new value to all of the existing responses of the original name or value.

In the **Movies.jmp** data table, we want to replace PG-13 movies with a PG rating because many movies made before 1985 only contain ratings of G, PG, and R.

1. Select the column that you want to recode, **Rating**. Select **Cols ▶ Recode**. (See Figure 2.40.)

Figure 2.40 The Recode Option

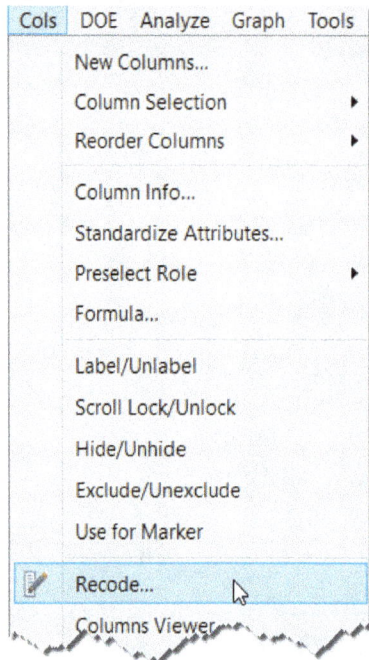

| Cols | DOE | Analyze | Graph | Tools |

New Columns...

Column Selection ▸

Reorder Columns ▸

Column Info...

Standardize Attributes...

Preselect Role ▸

Formula...

Label/Unlabel

Scroll Lock/Unlock

Hide/Unhide

Exclude/Unexclude

Use for Marker

Recode...

Columns Viewer

2. This command generates an input window of current and unique responses, with an area to the right to specify a new value. In the box to the right of PG-13, type **PG**, click out of the edited box, then click **Recode**. (See Figure 2.41.)

Figure 2.41 The Recode Dialog Box

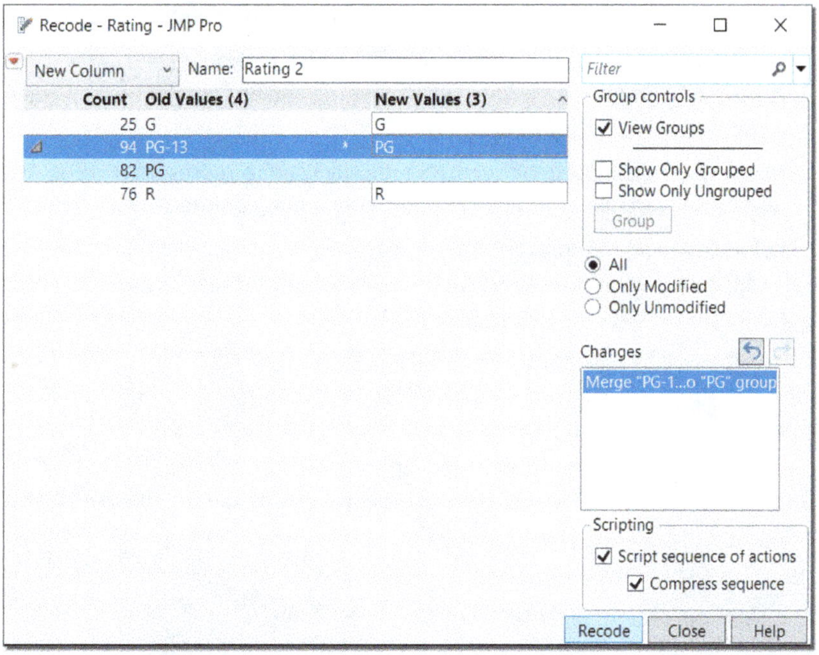

Once you have selected a new value, you can replace that value in the same column, create a new column with these values, or even create a formula column. Be careful! If you select **In Place**, these values cannot be changed back because the Recode command replaces values in that column. The default in JMP 15 places recoded values in a new column.

2.5 Selecting, Highlighting, and Filtering Data: Row States

Thus far, we have focused on column properties. Let's now look at rows or the observations in your data. In the process of exploring or analyzing data, it is often valuable to *drill-down* or to see and compare subsets of your data or rows. JMP makes this task seamless and simple through a concept called Row State that assigns one or more of the following six conditions to one or more rows (or sets of observations) within a data table: Selected, Hidden, Excluded, Labeled, Marked, or Colored (Figure 2.42).

- **Selected** will appear as a highlighted row(s) that will correspond to a highlighted point or area in any corresponding graph. You can also easily subset highlighted rows.

- **Hidden** means that the row will not appear in any graph. A blindfold icon appears next to hidden rows because you will not see that row in any graph.

- **Excluded** means that the row will not be used in any calculated result. A red circle with a line through it next to its row number indicates an excluded row.

- **Labels** provide the columns value when the observation is selected within an appropriate graph. Labels look like a price tag and appear next to the column in the column panel.

- **Markers** are distinguishing symbols typically used to represent a group within a categorical variable or to highlight particular points of interest.

- **Colors** are used to represent different groups in a categorical variable or as a gradient in a continuous variable within graphs. Both Colors and Markers are covered in the next section.

Figure 2.42 Row States Indicated in a Data Table

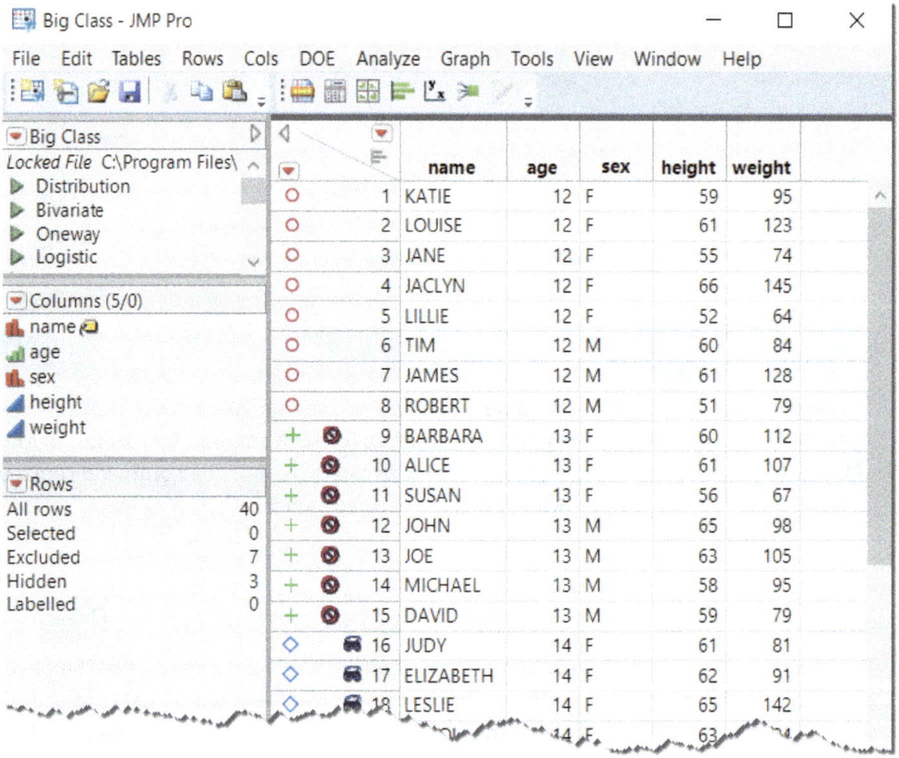

Hiding and Excluding Data: Using Data Filter

Hiding a row prevents that data from appearing in any graph (but is not excluded from any analysis). Conversely, excluding a row will remove the row from any calculated result but will still show the point in a graph. If you prefer to both hide and exclude a row, you can directly select a row or rows in your data table with your mouse, then go to **Rows ▶ Hide and Exclude**. This is an effective approach when you are dealing with few specific points like outliers.

When exploring data, it can be more efficient to hide and/or exclude entire groups or ranges within your data table, and rather than thinking about what you DON'T want to see or analyze, it is more natural to think about what you DO want to see. That is, to Show (rather than Hide) and Include (rather than Exclude).

JMP's Data Filter tool (**Rows ▶ Data Filter**) provides this capability and can be applied to any graph or analysis platform in the active data table. The Data Filter enables you to dynamically show, include, or select groups within a column and toggle between them, or to specify a custom range within a continuous variable and create a slider to filter the graph or analysis. The Data Filter automatically hides and/or excludes values that are not selected.

In Figure 2.43, we have launched and asked the Data Filter to "Show" and "Include" the male students in our Big Class data graph. Notice that it has automatically hidden and excluded the female students in data table in the background. The Data Filter will be illustrated in Chapter 6.

Figure 2.43 Using the Data Filter to Select a Group

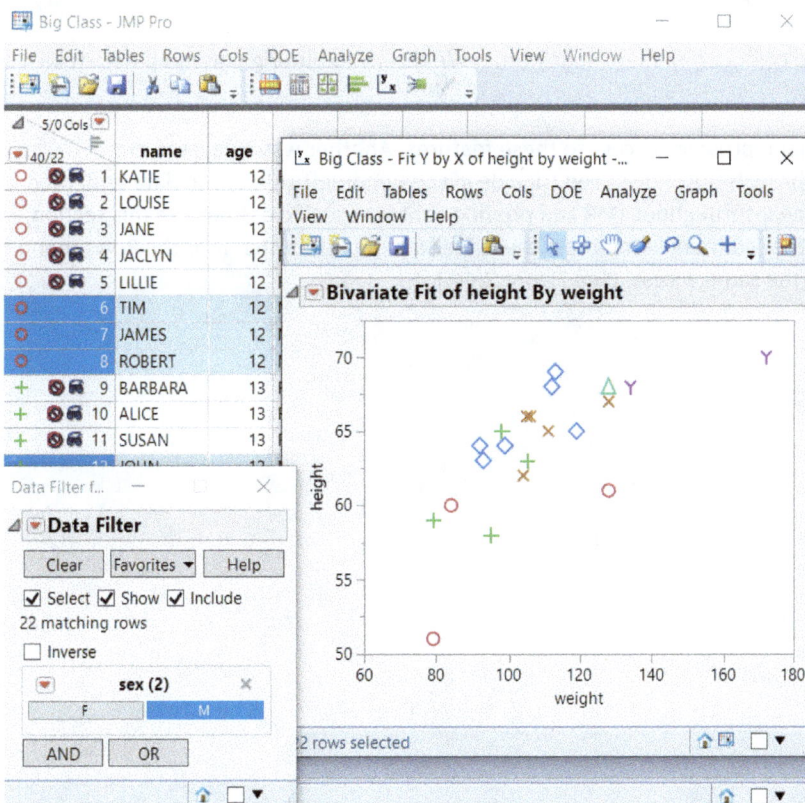

There are two types of Data Filters in JMP: the Global Data Filter and the Local Data Filter. The Global Data Filter, which we explained here is accessed in the Rows menu, is applied to the data table and all analyses and graphs for that data table. The Local Data Filter is found within individual output windows in JMP (under the red triangle) and can be applied to filter values within that particular output window only. Examples of the Local Data Filter will be illustrated later in the book.

2.6 Adding Visual Dimension to Your Data

JMP is designed to be visual. Its many useful tools help you visualize or communicate your data effectively. For example, you can use colors or unique markers to signify a range or value of another column in any appropriate graph. Any color or marker assigned to your data can be saved and used in any number of graphs.

You can change these colors or markers at any time. We return to our **Big Class.jmp** data table to illustrate this feature:

1. The **Rows** menu provides access to these features. Another way to access these features is through the "Rows" red triangle in the upper left side of the data grid. (Red triangles appear throughout JMP and provide context-specific options.) In this section, we will use the red triangle to access these features, but note that it or the Rows menu will provide the same access. (See Figure 2.44.)

Figure 2.44 Color or Mark by Column

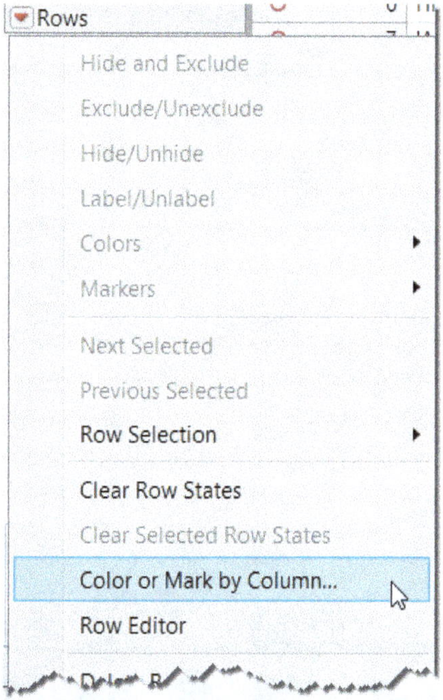

2. Select **Color or Mark by Column**. Select the column that you would like to distinguish with color (sex, in this example). You can see how JMP will express these values in color on the right side of the window. (See Figure 2.45.) Once you are satisfied, click **OK** and you will see colored markers preceding the row numbers in the data table.

Figure 2.45 Color by a Column

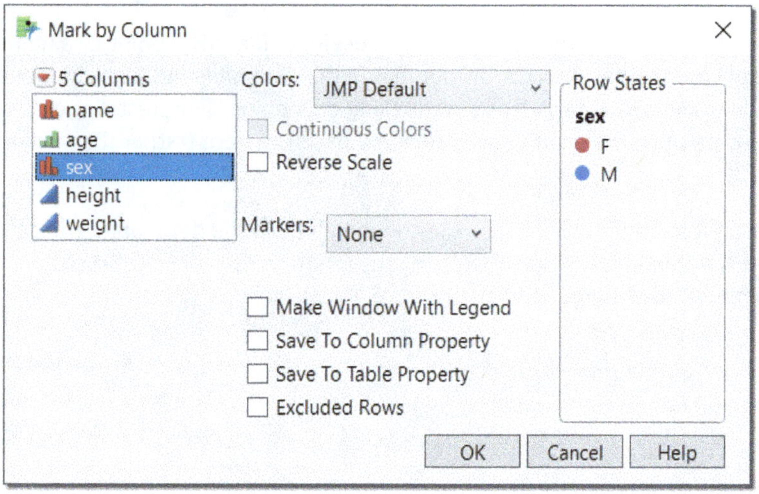

3. Alternatively, in the same window, you can also distinguish points by using unique markers (for example, symbols). Like colors, unique markers can be assigned to

categorical or continuous columns. Click the **Markers** drop-down menu. (See Figure 2.46.) JMP provides many different marker types, and a submenu enables you to view and select the desired type.

Figure 2.46 Mark by a Column

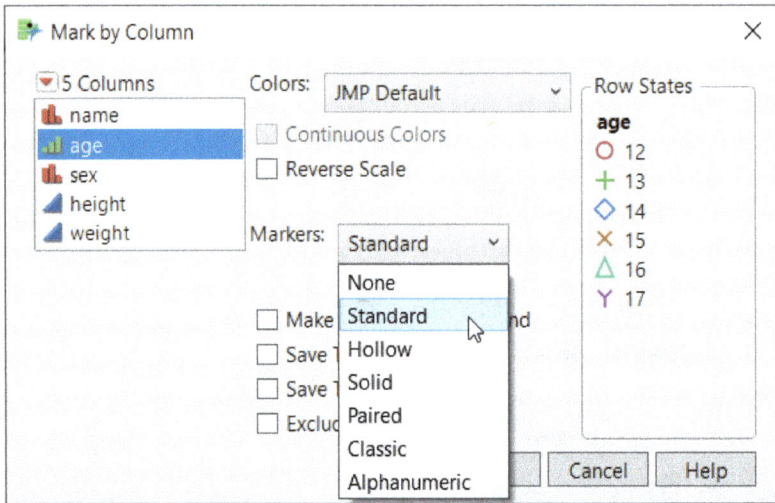

Adding Labels to Data

Sometimes in the process of exploring your data, it is useful to identify a point by a name, territory, or product type rather than its row number in the data table. Adding labels enables you to see these identifiers in a graph by simply clicking on a point of interest. For example, in the **Big Class.jmp** data table, we want to see the name of the student (rather than a row number) in a graph:

1. First, select a column by clicking on the column name in the data table. This selection activates the **Label option** when selecting the Columns red triangle. Select **Label/Unlabel**. (See Figure 2.47.)

Figure 2.47 Adding Labels to a Column

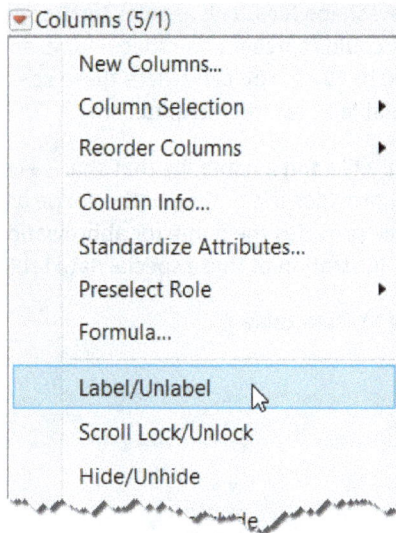

2. You then see a label or what might look like a price tag next to that column in the Columns panel of the data table. (See Figure 2.48.) When creating a graph, labels with the name of the student (rather than the row number) are displayed when they are selected with the mouse.

3. Figure 2.48 A Column to Display Labels in Graphs

Note

You can add labels for more than one column. For example, we might want to have a label containing both the name and age of the student. Simply highlight all the columns you would like to label first or repeat this process and add a label to "age."

2.7 Shape Files and Background Maps

Creating thematic maps is easy to do and is explained step-by-step in Chapters 3 and 4. In this section, we will describe a special type of data table required to create thematic maps called shape files.

JMP includes 12 common shape files such as States and Counties of the US, Countries of the World, and others. You can import or create new shape files such as sales territories, and these do not need to be geographic maps. Shape files could represent any space: for example, a football stadium, an assembly plant, or an office building. You can create these special purpose maps with the Custom Map Creator add-in available at community.jmp.com.

Shape files consist of two data tables: a boundary file and a name file that share a common Shape IDs column. The boundary file (or XY file) provides the outline of the shape as a polygon that corresponds to each Shape ID. The name file provides the name (or abbreviations of the name) of the shape ID. Figure 2.49 provides an illustration of these special data tables.

Figure 2.49 Shape Files Contain Both a Name and XY Data Table

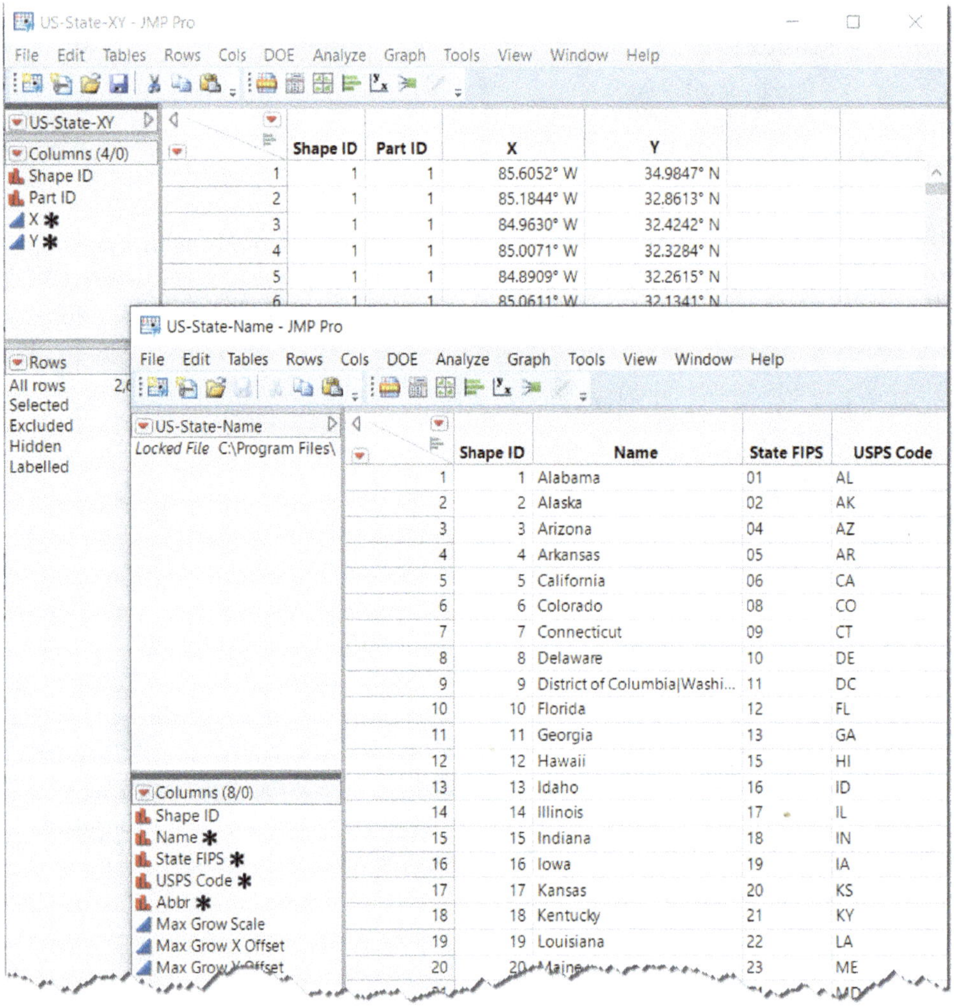

Built-in shape files are included in the Maps folder: **C:** ▶ **Program Files** ▶ **SAS** ▶ **JMP** ▶ **15** ▶ **Maps.** Should you want to create or import new shape files, they must be placed in this same Maps folder.

When a column containing names (corresponding with a Shape ID) is dragged into the Shape/Map box in Graph Builder, it will render the boundaries/shapes. (See Figure 2.50.)

Figure 2.50 Using Shape to Create Thematic Maps

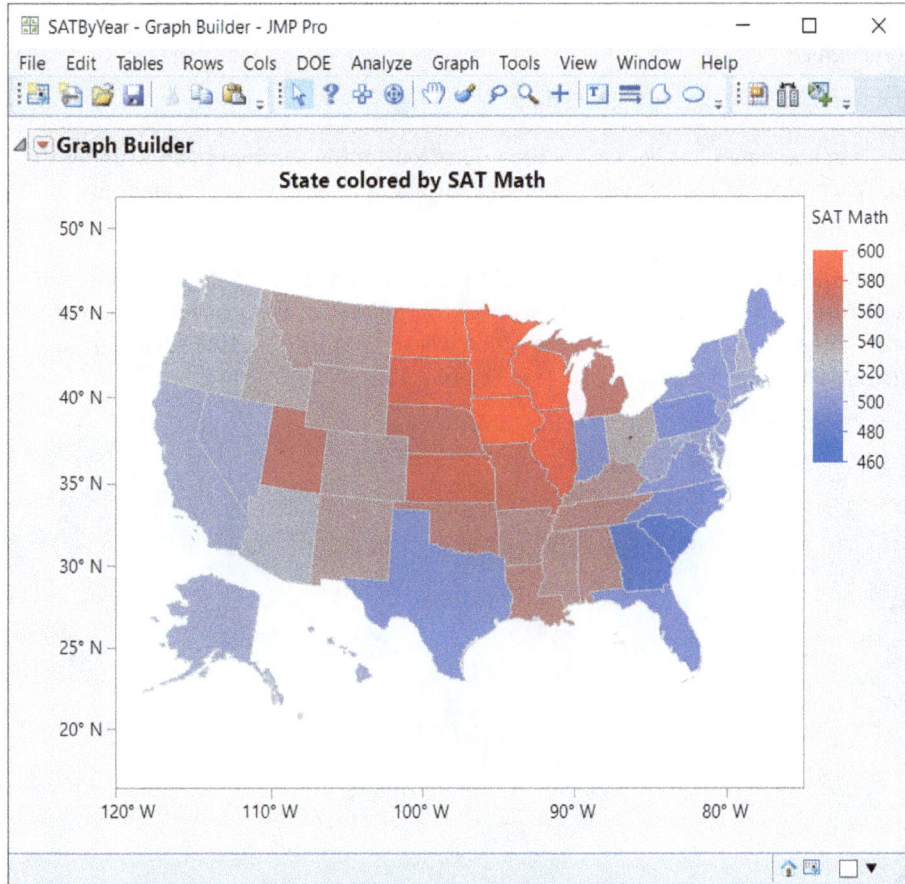

Note

In addition to shape files, JMP contains a variety of background or reference maps that provide the base map that you can plot data upon, provided that the points you want to plot on the background contain latitude and longitude information. OpenStreet Maps for example, enables you to show data on a map at the street address level. (See Section 4.2.)

2.8 The Tables Menu

The Tables menu is a collection of JMP tools that you will need to manage your data, whether you are sorting it, transposing it, or joining multiple data tables. Put another way, if your data is not structured in a manner that fits the JMP analysis framework, you need to use these commands to improve the structure. To keep things simple, we will cover just a few of these features, including sorting, joining, and dealing with missing data. In this section, we learn:

- How to structure your imported data into a form that you would like to see or that JMP will recognize.

- What to do when you have missing data.

Using the **Big Class.jmp** data table, let's first take a quick look at the **Summary** option under the **Tables** menu. This command enables you to obtain a variety of summary statistics for any column.

1. Select **Tables ▶ Summary**. (See Figure 2.51.) Choose **height** from **Select Columns** and **Mean** from the **Statistics** menu. (See Figure 2.52.) Select **age** from **Select Columns** and place in the **Group** area, Click **OK.** This action will generate a new data table, *Big Class By (age)*, with mean heights by age. (See Figure 2.53.)

Figure 2.51 Summary Platform from the Tables Menu

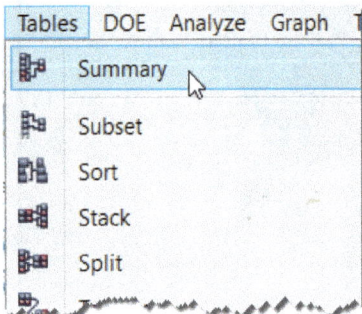

Figure 2.52 Summary Dialog Box

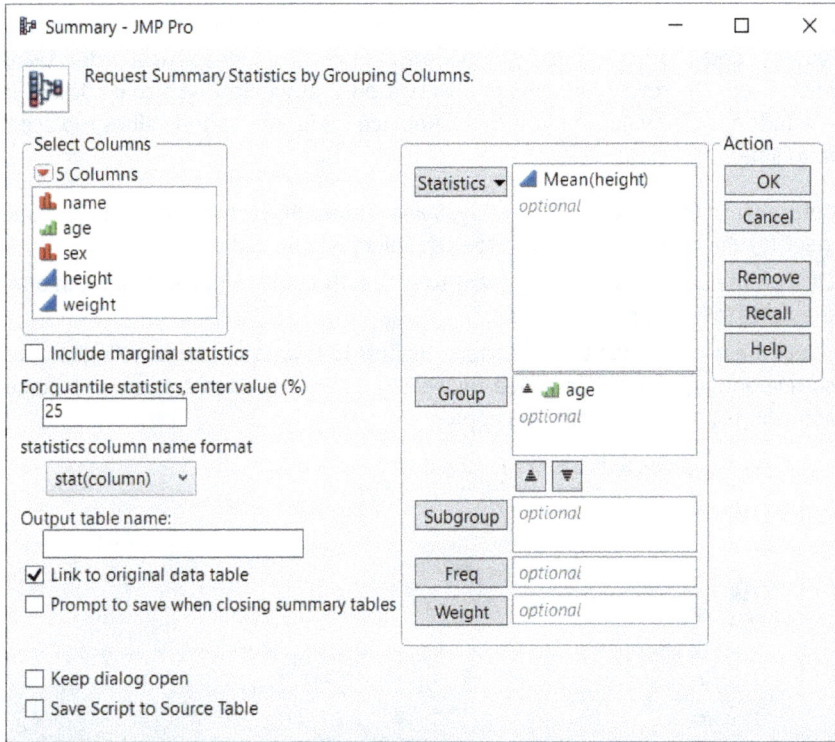

Figure 2.53 Summary Data Table

	age	N Rows	Mean(height)
1	12	8	58.1
2	13	7	60.3
3	14	12	64.2
4	15	7	64.6
5	16	3	64.3
6	17	3	66.7

Sorting

You can sort numeric columns from highest to lowest or lowest to highest. With character columns, you can sort character data by alphabetical or reverse alphabetical order. Using JMP's sorting option keeps the rows (your sets of observations) intact. Sorting also creates a new JMP data table with the sorted values (if you check **Replace table**, the sorted values replace the existing data table).

1. Returning to the Big Class data table, Select **Tables ▶ Sort**. In the resulting window, identify which column you want to sort. Select **height** and click **By.**
2. Click on the column(s) that you want to sort in the right window (height, in our example) to highlight the column.
3. Select the way you want to sort them, highest to lowest or lowest to highest, using the corresponding triangle icon. (See Figure 2.54.) Click **OK.** Each entire row is sorted according to the conditions you apply.

Figure 2.54 The Sort Launch Dialog Box

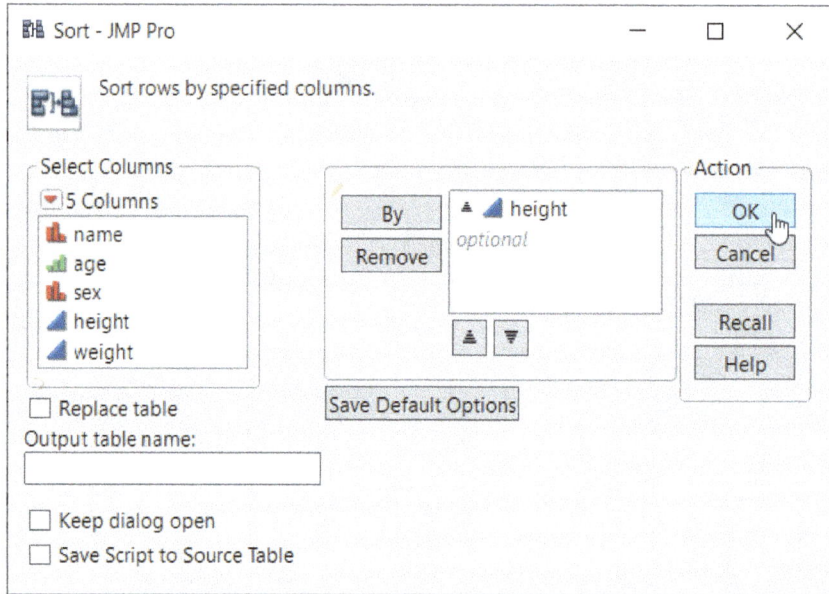

More information about sorting is available in **Help ▶ JMP Documentation Library ▶ Using JMP ▶ Ch. 6 Reshape Data**.

Joining

The **Join** option from the **Tables** menu enables you to combine or merge two or more different data tables into one. If some of the columns in your original data have the same name and type, this is a simple process. If not, there are some handy JMP tools to help you specify how two different data tables can be joined. Let's look at a simple example:

1. First open the data tables that you would like to join. Select **Tables ▶ Join**.
   We will use **Trial1.jmp** and **Trial2.jmp** for this example, which can be found at **Help ▶ Sample Data ▶ Open the Sample Data Directory ▶ Trial1.jmp**. Repeat for **Trial2.jmp**.

2. The window indicates your active data table (Trial1.jmp) and prompts you to select another data table that you want to merge or join (Trial2.jmp). Select the data table(s) you want to join. (See Figure 2.55.) The column headings of each appear in the Source Columns windows.

Figure 2.55 The Join Launch Dialog Box

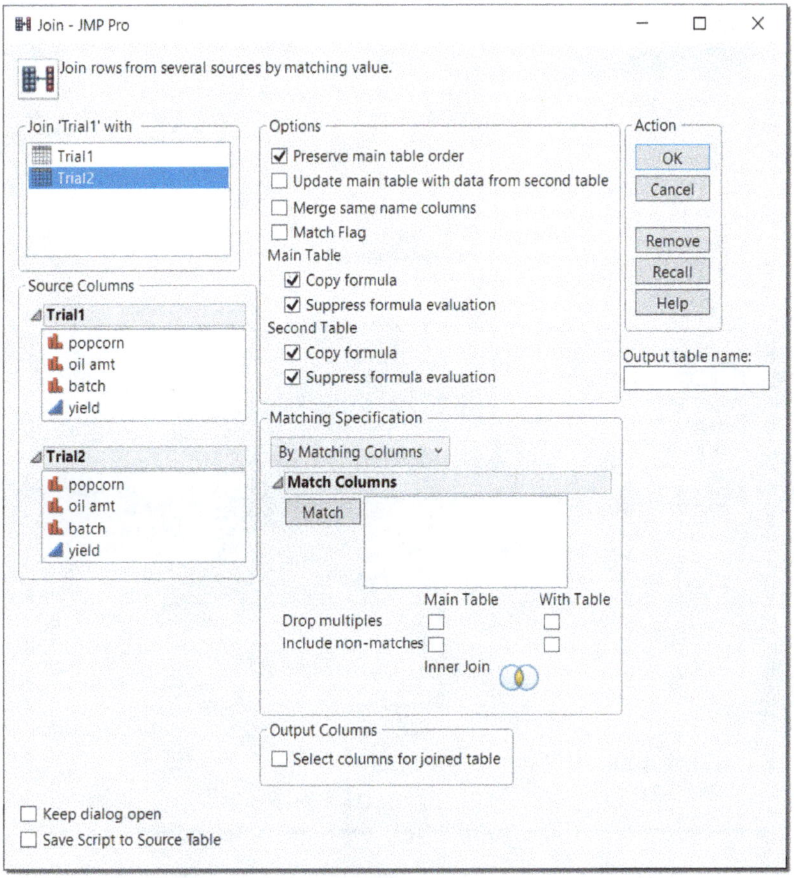

3. Decide how you want to join the data under the **Matching Specification** drop-down menu.

   a. **By Row Number** joins your data side-by-side by its row number.

   b. If your data has different column headings or you want to select specific columns to match, use **By Matching Columns**. Click on a column from each of the **Source Columns** windows that you would like to match and click **Match**. You now see each of those selected columns in the Match columns window with an "=" symbol between them.

4. If you want to name the new data table, enter the name in the **Output** table name box (otherwise, it will be named Untitled), and click **OK**. (See Figure 2.56.) A new data table appears. (See Figure 2.57.)

Figure 2.56 Join by Row Number

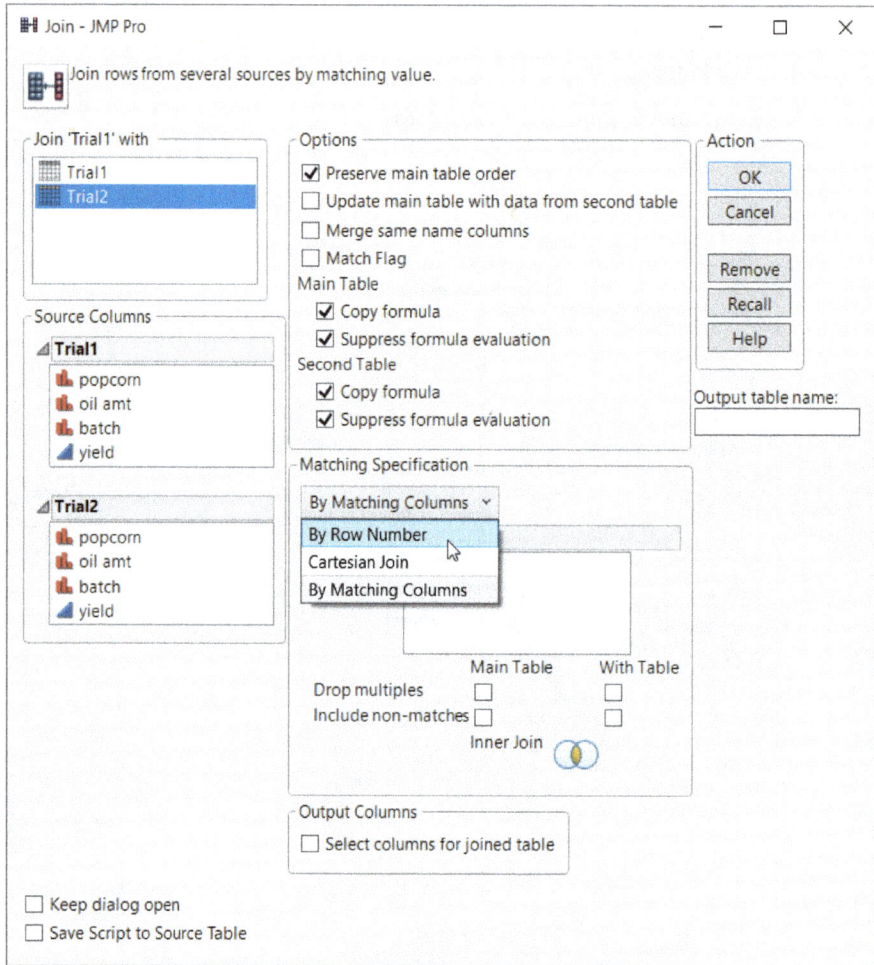

Figure 2.57 The Joined Data Table

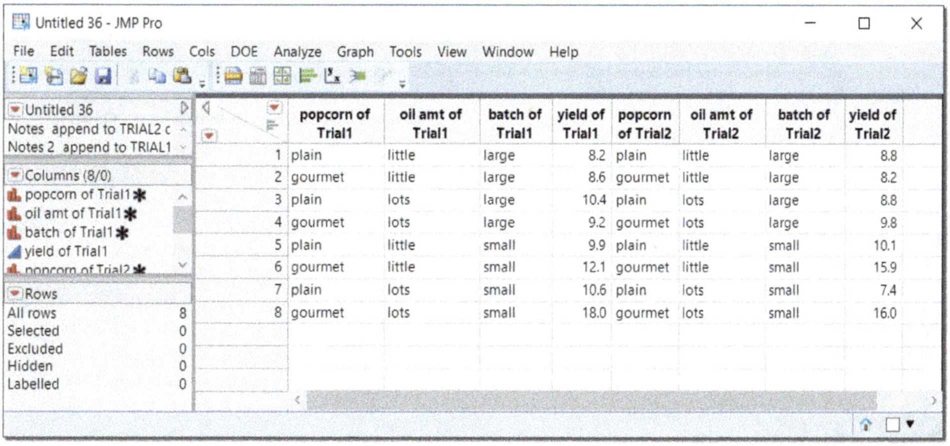

Join by Matching Columns

While joining by row number is very easy to do, it relies on the data in the two data tables being in the same row order with no items missing. This might not always be the case. Examine the data tables **Candy Bar Sizes.jmp** and **Candy Nutrition Information.jmp** shown below in Figure 2.58. These data tables are available at the author's book page. As an analyst, you might want to have all this information for the candy bars in a single data table.

Figure 2.58 Data Tables to Be Joined by Matching Columns

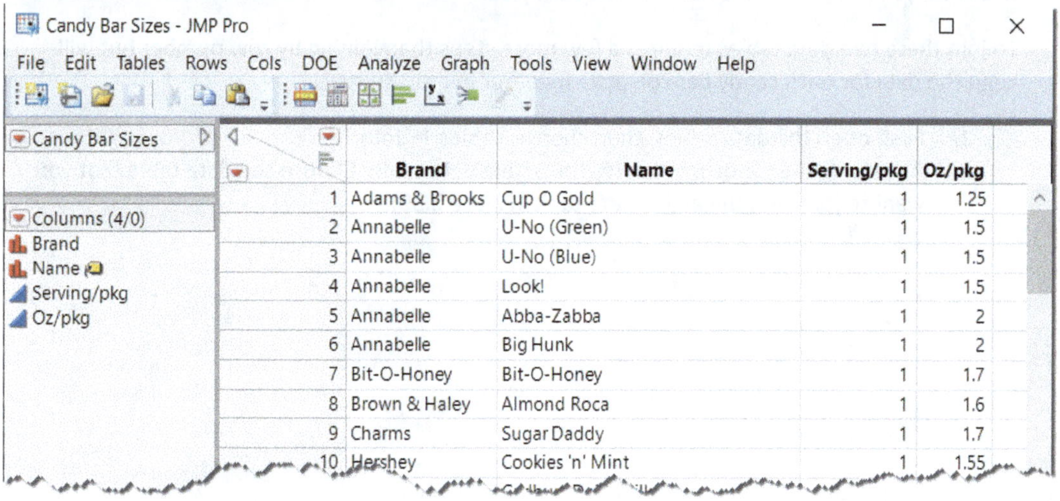

A quick look at these data tables reveals that the data is sorted differently in the two tables. For example, Bit-O-Honey candy bar appears in row 7 in the Candy Bar Sizes data table, but in row 14 in the Candy Nutrition Information data table. Joining these data tables by row number would result in incorrect information. In situations such as these, you can join by Matching Columns. In this case, you can join by the name of the candy bar.

To join these two data tables requires a few more steps than joining by row number but will keep the data for each candy bar correctly together in a single row.

1. First, open the data tables, then choose **Tables ▶ Join**.
2. As before, the window indicates the active data table. Choose the data table that you want to join with the active data table (Figure 2.59).

Figure 2.59 Joining by Matching Columns

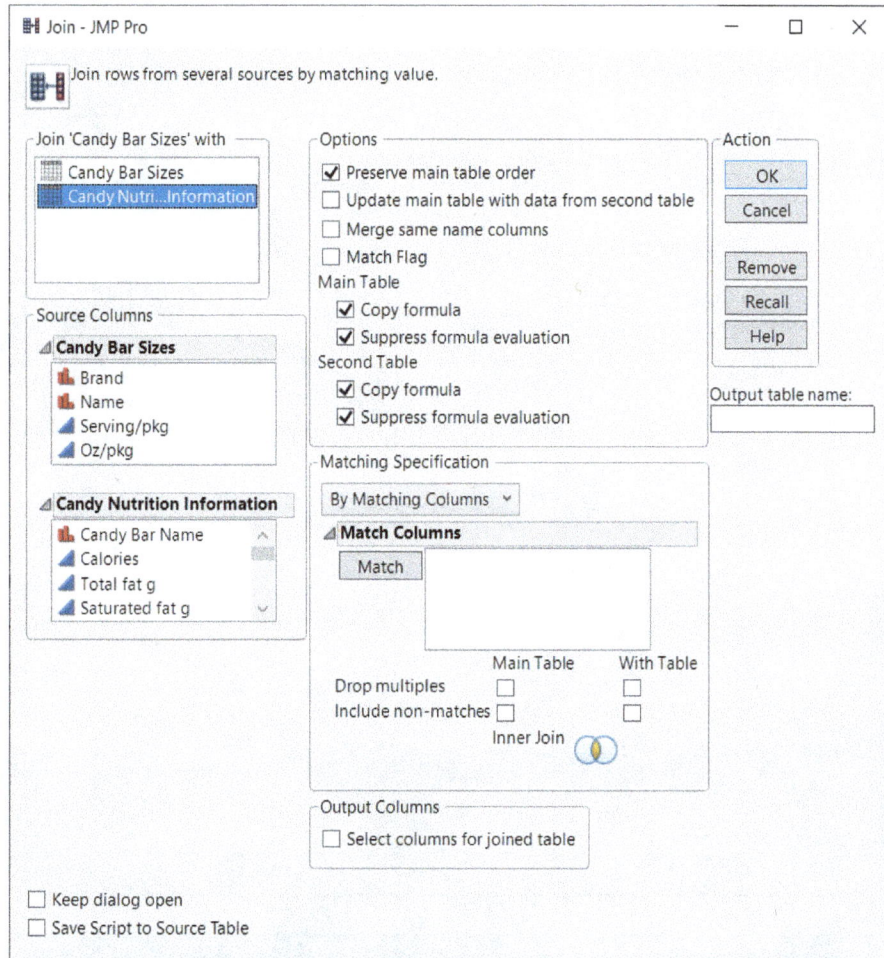

3. Be sure **By Matching Columns** is selected, then choose **Name** from the **Candy Bar Sizes** table columns and **Candy Bar Name** from the **Candy Nutrition Information** table columns.
4. Click the **Match** button. (See Figure 2.60.)

Figure 2.60 Join Dialog Box with Matching Columns Assigned

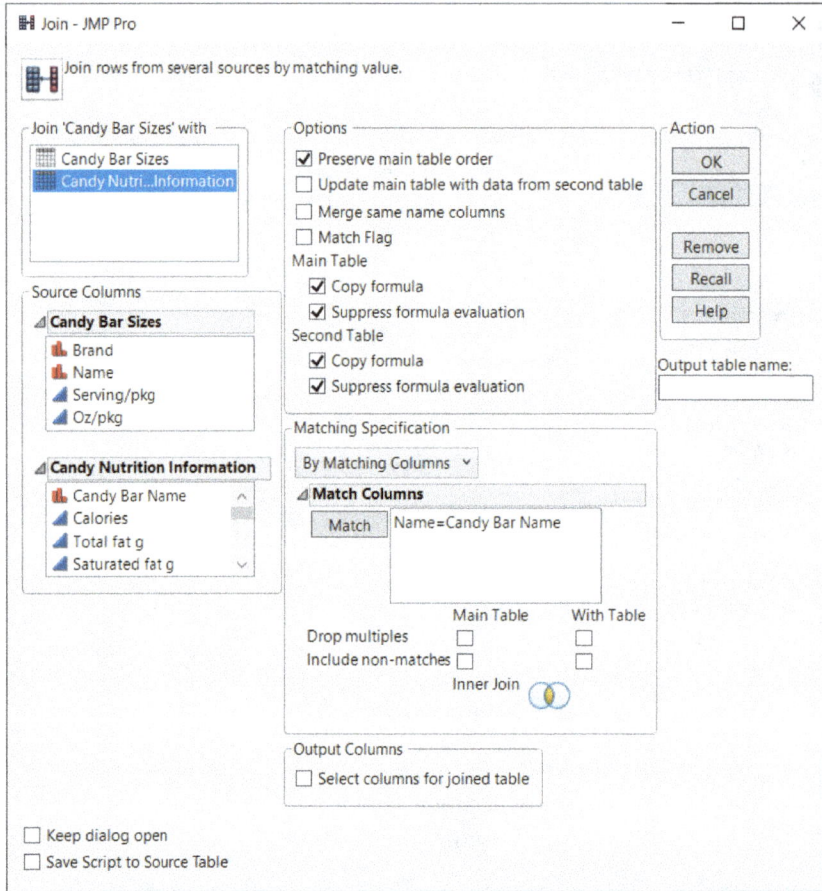

> **Note**
>
> If the two data tables do not contain the same number of candy bars or if there may be multiple rows for the same candy bar, you might need to choose whether to drop multiple rows or include non-matches in the resulting data table.

5. Click **OK**.

The final data table has all of the information for each candy bar. (See Figure 2.61.) You might want to delete the second column of candy bar names in the new data table.

Figure 2.61 Joined Data Table

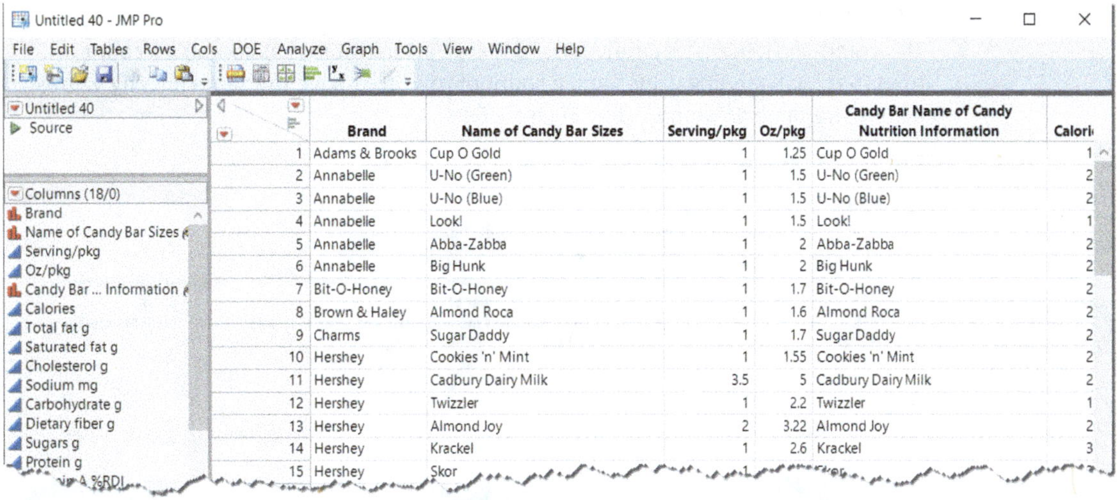

Missing Data

The Missing Data Pattern window can help you identify the quantity of missing data or whether any patterns exist due to non- response, data importing, or data entry errors. The **Missing Data Pattern** feature under the **Tables** menu searches your specified columns and summarizes the frequencies of missing data. To explore this feature, some values from the **Big Class.jmp** data table have been removed.

1. Select **Tables ▶ Missing Data Pattern**. This generates the window on the right. (See Figure 2.62.)

   Figure 2.62 Missing Data Pattern Dialog Box

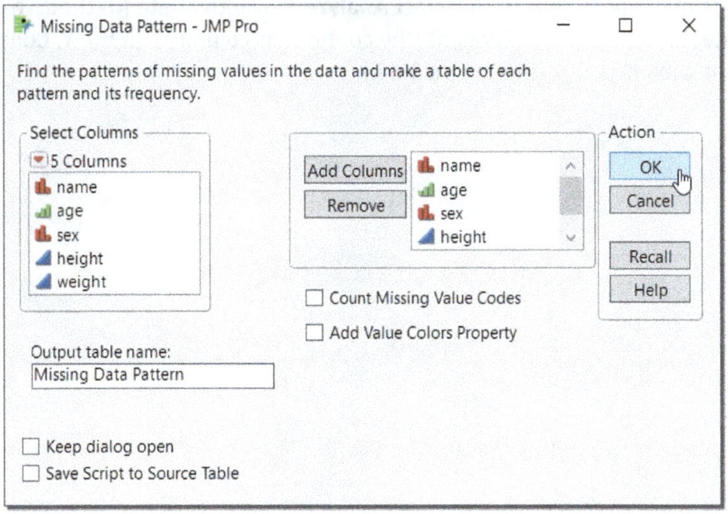

2. Select the columns in the left panel that you want to search. Select **Add Columns** and then click **OK**.
3. This command generates a new Missing Data Pattern table that contains a count of rows that have missing values and a count of rows that have the same missing values among the same column(s). (See Figure 2.63.)

Figure 2.63 Missing Data Pattern Results

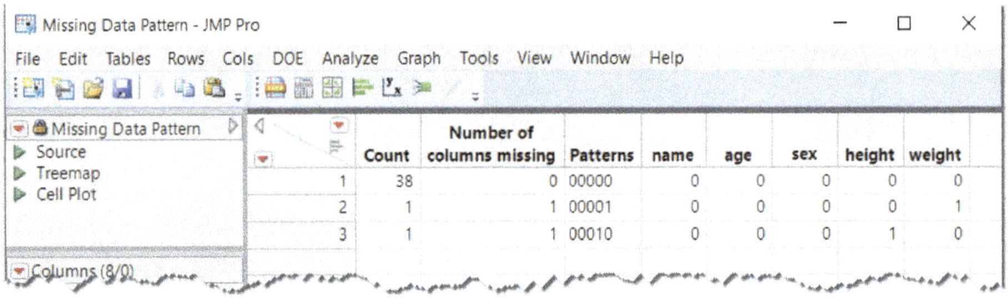

	Count	Number of columns missing	Patterns	name	age	sex	height	weight
1	38	0	00000	0	0	0	0	0
2	1	1	00001	0	0	0	0	1
3	1	1	00010	0	0	0	1	0

Note

You can proceed without addressing missing values, but JMP will by default ignore (or exclude) any rows containing missing values in most analysis platforms. JMP includes "Informative Missing" options in a few platforms that use rows that would otherwise be ignored (JMP Pro has this feature in several platforms). This is important if you have lots of missing data because more usable observations will generally lead to better statistical models.

One solution to missing continuous data is to impute them. *Imputing* analyzes similar values in other columns and rows to estimate the missing value. JMP has an imputation feature under the red triangle within the Multivariate window. To illustrate, some values from the height and weight columns in **Big Class.jmp** have been removed:

1. First, run the multivariate platform. Select **Analyze ▶ Multivariate Methods ▶ Multivariate**. Select **height** and **weight** (the continuous columns), in the **Y, Columns** window, and then click **OK**. (See Figure 2.64.)

Figure 2.64 The Multivariate Platform Offers Imputation

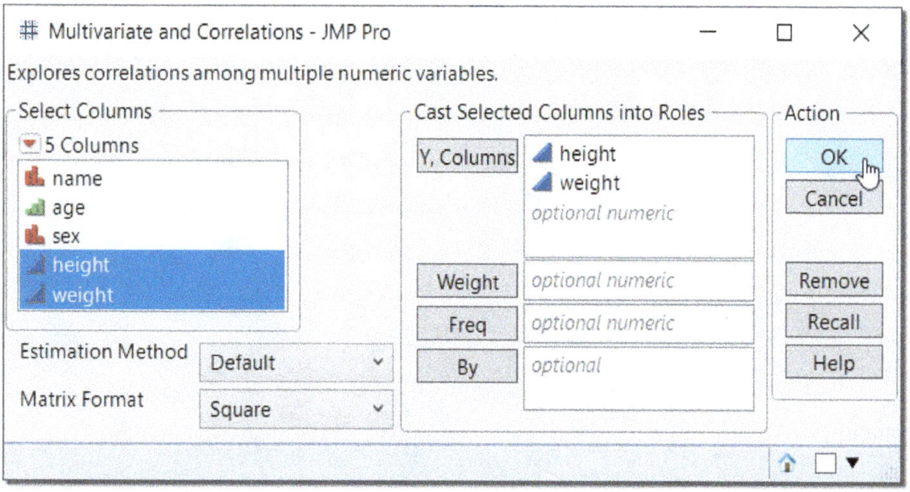

2. Click the red triangle next to **Multivariate** and select **Impute Missing Data**. (See Figure 2.65.) **Note:** Like many menu options in JMP, Impute Missing Data only appears in the menu when appropriate – in this case, when data is missing. A new data table is generated with the estimated missing values in place.

Figure 2.65 Launching the Imputation Option

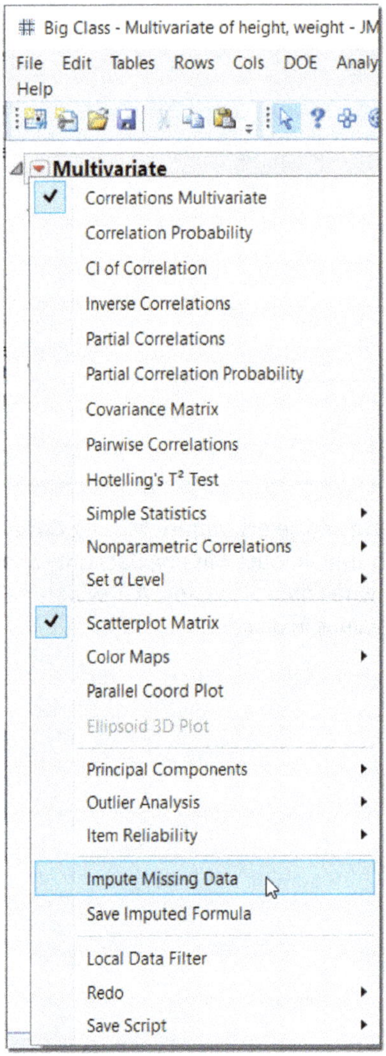

3. Because you can only impute continuous values, cut and paste these columns into your original data table, which might contain other data types. (Alternatively, use **Update** from the **Tables** menu.)

Note

JMP provides several methods of imputing missing values. Additional functionality is available in the Analyze menu under **Analyze ▶ Screening ▶ Explore Missing Values**. Methods of handling missing values are best selected with the help of an expert.

2.9 Summary

In this chapter, we covered a wide range of topics on getting your data into JMP and learning how to manage it. Because the data table not only stores your data but also stores key information that drives the appropriate analysis and graphs, it serves as the critical starting point for all exploration and visualization within JMP.

Analyzing and visualizing data often requires special features, and there are many advanced features in JMP that we did not address. As we have indicated, your copy of JMP includes extensive documentation, which you can access through the **JMP Documentation Library** under the **Help** menu. We recommend the *Using JMP* book for a complete discussion of data, data tables, and the **Tables** menu.

Chapter 3: Index of Graphs

This chapter is a quick reference to commonly used graphs in JMP. The format of this chapter differs from other chapters for good reason. In this chapter, you can peruse the graphs much like you would look through a cookbook – find what you want and follow the recipe. A picture of the graph, brief description, required data conditions, usage description, and the steps required to generate the graph immediately follow. This chapter is not intended to be a complete index of graphs available in JMP, but we have tried to choose those that we see used most often. This chapter is for users who know what graph they want and can select it by how it looks or what it's called.

You will find that many graph windows have additional options that enable you to further enhance your graphical result. However, we have focused on the steps to generate the base case of each graph, which is illustrated in the figure that accompanies each graph.

Some of the graphs illustrated in this chapter are accessed from the **Graphs** menu, while others are accessed from the **Analyze** menu.

Statistical output is provided with graphs generated from the **Analyze** menu. For instructions on sharing or printing graphs and on surfacing graphs into other applications such as Microsoft Word or PowerPoint, interactive HTML, JMP Public, or JMP Live, see Chapter 7.

There is more than one way to generate many of the graphs in this chapter. A selection of graphing methods will be presented including methods available in the more intuitive Graph Builder and Control Chart Builder platforms. You can decide which method works best for you. The preferred method is presented first.

You can customize the appearance of any graph (including colors, markers, axes, legends, and fonts) by interacting with simple palettes and controls or by simply right-clicking on the area or item that you want to change. See Chapter 2, sections 5 and 6, for more details about customizing the appearance of your graphs.

3.1 Basic Charts

The Graph Builder platform produces dozens of graphs for general purposes. The platform responds differently depending on the modeling types of the data and the roles of the columns that you select. In the context of those modeling types, it visually alerts you to charting options that come with your selections. This section is designed to help you understand some of the more commonly used charts that can be produced with this multi-purpose platform and how to produce them.

Because the type of plot generated depends on the columns assigned to the drop zones, a clear understanding of those zones is helpful.

Figure 3.1 Graph Builder Drop Zones

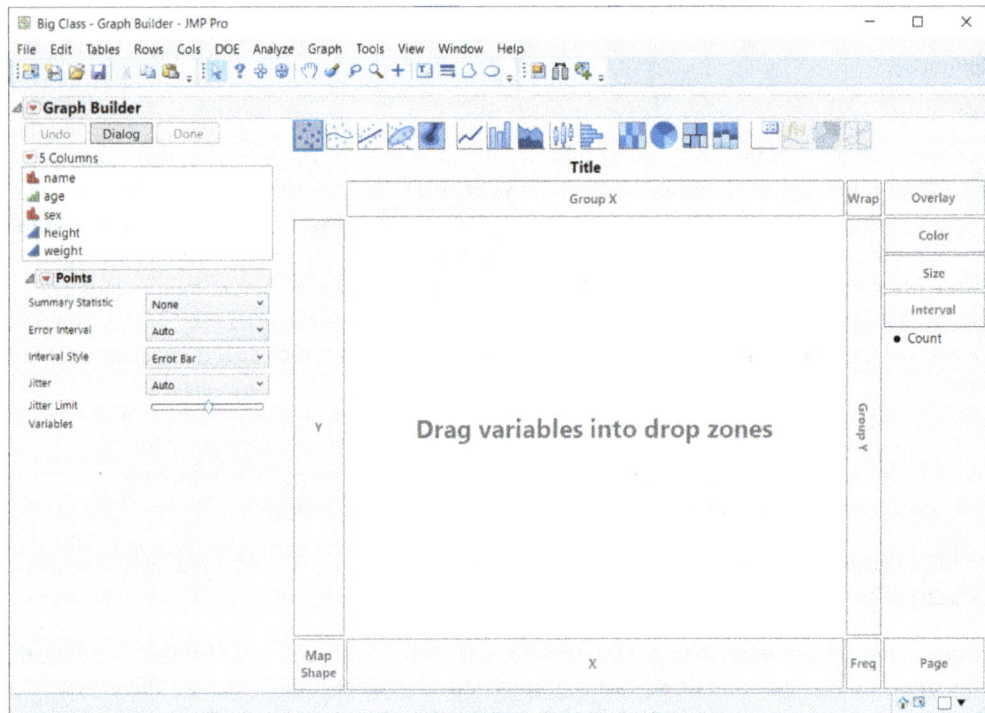

Graph Builder has 12 distinct drop zones (see Figure 3.1):

- **Y**: A column placed in this drop zone will be graphed on the Y axis.

- **Group Y**: A column placed in the drop zone will be used to create separate graphs along the vertical axis. There will be one graph for each level of a nominal or ordinal column. A continuous column is grouped when used in this zone.

- **X**: A column placed in this drop zone will be graphed on the X axis.

- **Group X**: A column placed in the drop zone will be used to create separate graphs along the horizontal axis. There will be one graph for each level of a nominal or ordinal column. A continuous column is grouped when used in this zone.

- **Wrap**: Has the same effect as a Group X or Group Y column except the resulting graphs are presented in a grid or trellis rather than strictly by the horizontal or vertical direction.

- **Freq**: Enables you to assign a frequency column. This is used in instances where the data is summarized. That is, each row in the data table represents more than one observation.

- **Map Shape**: Used to create a map, such as states or counties. This is discussed further in the section on maps.

- **Overlay**: Similar to the Group X or Group Y, a column in this role creates separate graphs for each level of a nominal or ordinal column or for groups of levels for a continuous column. However, the graphs are all placed on the same set of axes with colors used to identify the different values of the overlay column.

- **Color**: The color role is used to color any points on the graph according to the level of a nominal or ordinal variable assigned. If a continuous variable is assigned, a gradient of colors is used.

- **Size**: Used to vary the size of graph elements.

- **Interval**: Columns to be used to determine the numeric value to be used when error bars are drawn on the graph.

- **Page**: Used to specify a nominal or ordinal column whose levels are used to create separate graphs with the same features.

The Graph Builder platform is accessed from the Graph menu. (See Figure 3.2.)

Figure 3.2 Graph Builder Menu

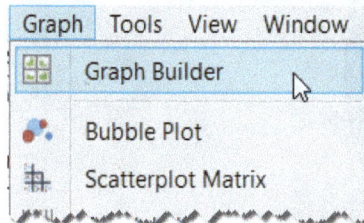

Graphs appear in the center area framed by X and Y drop zones. When columns are selected, dragged, and dropped into zones, graphs instantly appear depending on the data and modeling types of the columns. Additional options become enabled. If column modeling types like continuous and nominal and ordinal are unfamiliar to you, see Section 2.3.

By default, a point chart is generated for any column type when dragged to the X or Y drop zones. Charting choices are revealed in the element icon palette at the top of the Graph Builder window. (See Figure 3.3.)

Figure 3.3 Element Icon Palette

As columns are dragged to drop zones, graphing options become enabled and their icons are highlighted in the palette. Graphing options respond to the data types and modeling types of the columns and the drop zones selected. Unavailable graphing options for the column combinations selected are automatically disabled and appear to be grayed out. You can use the Shift key to apply multiple elements at once. If you are unfamiliar with the graphs depicted in the element type icon palette, experiment by clicking on them when they are enabled. You can also click **Undo** if you don't like the result.

Pie Chart

A *pie chart* is a circular chart divided into areas proportional to the percentages of the whole or total.

Figure 3.4 Pie Chart

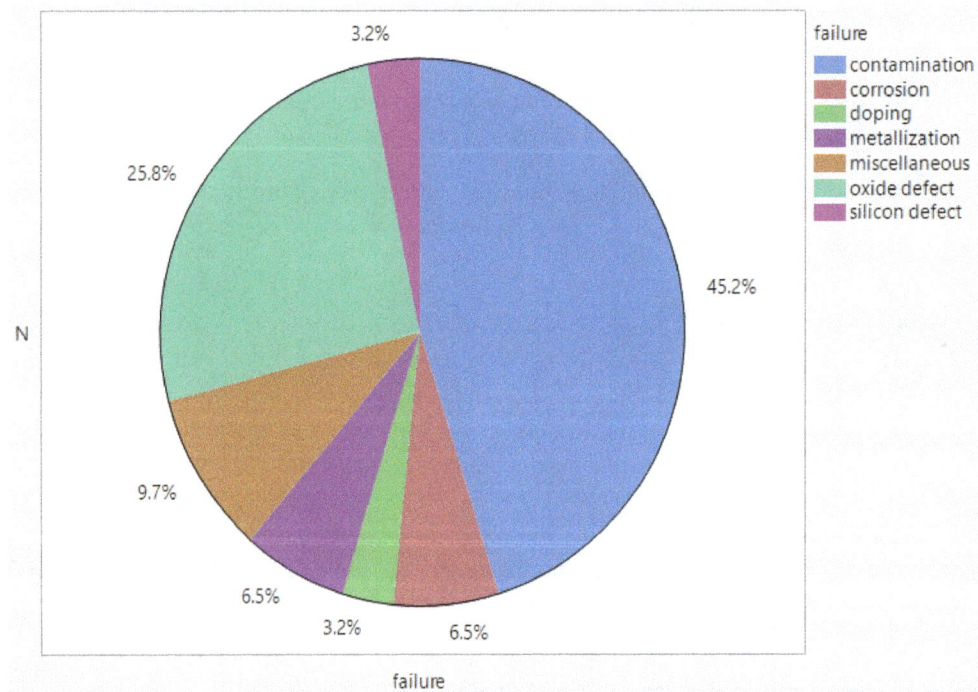

Data Table Access

To access the data table used for the above pie chart example, follow **Help ▶ Sample Data Library ▶ Quality Control ▶ Failure.jmp**.

Usage

Used when representing proportions, percentages, or fractions of any measured quantity. Some examples are market share, customer preferences, and percent of any type of category or group. As shown, pie charts can display percentages of the different failure types. (See Figure 3.4.)

Required

One ordinal or nominal column for labels and a continuous column to define the size of the pie chart sections.

Other column combinations are supported. See **Help ▶ JMP Help**. Type Pie Chart in the search field.

Select **Graph ▶ Graph Builder**. Drag a continuous column with count data to the Y drop zone and drag a nominal variable with the group identifier to the X drop zone. From the element palette, choose the pie element 🥧. To add percentages to the pie slices as depicted in Figure 3.4, choose **Label by Percent of Total Values** from the **Control Panel** on the left.

Dot Plot

While the Pie Chart might be considered the simplest graph to depict the values of categorical data, a *Dot Plot* is often used to depict the values of continuous data when the data set is fairly small. As with the pie chart, a dot plot is used to visually see the distribution of the data.

Figure 3.5 Dot Plot

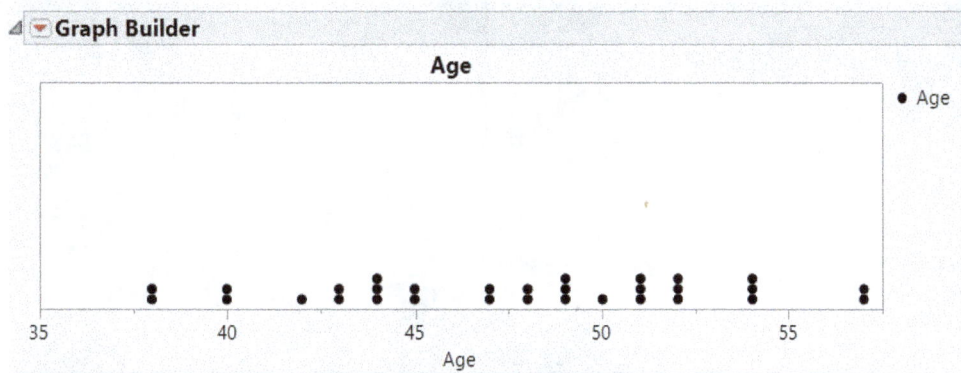

In a dot plot, data values are displayed along the axis. Each dot represents a single data point at the point's value along the axis. From the graph shown in Figure 3.5, you can see that the youngest people in this data table are two individuals who are 38 years old. The oldest are 57 years old. No more than 3 people are the same age as one another.

Data Table Access

To access the data table used for the above dot plot example, follow **Help ▶ Sample Data Library ▶ Fitness.jmp**.

Usage

To view the properties of a continuous distribution such as shape and range.

Required

A numeric column for either the X or Y role. In the graph in Figure 3.5, the column is placed in the X role yielding a dot plot that goes from low to high as you move from right to left. If the column were placed in the Y role, the plot would go from low to high as you move from bottom to top.

To generate the graph shown, select **Graph ▶ Graph Builder**. Drag **Age** to the **X** drop zone at the bottom of the graphing area. In the Control Panel, change the Jitter setting to **Positive Grid.** (See Figure 3.6.) In the plot shown, the Control Panel is hidden by clicking on the **Done** button, and the graph was resized by clicking on and dragging the bottom right corner of the plot.

Figure 3.6 Jitter Setting for Dot Plot

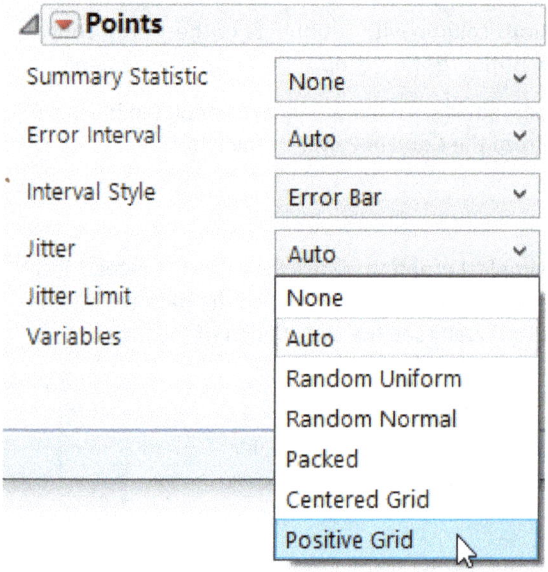

◢ ▼ Points		
Summary Statistic	None	⌄
Error Interval	Auto	⌄
Interval Style	Error Bar	⌄
Jitter	Auto	⌄
Jitter Limit	None	
Variables	Auto	
	Random Uniform	
	Random Normal	
	Packed	
	Centered Grid	
	Positive Grid	

Bar Chart, Line Chart, and Point (or Scatter) Chart

These charts use bars, lines, and points to show lengths or positions proportional to quantities.

Figure 3.7a Bar Chart with Values

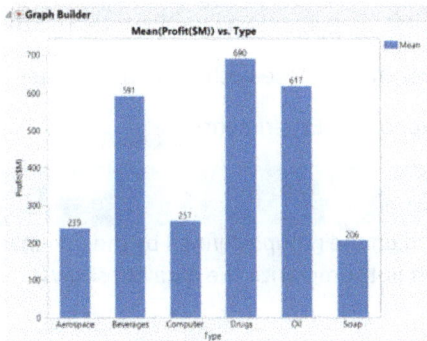

Figure 3.7b Line Chart

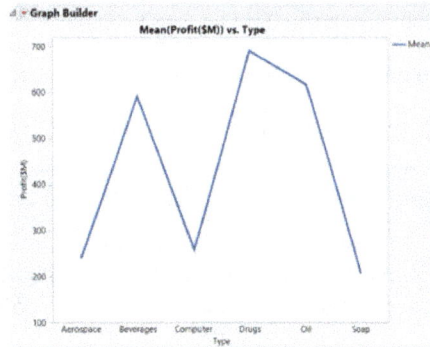

Figure 3.7c Point Chart

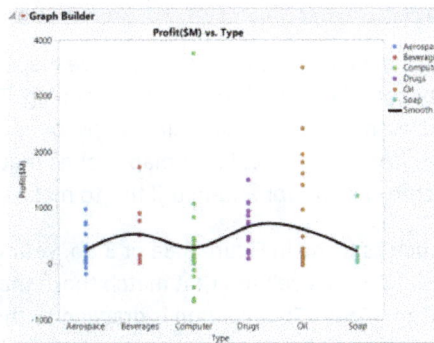

Data Table Access

> To access the data table used for the above chart examples, follow **Help ▶ Sample Data ▶ Business and Demographic ▶ Financial.jmp**.

Usage

> Similar to pie charts representing individual values, proportions, percentages, or depictions of any measured quantity. Some examples are market share proportions and customer preferences expressed as percentages of any type of category or group. As shown, these charts display profits for six company types in three different chart styles. (See Figures 3.7a, 3.7b and 3.7c.)

Required

> One continuous column for the Y drop zone and one nominal or ordinal column for the X drop zone.

Select **Graph ▶ Graph Builder**. Then drag a continuous column to the Y drop zone and drag a nominal or ordinal column to X drop zone. For a bar chart, choose the bar element ▦ . To add value totals to the bars as depicted, choose **Label ▶ Value** from the Bar Element Properties Panel on the left. Additional elements like Summary Statistic, Error Bars, and labeling are available from the Bar Element Properties Panel on the left. To generate a line chart, choose the line element ◠ . To generate a scatter chart with a smoother, choose both elements (scatter chart and smoother) ⬚⬚ . Many other charts are supported. Experiment!

> **Note**
>
> The bar chart and line chart summarize your data based on the groups defined by the nominal column placed in the X drop zone. The point chart does not summarize the data but shows each row in the data table as a separate point.

3.2 Thematic Maps

The Graph Builder platform can be used to create interactive maps with boundaries such as states, provinces, or county boundaries and street-level mapping. These mapping tools are included in JMP and stored as shape files, background maps, or as a link to a street map server. Other sources of shape files (for example, ESRI) or map shapes that you create yourself can be used in JMP. See the JMP.com website for add-in utilities to make your own map shapes.

To create thematic maps such as those in Figure 3.8a or 3.8b, your data needs to contain boundary names or abbreviations in a column that match those that appear in the shape file, for example, "California", "CA" or "Calif". That column is dragged to the **Map Shape** drop zone. (See Figure 3.8d.) If you have latitude and longitude columns, you can plot points on a background map as those in Figure 3.8c. (See Section 2.7 for more information about these file types.)

Figure 3.8a SAT Math Score Map

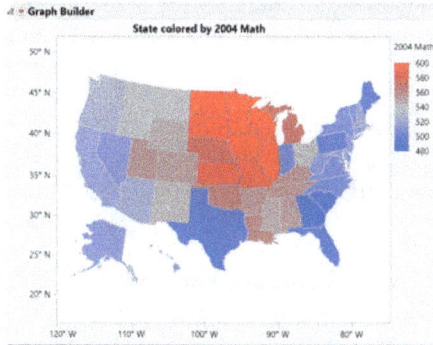

Figure 3.8b SAT Math Score Map with Percent Taking Overlay

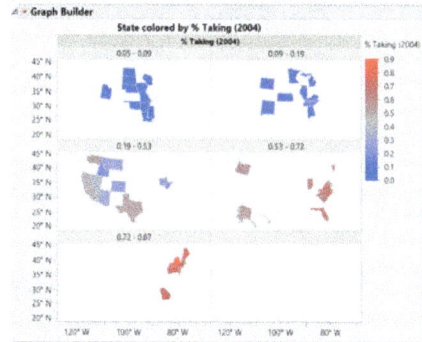

Figure 3.8c San Francisco Crime Map

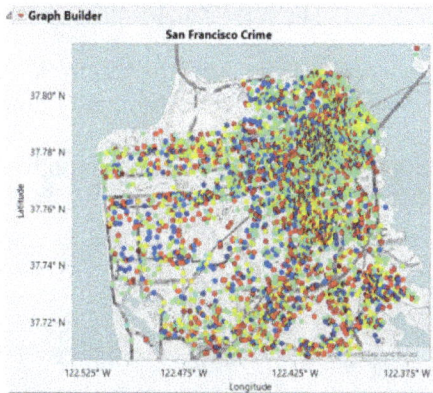

Figure 3.8d Graph Builder showing Map Shape Drop Zone

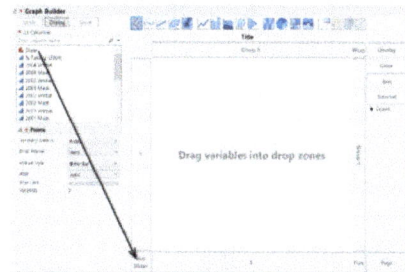

Data Table Access

To access the data table that uses a map shape file, select **Help ▶ Sample Data Library ▶ SAT.jmp**.

To access the data table that uses latitude and longitude, select **Help ▶ Sample Data Library ▶ San Francisco Crime.jmp**.

Usage

A method to visualize data in a spatial system. Some examples are socio-economic indicators overlaid on political boundaries, crime incidents overlaid on street maps, or where boundaries can be defined in two dimensions.

Required for thematic maps (with a shape file)

One column recognized as a map shape to provide map boundaries and one continuous column to provide a color range.

Optional

A continuous, nominal, or ordinal column for the additional visualizations. See Appendix B for descriptions of these terms. As shown in Figures 3.8a and 3.8b, the maps display US State boundaries for 2004 SAT Scores for Math and percentage of students taking the SAT in that year.

Select **Graph ▶ Graph Builder**. Then, drag a map shape column from the **Variables** list (for example, State) to the map shape drop zone (Figure 3.8d). The states can be whole names or two-letter abbreviations. Drag a continuous column (for example, 2004 Math) to the color drop zone.

Required for plotting points on a background map (with latitude and longitude)

One column with latitude coordinates in the **Y** drop zone and one column with longitude coordinates in the **X** drop zone. **Optional:** A continuous, nominal, or ordinal column for the additional visualizations. As shown in Figure 3.8c, San Francisco crime incidents are displayed by type of crime overlaid upon a street level map.

Select **Graph ▶ Graph Builder**. Then, drag the longitude column from the **Variables** list to the **X** drop zone. Then, drag the latitude column from the **Variables** list to the **Y** drop zone. Assuming you have an internet connection, right-click in the map and select **Graph ▶ Background Map ▶ Street Map Service ▶ OK**. Finally, remove the smoother from the graph.

3.3 Control Charts, Pareto, Variability, and Overlay Plots

This section groups typical graphs used to measure product or service quality. The graphs are common in quality improvement scenarios.

Control charts are graphical and analytic tools for deciding whether a process is in a state of statistical control and for monitoring an in-control process. Control charts help determine whether variations in measurement of a product are caused by small, normal variations that cannot be controlled or by some larger, special cause that can be controlled. The type of chart to use is based on the nature of the data.

Control charts are broadly classified into control charts for variables and control charts for attributes. Control charts for variables and control charts for attributes come in several varieties with names or letters attached to them. Control charts can be generated from the **Control Chart Builder** under **Analyze ▶ Quality and Process ▶ Control Chart Builder** or **Control Chart.** (See Figure 3.9.) This chapter focuses on use of the Control Chart menu item, which enables you to specify in advance the type of control chart you want to create. In Chapter 4, we will examine the Control Chart Builder menu choice, which provides a more interactive way to explore your data.

Figure 3.9 Quality and Process Control Chart Builder and Control Chart

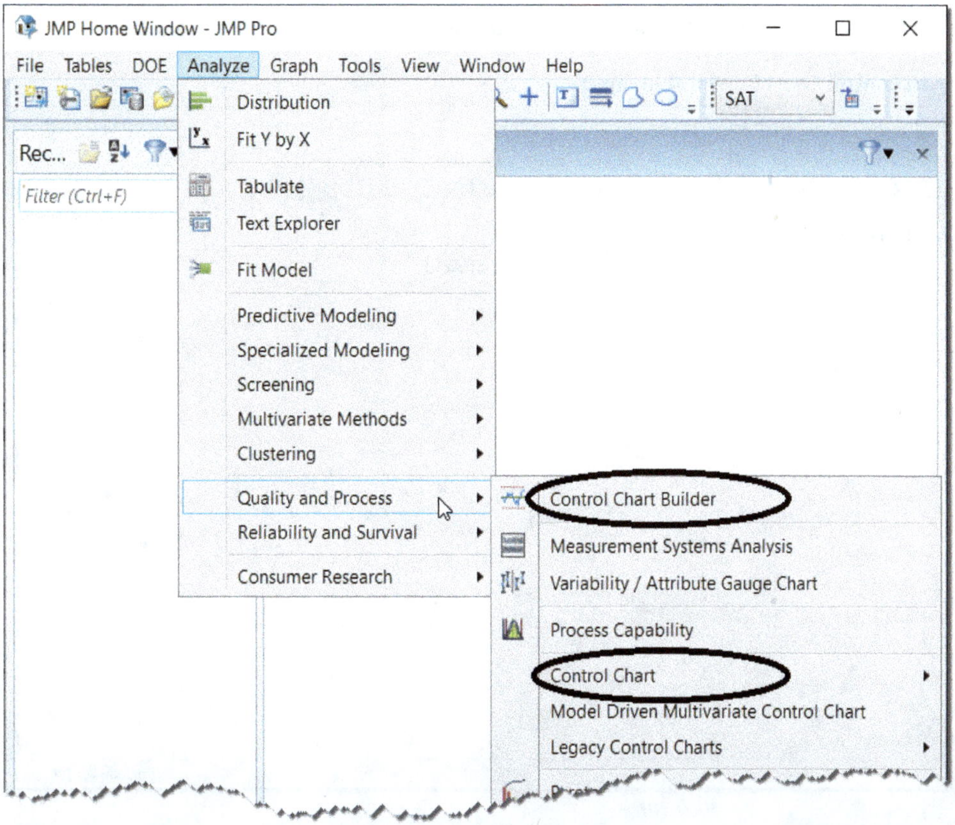

Note

In a few of the graph instructions that follow, we use the term "numeric," which refers to the nature of the data. This term is used in place of modeling type (continuous, nominal, ordinal).

Run Chart

A *run chart* displays a column of data as a connected series of points. (See Figure 3.10.)

Figure 3.10 Control Chart Builder Run Chart

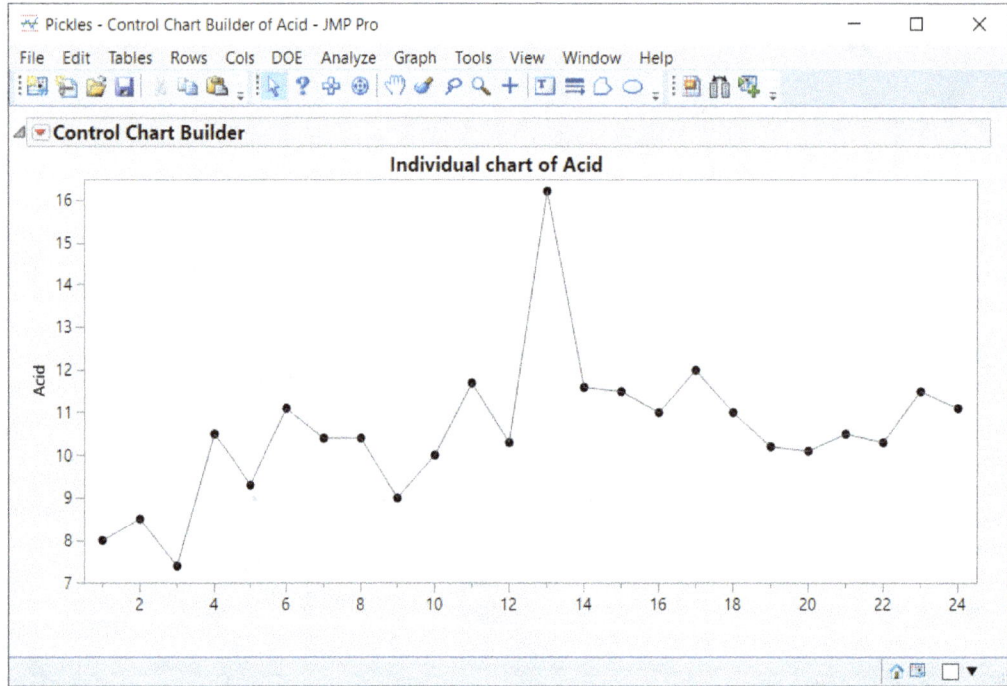

Data Table Access

To access the data table used for this example, follow **Help ▶ Sample Data Library ▶ Quality Control ▶ Pickles.jmp**.

Usage

Displays data from a column. Frequently used as a first visualization of quality data and to assess ranges of variability. Examples include delay times, process yield, or any other continuous measurement, generally over time. As shown, the run chart displays Acid measurements over 24 Pickle Vats.

Required

One or more continuous columns for the **Y** role.

Optional

Nominal, ordinal, or continuous columns for the **By** role.

If data in a column is sorted by ascending values of time, then the X axis will be displayed in the time-sorted order.

Select **Analyze ▶ Quality and Process ▶ Control Chart ▶ Run Chart**. Assign a continuous column to the **Y** role.

Individual & Moving Range Chart

A process behavior chart that displays individual measurements, an *individual measurements chart* displays a single measurement with each point. This type of chart is appropriate when there is no logical subgrouping of the data values collected. That is, only one measurement is available for each sample.

The accompanying *moving range chart* displays moving ranges of two or more successive measurements. Moving ranges are computed using the number of consecutive measurements entered in the Range Span box. The default range span is 2.

Figure 3.11 Control Chart Builder IR Chart with Phases

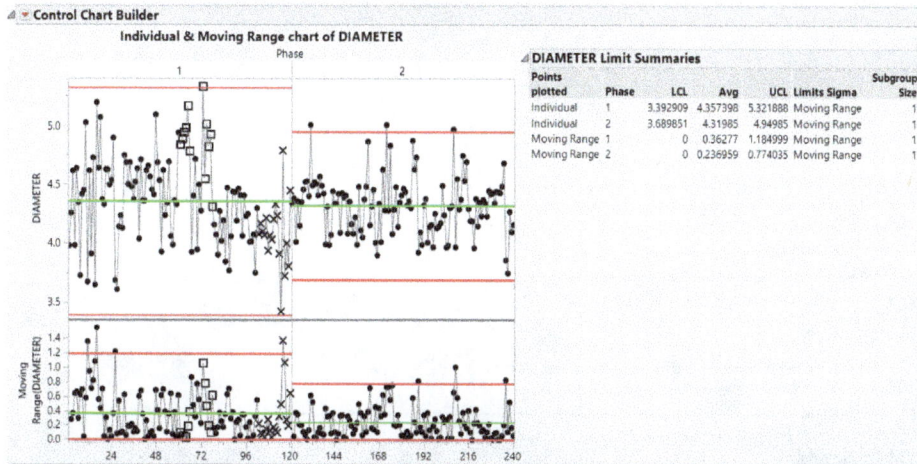

Data Table Access

> To access the data table used for the above chart examples, follow **Help ▶ Sample Data Library ▶ Quality Control ▶ Diameter.jmp**.

Usage

> Used when only one measurement is available for each subgroup. An example would include measurements of the diameter of one part from each of various lots. Individual & moving range charts are efficient at detecting relatively large shifts in the process average. As shown, the charts display the contrasting control limits and process performance before and after a quality improvement as phases 1 and 2. (See Figure 3.11.)

Required

> One or more numeric columns for the **Process** or **Y** role.

Optional

> Continuous, nominal, or ordinal columns for the Sample Label, Phase, and By roles.

Select **Analyze ▶ Quality and Process ▶ Control Chart ▶I/MR Control Chart**. Assign a continuous column to the **Y** role. To add phases as shown, assign a nominal or ordinal column to the **Phase** role (see Figure 3.11)

XBar & R Chart

An *XBar and R chart* displays quality characteristics measured on a continuous scale. A typical analysis shows both the process mean and its variability, aligned above a corresponding range or standard deviation chart, respectively.

Figure 3.12 Control Chart Builder XBar and R Chart

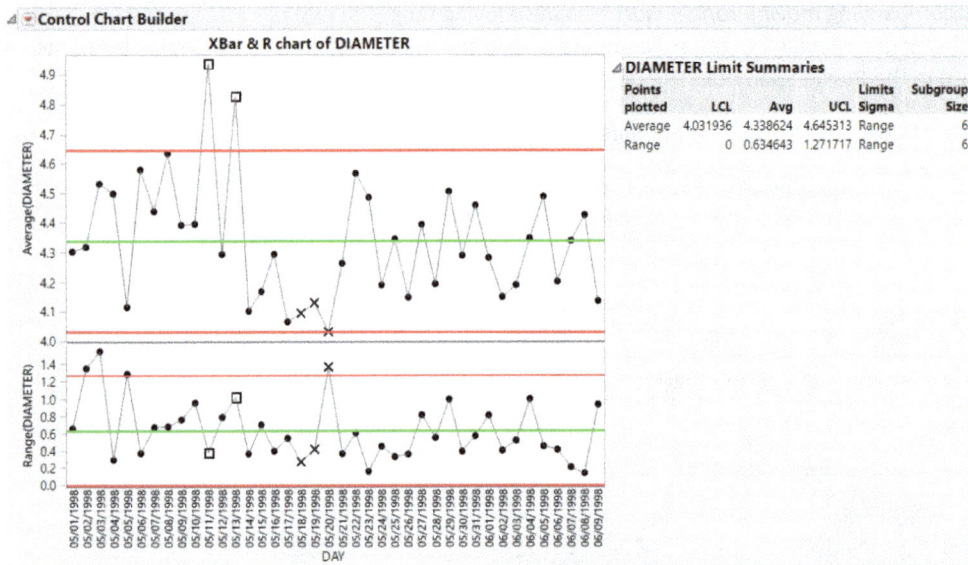

Data Table Access

To access the data table used for these chart examples, follow **Help ▶ Sample Data Library ▶ Quality Control ▶ Diameter.jmp**.

Usage

Normally used for numeric data that is recorded in subgroups in some logical manner (for example, three production parts measured every hour). A special cause, such as a broken tool, will then appear as an abnormal pattern of points on the chart. As shown, the chart displays several sample mean diameters outside of the control limits. (See Figure 3.12.)

Required

One or more numeric columns for the **Process** (or **Y**) role.

Optional

Nominal, ordinal, or continuous columns for the **Sample Label** and **By** roles.

Select **Analyze ▶ Quality and Process ▶ Control Chart ▶ XBar Control Chart**. Assign a continuous column to the **Y** role. Assign a column to the **Subgroup** (or **X**) role. (See Figure 3.12.)

P Chart

A *P chart* is an attribute chart that displays the proportion of nonconforming (defective) items in subgroup samples, which can vary in size. Because each subgroup for a P chart consists of N items, and an item is judged as either conforming or nonconforming, the maximum number of nonconforming items in a subgroup is N. The binomial distribution is used to calculate P chart control limits.

Figure 3.13 Control Chart Builder P Chart of Defective

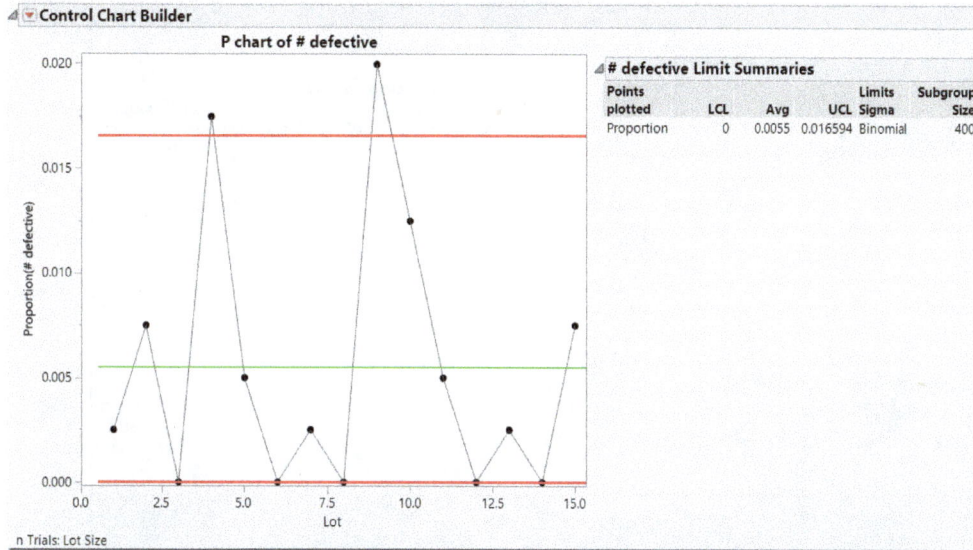

Data Table Access

To access the data table used for these chart examples, follow **Help ▶ Sample Data ▶ Control Charts ▶ Washers.jmp**.

Usage

Used when each sample is assessed as good or bad, that is, it passes or fails. The proportion is then the number of good items divided by the total number of items in the sample, where the sample sizes can vary. As shown, the chart displays the proportion of defective washers across many subgroups of washers within and outside the control limits. (See Figure 3.13.)

Required

One or more numeric columns for the **Process** (or **Y**) role. A constant or variable sample size can be specified and must be numeric.

Optional

Continuous, nominal, or ordinal columns for the **Subgroup**, **Phase**, and **By** roles.

Select **Analyze ▶ Quality and Process ▶ Control Chart ▶ P Control Chart**. Assign a continuous, numeric column to the Y role and, optionally, a numeric column to the n Trials role.

NP Chart

An *NP chart* is an attribute chart that displays the number of nonconforming (defective) items in fixed-sized subgroup samples. Because each subgroup for an NP chart consists of N_i items, and an item is judged as either conforming or nonconforming, the maximum number of nonconforming items in subgroup i is N_i. The binomial distribution is used to calculate the control limits of an NP Chart.

Figure 3.14 Control Chart Builder NP Chart of Defective

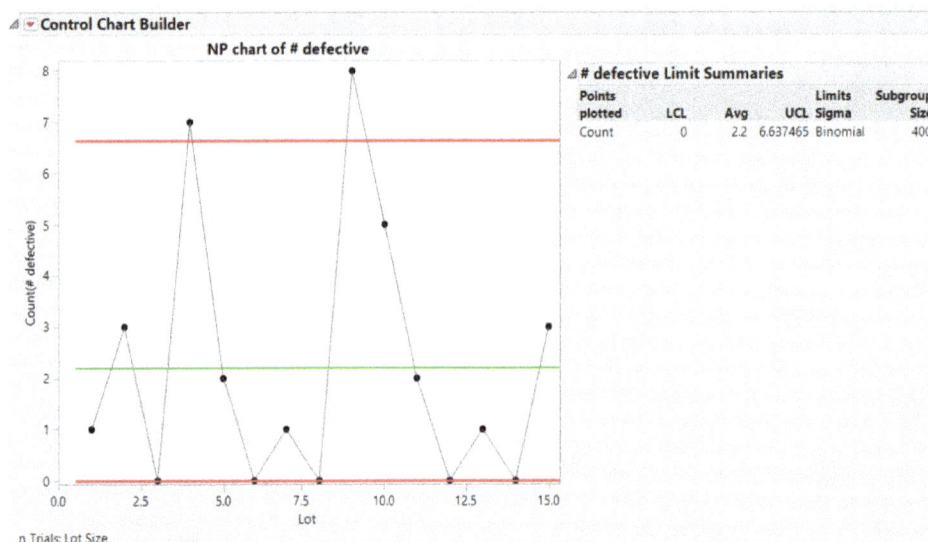

Data Table Access

To access the data table used for the above chart examples, follow **Help ▶ Sample Data Library ▶ Quality Control ▶ Washers.jmp**.

Usage

A fixed sample is taken from an established number of transactions or manufactured items each month. From this sample, the number of transactions or items that had one or more errors is counted. The control chart then tracks the number of items with errors per group or lot. As shown, the chart displays the number of defective washers across many lots of washers within and outside the control limits. (See Figure 3.14.)

Required

One or more numeric columns for the **Y** (or **Process**) role. A numeric Sample Size must be specified for each subgroup in the **n Trials** role.

Optional

Continuous, nominal, or ordinal columns for the **Subgroup**, **Phase**, and **By** roles.

Select **Analyze ▶ Quality and Process ▶ Control Chart ▶ NP Control Chart**. Assign a continuous column to the **Y** role and a column representing lot size to the **n Trials** role. Optionally, assign a column representing subgroups to the **Subgroup** role.

C Chart

A *C chart* is an attribute chart that displays the number of nonconformities (defects) in a subgroup. The Poisson distribution is used to calculate the control limits in a C chart.

Figure 3.15 Control Chart Builder C Chart of Flaws

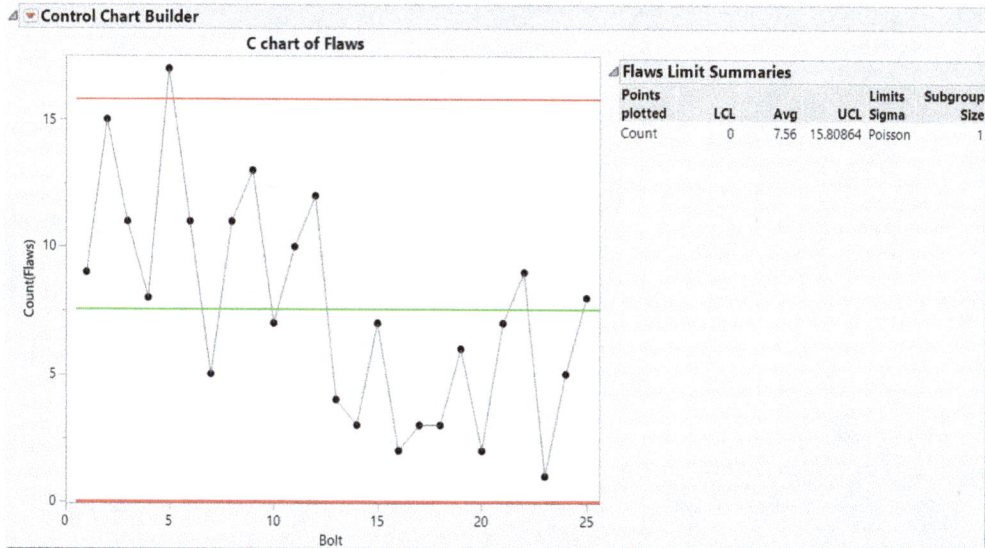

Data Table Access

> To access the data table used for the above chart examples, follow: **Help ▶ Sample Data Library ▶ Quality Control ▶ Fabric.jmp**.

Usage

> Used when one or more errors might propagate within the same sample, such as flaws on a DVD. As shown, the chart displays the number of flaws in each bolt of fabric within and outside the control limits. (See Figure 3.15.)

Required

> One or more numeric columns for the **Y** (or **Process**) role.

Optional

> Continuous, nominal, or ordinal columns for the **Subgroup**, **Phase**, and **By** roles. A constant or variable sample size can be specified in the **n Trials** role and must be numeric.

Select **Analyze ▶ Quality and Process ▶ Control Chart ▶ C Control Chart**. Assign a continuous column representing defects to the **Y** role. Assign a column representing the subgroups to the **Subgroup** role.

U Chart

A *U chart* is an attribute chart that displays the number of nonconformities (defects) per unit in subgroup samples that can have a varying number of inspection units. The Poisson distribution is used to calculate the control limits in a U chart.

Figure 3.16 Control Chart Builder U Chart of Defects

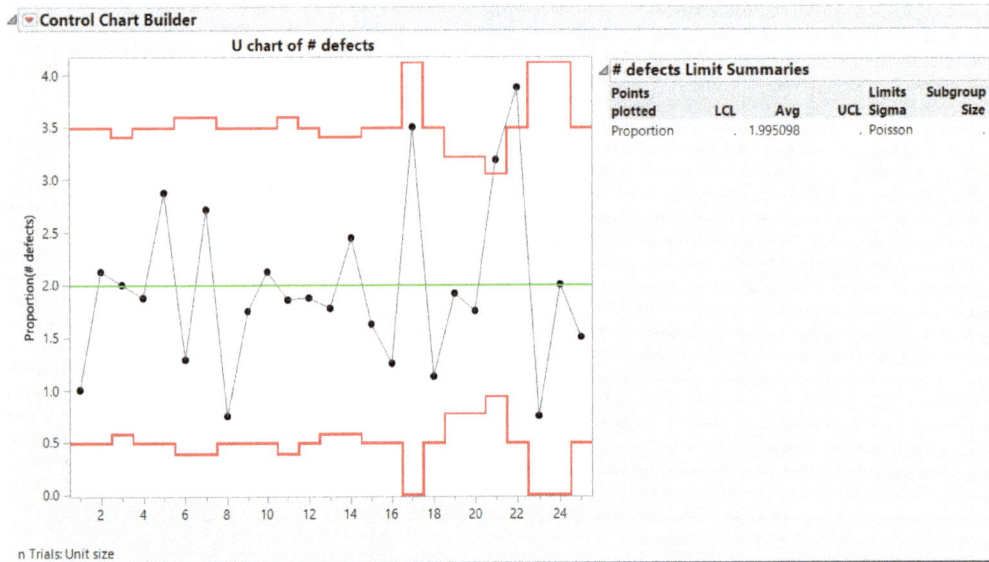

Data Table Access

> To access the data table used for the above chart examples, follow **Help ▶ Sample Data Library ▶ Quality Control ▶ Braces.jmp**.

Usage

> To count the number of defects in subgroups of varying numbers. As shown, the chart counts the number of defects on braces in groups of braces of varying size on specific dates and indicates whether the count is within or outside of the control limits. (See Figure 3.16.)

Required

> One or more numeric columns for the **Y** (or **Process**) role. The column assigned to the **n Trials** role must be numeric.

Optional

> Continuous, nominal, or ordinal columns for the **Subgroup**, **Phase**, and **By** roles.

Select **Analyze ▶ Quality and Process ▶ Control Chart ▶ U Control Chart**. Assign a continuous column representing defects to the **Y** role and a continuous column representing the number of units to the **n Trials** role.

Variability Chart

A *variability chart* illustrates how numeric values vary across a categories or subgroups. Along with the data, you can view the mean, range, and standard deviation of the data in each category. The analysis options assume that the primary interest is in how the mean, range, and variance change across the categories. One use of this graph is in assessing the variability caused by the measurement system itself. That is, how much of the variability is in the product and how much is caused by the way we measure it (for example, the instruments or operators).

Figure 3.17a Graph Builder Variability Chart of Diameter

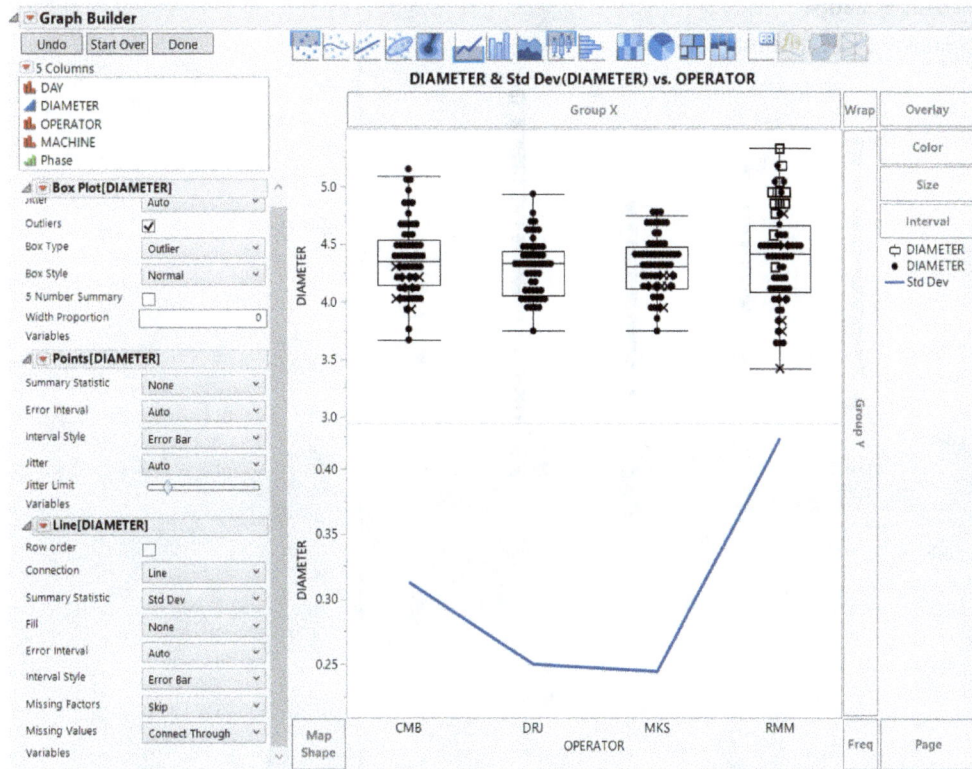

Data Table Access

> To access the data table used for these chart examples, follow **Help ▶ Sample Data Library ▶ Quality Control ▶ Diameter.jmp**.

Usage

> For viewing the ranges, standard deviation, and means of a measured column across groups and subgroups. As shown, part diameter variability is displayed across different operators. (See Figure 3.17a).

Required

> At least one numeric column for the **Y, Response** role and at least one nominal, ordinal, or continuous column for the **X, Grouping** role.

Using the Graph Builder method, select **Graph ▶ Graph Builder.** Drag a continuous column to the **Y** drop zone. Drag one or more nominal or ordinal columns to the **X** drop zone for groups or subgroups. Use selections on the element palette and line and point controls to modify the graph.

To produce a variability chart for a variability study, select **Analyze ▶ Quality and Process ▶ Variability / Attribute Gauge Chart.** (See Figure 3.17b.)

Figure 3.17b Classic Variability Chart

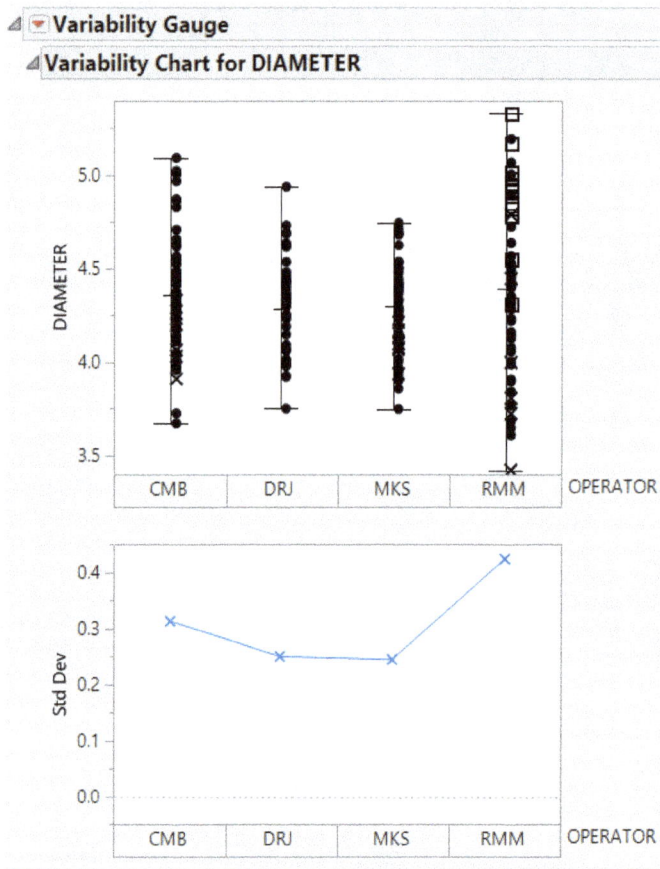

Select one column for the **Y, Response** role and one or more columns for the **X, Grouping** role. Multiple nominal columns for the **X, Grouping** role produce horizontally nested results by the subgroup, overlaid. Click the **Help** button for many additional options.

Overlay Plots

Like the variability chart presented earlier, the *overlay plot* enables you to visualize values over any specified times or groups. A key difference, however, is that this plot enables you to specify multiple Y columns and group values in a meaningful way. An example of an overlay plot with more than one Y value and more than one X value is shown for reference here. (See Figure 3.18a.)

Figure 3.18a Overlay Plot Example

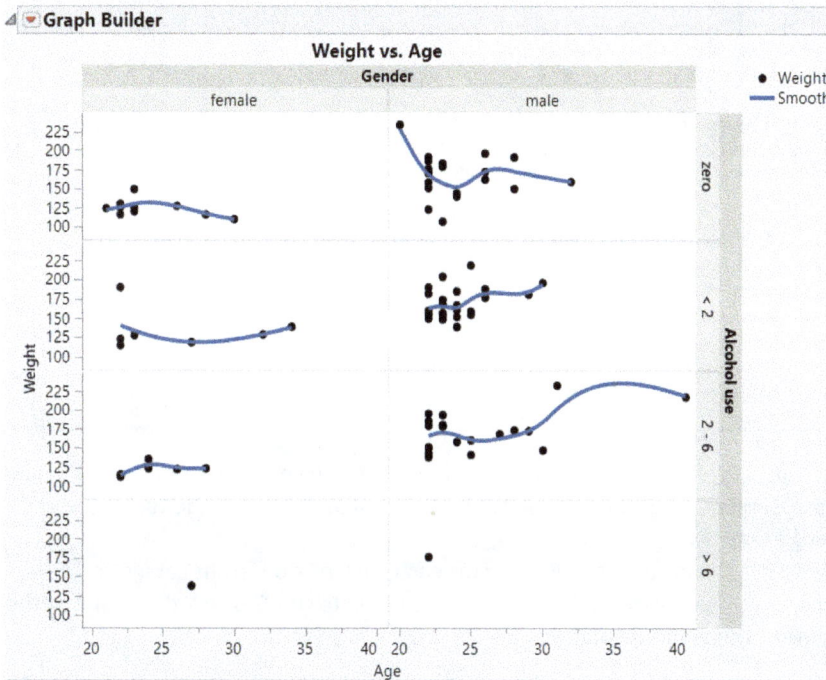

Data Table Access

To access the data table used for Figure 3.18a, follow **Help ▶ Sample Data Library ▶ Lipid Data.jmp**.

The overlay plot pictured is a scatter plot of weight by age, with separate plots for each Gender and Alcohol Use combination. To generate an overlay plot like the one pictured:

1. Select **Graph ▶ Graph Builder**.
2. Drag a continuous column of interest, in this case **Weight**, to the **Y** drop zone on the left side of the graph area.
3. Drag a second continuous column of interest, in this case **Age**, to the **X** drop zone at the bottom of the graph area. You now have a scatter plot of Weight by Age as shown in Figure 3.18b.

Figure 3.18b Scatter Plot of Weight by Age

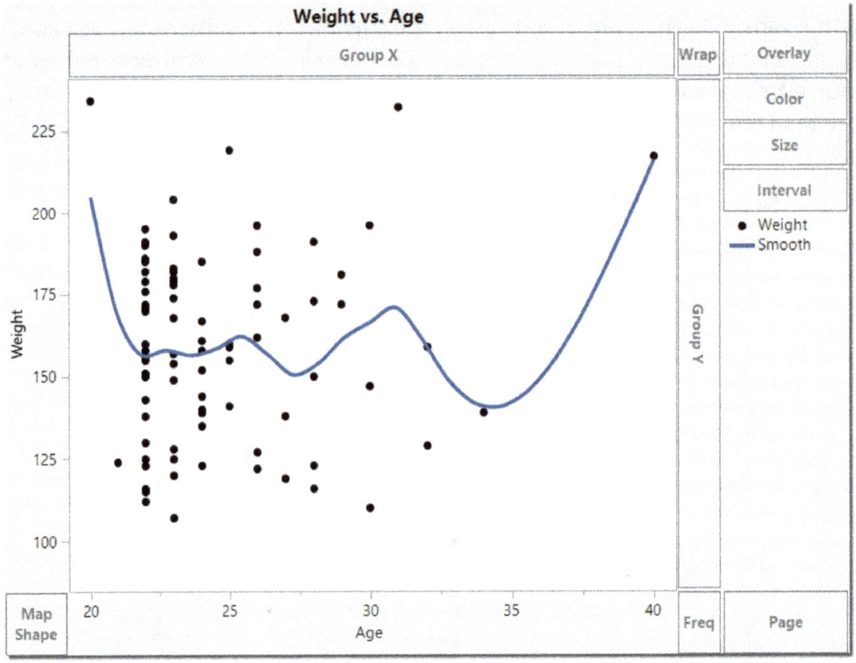

4. To get separate, side-by-side scatter plots for each level of a categorical variable, in this case **Gender**, drag that column to the **Group X** drop zone at the top of the graphing area.
5. Finally, to get separate plots vertically for each level of a different categorical variable, in this case **Alcohol Use**, drag that column to the **Group Y** drop zone at the right side of the graphing area.

Alternate Overlay Plots

Another type of overlay plot produces overlays of columns or groups on a single bivariate plot.

Figure 3.19a Alternate Overlay Plot

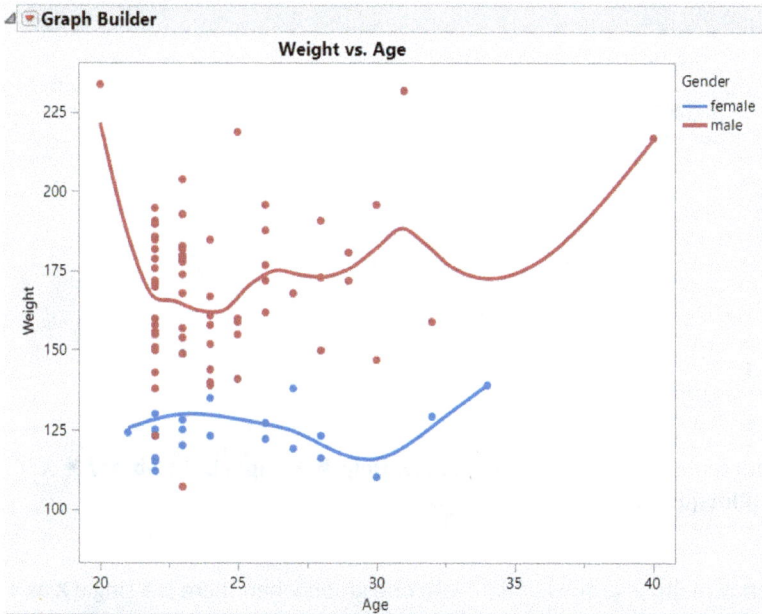

Usage

You want to overlay multiple groups of data into one graph when there is a single X and a single Y axis. This graph shows a scatter plot of weight by age with contrasting colors for each gender (see Figure 3.19a).

Required

Two numeric columns, one for the **Y** role and one for the **X** role. A nominal or ordinal column for the **Overlay** role.

1. Select **Graph ▶ Graph Builder**.
2. Drag a continuous column of interest, in this case **Weight**, to the **Y** drop zone on the left side of the graph area.
3. Drag a second continuous column of interest, in this case **Age**, to the **X** drop zone at the bottom of the graph area. As before, you now have a scatter plot of Weight by Age.
4. Drag a nominal or ordinal column, in this case **Gender** to the **Overlay** drop zone at the top right corner

A final type of overlay plot enables you to use multiple continuous columns in the Y role with a single X column.

Figure 3.19b Overlay Plot with Multiple Y Columns

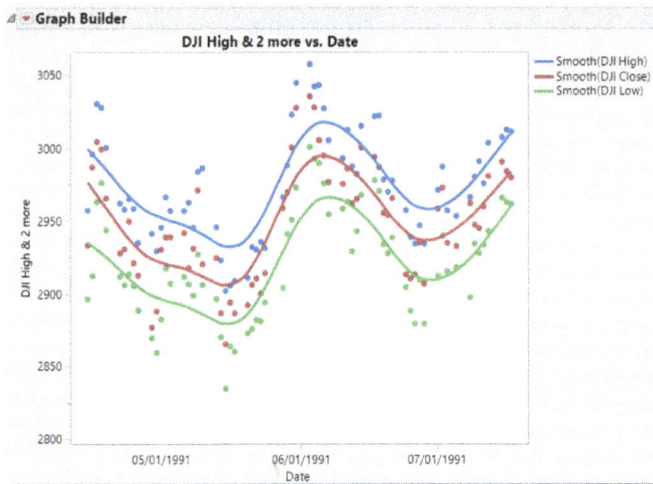

Data Table Access

To access the data table used for this graph, follow **Help ▶ Sample Data Library ▶ XYZ Stock Averages (Plots).jmp.**

Usage

You want to overlay multiple groups of data into one graph when there is a single X axis, but multiple Y columns. The graph shows the Dow Jones High, Low, and Close index for a period of time with contrasting colors for High, Low, and Close (see Figure 3.19b).

Required

At least two numeric columns for the **Y** role and one numeric column for the **X** role. **Optional:** Continuous, nominal, or ordinal columns for the **Grouping** role.

1. Select Graph ▶ Graph Builder.
2. Select at least two continuous columns simultaneously (in this case DJI High, DJI Low, and DJI Close) and drag them to the **Y** drop zone on the left side of the graph area.
3. Drag another continuous column of interest, in this case **Date**, to the **X** drop zone at the bottom of the graph area

> **Note**
>
> When using multiple continuous Y columns in this manner, they should be columns that have magnitudes close to one another. Columns with magnitudes that differ widely will create a graph that is difficult to read.

> **Note**
>
> When the overlay chart appears, additional customizations are available on the element icon palette and with a right mouse click inside the graph frame.

Pareto Plot

A chart often included as a quality metric for processes and products, the *Pareto plot* produces charts to display the relative frequency of problems in a quality-related process or operation. A Pareto plot is a bar chart that displays the classification of problem occurrences, arranged in decreasing order. The column with values that are the cause of a problem is displayed as **X** in the plot. An optional column with values assigning the frequencies is assigned as **Freq**. An optional column whose value holds a weighting value is assigned as **Weight**.

Figure 3.20 Pareto Plot of Failure

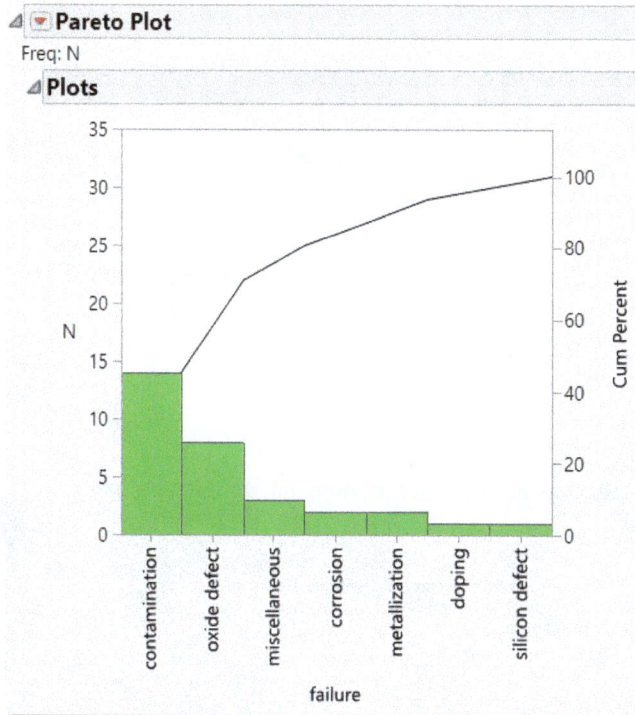

Data Table Access

To access the data table used for the above chart example, follow **Help ▶ Sample Data Library ▶ Quality Control ▶ Failure.jmp**.

Usage

For counts of defects by occurrence of defect causes. The plot can be used to target improvement efforts toward those failures that are most serious or common. As shown, the chart displays defect counts and cumulative percents of seven types of semiconductor defects. (See Figure 3.20.)

Required

At least one continuous or nominal column for the **Cause** role. Additional options for **Grouping, Frequency** and **Weight** roles.

To generate the plot, select **Analyze ▶ Quality and Process ▶ Pareto Plot**.

3.4 Graphs of One Column

Unlike the charts introduced so far, graphs in this section are accompanied by statistical results. These graphs depict the distribution of values for one column of data (also known as univariate) and provide appropriate tools to assess their properties.

These graphs help you understand the nature of a column, such as how widely the values vary or whether there are any curious qualities to the data.

Most of these graphs are found within the Distribution platform from the **Analyze** menu. We also briefly cover time series graphs in this section.

Many JMP graphs can be saved as interactive HTML and retain their interactivity when opened in a web browser.

> **Note**
>
> You can choose more than one column with these graphs, but each column will be graphed and independently analyzed side-by-side. When you are looking at more than one column, the graphs are linked, which enables you to click on any part of the graph to see and explore those values represented in the graphs of other selected columns. See Chapter 5 for more information about how graphs and data are linked.

Histograms and Bar Charts

The Distribution platform examines properties of a continuous, nominal, or ordinal column individually, or in a univariate fashion.

Figure 3.21a Distribution of Profit($M)

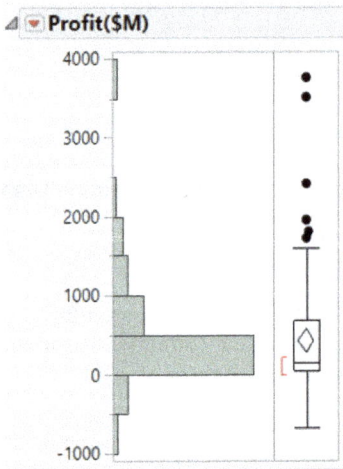

Figure 3.21b Distribution of Company Type

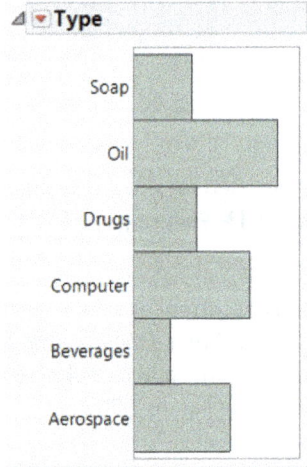

Data Table Access

To access the data table used for the above distribution chart examples, follow **Help ▶ Sample Data Library ▶ Financial.jmp**.

Continuous Usage

To view the properties of a continuous distribution such as shape, range, and data density. As shown in Figure 3.21a, the chart displays the profits (or losses) of a selection of technology companies from the late 1990s.

Continuous Distribution Requires

One or more continuous columns for the Y, Columns role.

Nominal, Ordinal Usage

The default plot is similar to a bar chart and enables you to view the properties of a frequency distribution such as the relative counts or percentages of fixed groups. As shown in Figure 3.21b, the chart displays the frequency of company types as bars for a selection of technology companies from the late 1990s. The nominal and ordinal distribution plots are related to the mosaic plot in the next section.

Frequency Distribution Requires

One or more nominal or ordinal columns for the **Y, Columns** role.

Select **Analyze ▶ Distribution**. Select a column and place it in the **Y, Columns** role, and click **OK**.

Optional

To generate a distribution plot like the ones pictured using **Graph Builder** select **Graph ▶ Graph Builder**. Drag a continuous or nominal column to the **Y** drop zone. Select the histogram element from the elements palette.

Outlier Box Plot

An *outlier box plot* is a chart for detecting extreme values and properties of a distribution, sometimes called a *Tukey box plot*. See Appendix B for a description of this term.

Figure 3.22 Tukey Box Plot of Profit

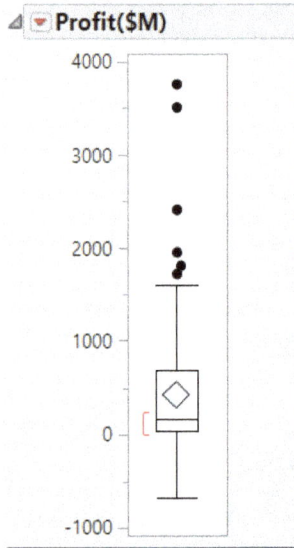

Data Table Access

To access the data table used for the above chart example, follow **Help ▶ Sample Data Library ▶ Financial.jmp**.

Usage

To view the properties of a continuous distribution such as quartiles, moments, and outliers. See Appendix B for a description of these terms.

As shown, the plot displays a few very profitable companies as points that are well beyond the main body of companies, the middle half of which are contained in the box. (See Figure 3.22.)

Required

One or more continuous columns for the **Y, Columns** role.

Select **Analyze ▶ Distribution**. Select a continuous column and place it in the **Y, Columns** role, and click **OK**. By default, the outlier box plot is displayed with the histogram. If it is not, click the red triangle and select **Outlier Box Plot**.

Optional

To generate an outlier box plot like the one pictured using **Graph Builder** select **Graph ▶ Graph Builder**. Drag a continuous column to the **Y** drop zone. Select the Box Plot element from the Elements palette.

Normal Quantile Plot

A *normal quantile plot* is a chart for visualizing the extent to which a column is consistent with a normal distribution. In a normal distribution, the points would fall about the solid red line in the display and not beyond the confidence curves.

Figure 3.23 Normal Quantile Plot of Profits with Box Plot

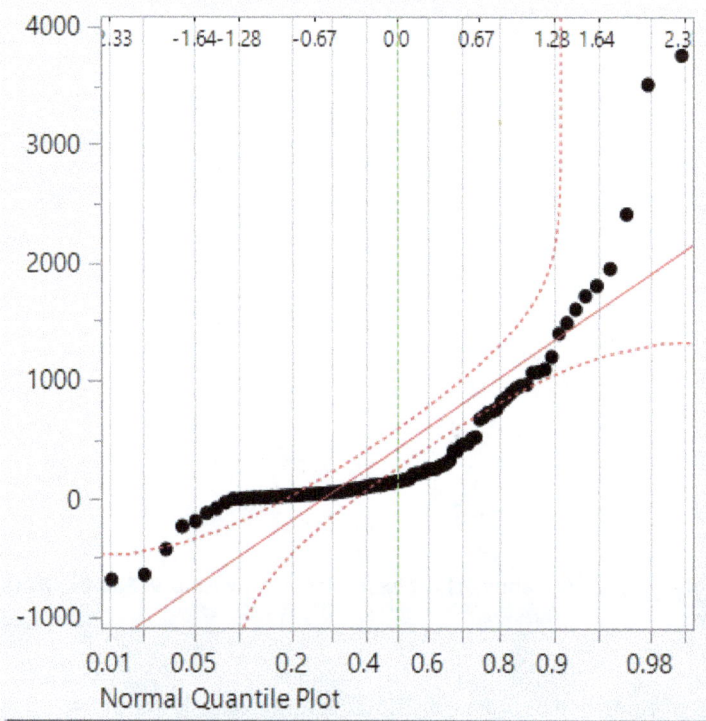

Data Table Access

To access the data table used for the above chart example, follow **Help ▶ Sample Data Library ▶ Financial.jmp**.

Usage

To view the properties and visually assess the extent to which the data is normally distributed. (See Figure 3.23.)

In this example, the plot displays the profits from a sample of companies. The data do not follow the solid red line; some fall beyond the dotted red confidence bands. The data are not consistent with a normal distribution.

Required

One or more continuous columns for the **Y, Columns** role.

Select **Analyze ▶ Distribution.** Drag a continuous column into the **Y, Columns** role, and click **OK**. Click the red triangle and select **Normal Quantile Plot**.

Mosaic Plot

A *mosaic plot* is a stacked bar chart where each segment is proportional to its group's frequency count.

Figure 3.24 Mosaic Plot of Type

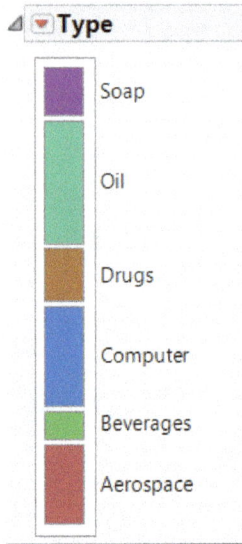

Data Table Access

To access the data table used for the above chart examples, follow **Help ▶ Sample Data Library▶ Financial.jmp**.

Usage

To view the properties of a nominal or ordinal distribution, or to visually assess the proportions of data that fall within each group. As shown, the chart displays the proportions or counts of each type of company from a stock portfolio. (See Figure 3.24.)

Required

One or more nominal or ordinal columns for the **Y,Columns** role.

Select **Analyze ▶ Distribution.** Drag a nominal or ordinal column into the **Y, Columns** role, and click **OK**. Click the red triangle and select **Mosaic Plot**.

Optional

To generate a mosaic plot like the one pictured using **Graph Builder,** select **Graph ▶ Graph Builder**. Drag a nominal column to the **Y** drop zone. Select the Mosaic element from the Elements palette.

Time Series

Time Series is a separate platform that generates a graph of a numeric value over time. It also serves as a platform to employ forecasting techniques and produces statistical results. For more information about these techniques, see the *JMP Statistics and Graphics Guide* (**Help > JMP Documentation Library > Predictive and Specialized Modeling**). The Time Series platform is available from the **Analyze** menu and the **Modeling** submenu. (See Figure 3.25a.)

Figure 3.25a Analyze Modeling Time Series

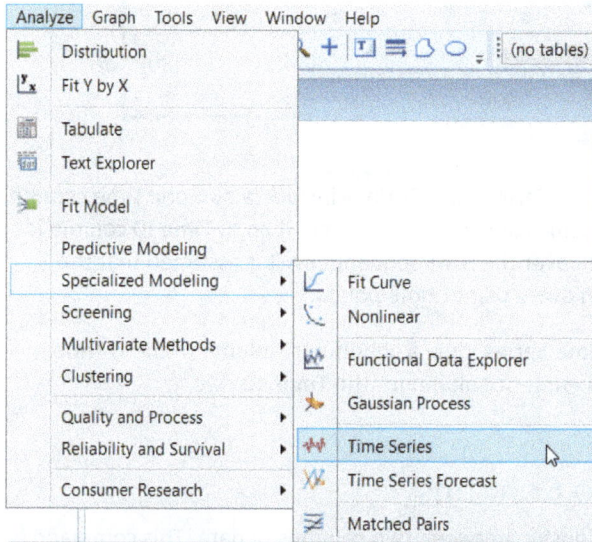

Data Table Access

To access the data table used for these chart examples, follow **Help ▶ Sample Data Library ▶ Time Series ▶ Raleigh Temps.jmp**.

Time Series Plot

A graph of numeric values, Y, over a time order, X.

Usage

To view and fit the variability and potential seasonality of a measured value over time. For example, the chart displays the average monthly temperatures with a clear seasonal trend in Raleigh, North Carolina, over a 130-month period. (See Figure 3.25b.)

Figure 3.25b Time Series

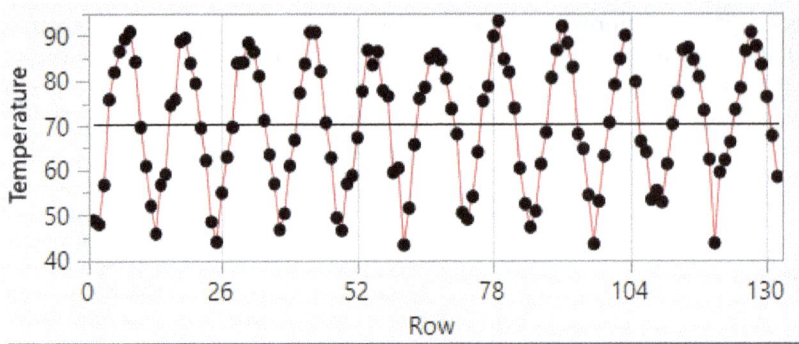

Required

One numeric column for the **Y, Time Series** role. Options include a numeric Time column (**X, Time ID**) with corresponding values and an input column. If an **X, Time ID** column is not specified, JMP orders the data over the rows sequentially. It is assumed that the time interval is constant between every pair of time points.

Select **Analyze ▶ Modeling ▶ Time Series**, drag a continuous column to the **Y, Time Series** role, and click **OK**. Select a numeric column for the **Time, ID** role.

3.5 Graphs Comparing Two Columns

The **Fit Y by X** command studies the relationship between two columns of data. This command is available from the **Analyze** menu and shows graphs with statistical results for each pair of x and y columns. The type of graph generated by JMP is determined by the modeling types (continuous, nominal, or ordinal) of the columns that are cast into the **X** and **Y** roles. When choosing a platform from the Analyze menu, JMP always creates the right graphs based on the modeling type. In important ways, **Fit Y by X** is four sets of graphs and analyses in one!

The matrix circled in the **Fit Y by X** window (see Figure 3.26) provides a visual preview of the graphs that will be generated depending on the modeling type of the Y (the vertical axis) and the X (the horizontal axis), which can be altered to obtain the desired analysis or plot.

> **Note**
>
> If the column modeling types nominal, ordinal, and continuous are unfamiliar to you, see Section 2.3.

Figure 3.26 Fit Y by X Contextual

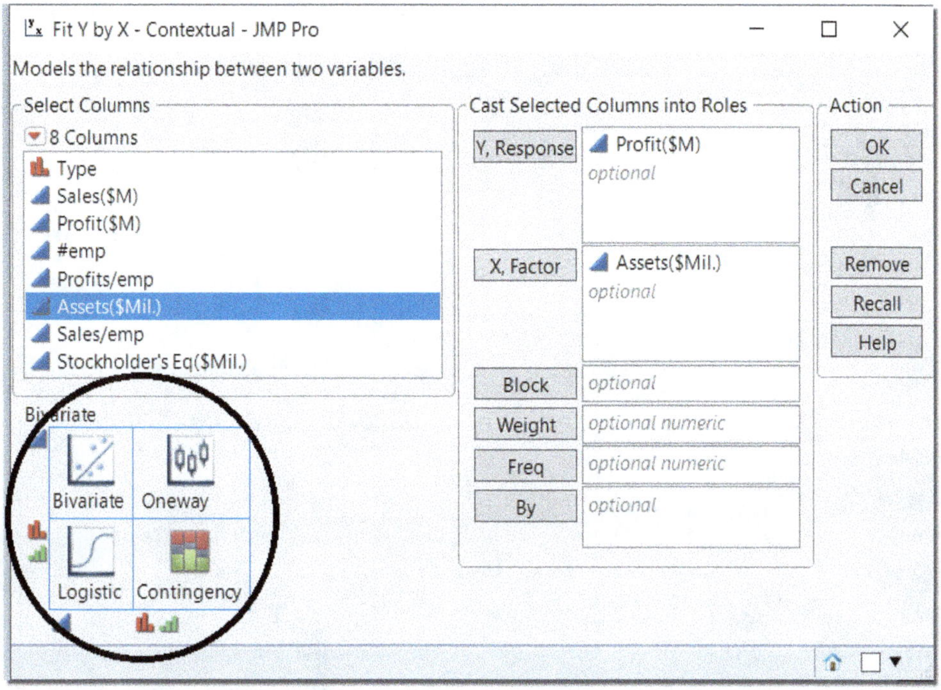

Scatterplot

A *scatterplot* is a graph of the continuous-by-continuous personality within the **Fit Y by X** command. The analysis begins as a scatterplot of points to which you can interactively add a linear fit and confidence curves.

Figure 3.27a Bivariate Profit by Assets

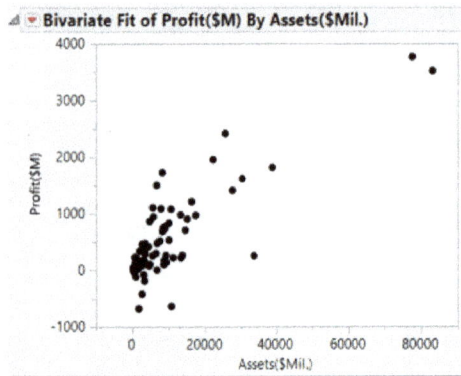

Figure 3.27b Bivariate Profit by Assets Fit Line

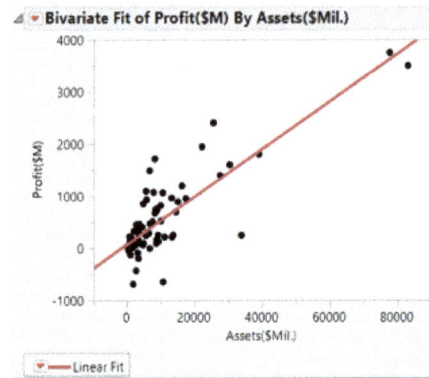

Figure 3.27c Fit Line menu

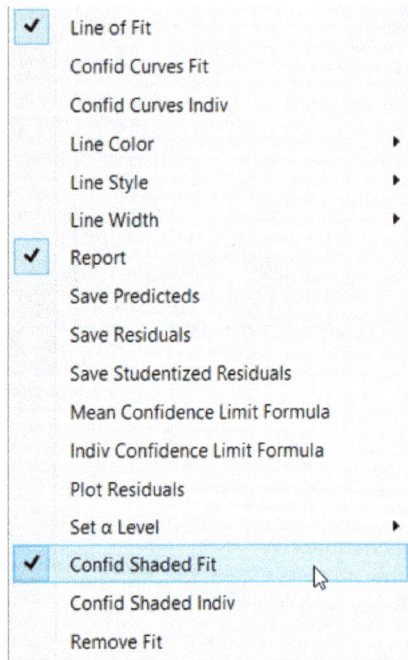

Figure 3.27d Bivariate Profit by Assets Fit Line Conf Shaded Fit

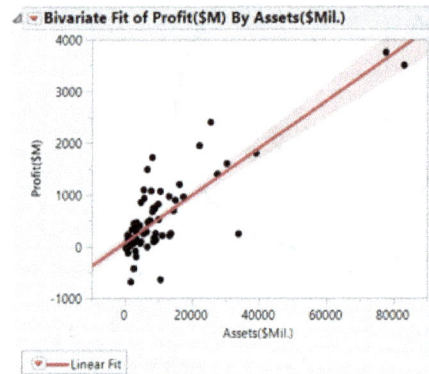

Data Table Access

To access the data table used for the above chart examples, follow **Help ▶ Sample Data Library ▶ Financial.jmp**.

Usage

To view the relationship of a continuous column to another continuous column. An example might be graphing the relationship of profits to assets for a selection of Fortune 500 companies and then fitting a regression line with 95% confidence curves, as shown. (See Figures 3.27a, 3.27b, 3.27c, and 3.27d.)

Required

One continuous column for the **Y, Response** role and one continuous column for the **X, Factor** role.

Select **Analyze ▶ Fit Y by X,** select a continuous **Y, Response** column and a continuous **X, Factor** column, and click **OK.** (See Figure 3.27a.)

To add the simple linear least squares fit: From the red triangle next to **Bivariate Fit**, select **Fit Line.** (See Figure 3.27b.)

To add the confidence shaded curves to the fit: From the red triangle on the **Linear Fit** item (see Figure 3.27c), select **Confid Shaded Fit.** (See Figure 3.27d.)

Optional

Using the Graph Builder method, select **Graph ▶ Graph Builder** and drag a continuous column to the Y drop zone and a continuous column to the X drop zone. Select the Line

Of Fit element from the elements palette.

Note
Scatterplots and bivariate fits are also supported in the Graph Builder Platform. These graphs can show colored markers. For more information about how to color or mark rows, see Section 2.6.

Scatterplot (with Polynomial Fit)

A *scatterplot with polynomial fit* is a graph that fits a polynomial curve to the degree you select from the **Fit Polynomial** submenu. After you select the polynomial degree, the curve is fit to the data points using least squares regression.

Figure 3.28 Bivariate 3rd Order Fit

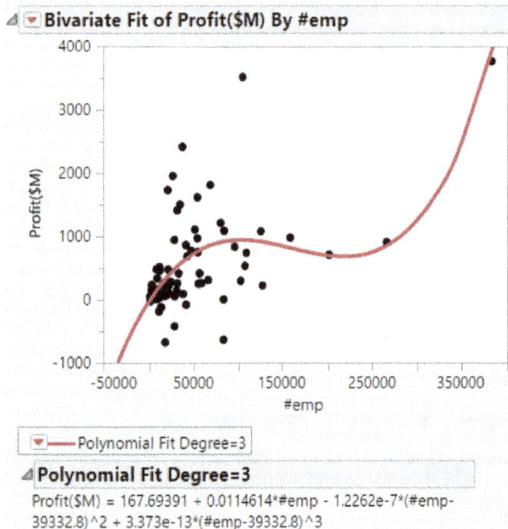

Data Access Table

To access the data table used for the above chart example, follow **Help ▶ Sample Data Library ▶ Financial.jmp**.

Usage

To view the relationship of a continuous column to another continuous column using a linear polynomial fit where curves produce the best fit of the data. The chart displays the third-order polynomial fit showing the relationship of profits to number of employees for a selection of companies. (See Figure 3.28.)

Required

One continuous column for the **Y, Response** column and one continuous column for the **X, Factor** role.

Select **Analyze ▶ Fit Y by X,** select a continuous **Y, Response** column and a continuous **X, Factor** column, and click **OK**. From the red triangle, select **Fit Polynomial** and select a degree number from the submenu.

Optional

Using the Graph Builder method, select **Graph ▶ Graph Builder** and drag a continuous column to the Y drop zone and a continuous column to the X drop zone. Select the Line Of Fit element from the elements palette, then adjust the degree of fit in the **Line Of Fit** panel.

Scatterplot (with Spline Fit)

A *scatterplot with spline fit* is a chart that fits a smoothing spline that varies in smoothness (or flexibility) according to a tuning parameter in the spline formula. You can use a spline of varying smoothness to highlight the overall trends in the data without using a linear function to describe the relationship. Larger values of the tuning parameter (lambda) smooth the relationship more, while smaller values fit the data more closely.

Figure 3.29 Bivariate Spline Fit

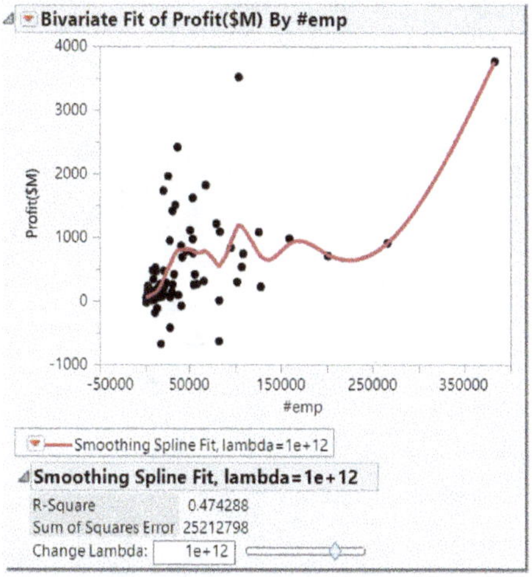

Data Table Access

To access the data table used for the above chart example, follow **Help ▶ Sample Data Library ▶ Financial.jmp**.

Usage

To view the relationship of a continuous column to another continuous column. For example, the chart illustrates that limited profit variation is present in companies with lower numbers of employees. The plot also shows that fewer companies have higher numbers of employees and higher profit variation. (See Figure 3.29.)

Required

One continuous column for the **Y, Response** role and one continuous column for the **X, Factor** role.

Select **Analyze ▶ Fit Y by X**. Select a continuous column and place it in the **Y, Response** role. Select a continuous column, place it in the **X, Factor** role, and click **OK**. From the red triangle, select **Flexible ▶ Fit Spline**, and from the submenu, select the degree of flexibility that you want in the spline fit by changing the lambda value.

Optional

Using the Graph Builder method, select **Graph ▶ Graph Builder** and drag a continuous column to the Y drop zone and drag a continuous column to the X drop zone. Adjust lambda to the value desired.

Oneway Plots

The *Oneway* platform analyzes how the distribution of a continuous Y column differs across groups defined by a categorical X column. Group means, as well as other statistics and tests, can be calculated and tested. The Oneway platform is the continuous (placed as Y) by nominal/ordinal (placed as X) personality of the **Fit Y by X** command.

Figure 3.30a Oneway Profits per Employee by Type

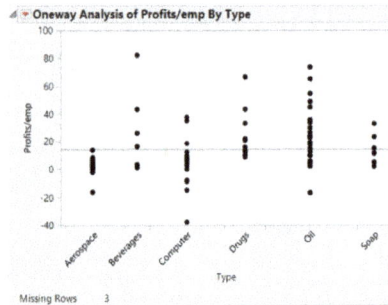

Figure 3.30b Oneway Profits per Employee by Type, ANOVA

Figure 3.30c Oneway Profits per Employee by Type, Quantiles

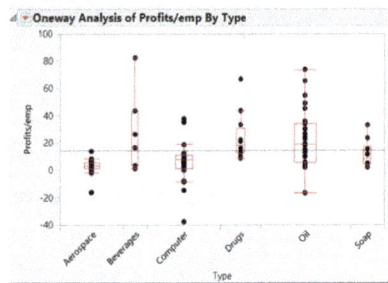

Data Table Access

To access the data table used for these chart examples, follow **Help ▶ Sample Data Library ▶ Financial.jmp**.

Usage

To compare the relationship of a continuous column across groups in a nominal or ordinal column. For example, the chart displays the difference in means and variation in profits per employees across six company types. (See Figures 3.30a, 3.30b, and 3.30c.)

Required

One continuous column for the **Y, Response** column and one nominal or ordinal column for the **X, Factor** role.

Select **Analyze ▶ Fit Y by X.** Select a continuous column for the **Y, Response** role and a nominal or ordinal column for the **X, Factor** role, then click **OK.** (See Figure 3.30a.)

From the red triangle, select **Means/Anova.** (See Figure 3.30b.) From the red triangle, select **Quantiles.** (See Figure 3.30c.)

Logistic Fit

The *Logistic* platform analyzes the probably of occurrence of one level of a categorical Y column across a continuous X column. Odds ratios and ROC curves are often reported from this type of analysis. The Logistic platform is the nominal/ordinal (placed as Y) by the continuous (placed as X) personality of the **Fit Y by X** command.

The Logistic platform displays a chart that estimates the probability of choosing one of the Y response levels as a smooth function of the X factor. The fitted probabilities will be between 0 and 1 and will sum to 1 across the response levels for a given factor value.

In a logistic probability plot, the Y axis represents probability.

Figure 3.31 Fit Y by X Logistic

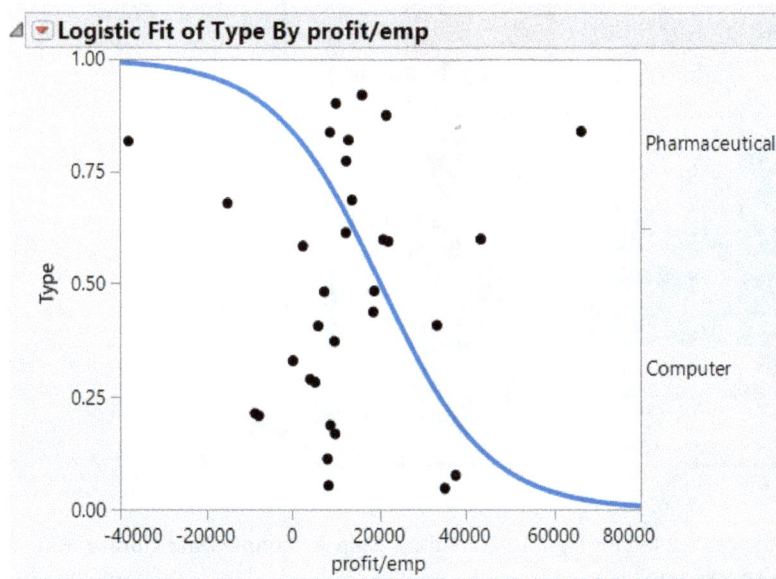

Logistic Fit of Type By profit/emp

Data Table Access
To access the data table used for Figure 3.31, follow **Help ▶ Sample Data Library ▶ Companies.jmp**.

Usage
To predict a group or groups by some continuous column. The chart displays a prediction (as a probability/percentage) of the type of company (Computer or Pharmaceutical) by its profits per employee. (See Figure 3.31.)

Required
One nominal or ordinal column for the **Y, Response** column and one continuous column for the **X, Factor** role.

Select **Analyze ▶ Fit Y by X**. Select a nominal or ordinal column for the **Y, Response** role and a continuous column for the **X, Factor** role, then click **OK**.

Mosaic Plot

A *mosaic plot* is a chart that is divided into small rectangles such that the area of each rectangle is proportional to a frequency count of interest.

The mosaic plot appears in the Contingency Platform and is the personality of the **Fit Y by X** command when both the Y and X columns are nominal or ordinal. Mosaic examines the distribution of a categorical Y column by the values of a categorical X column.

Figure 3.32 Fit Y by X Contingency Mosaic Plot

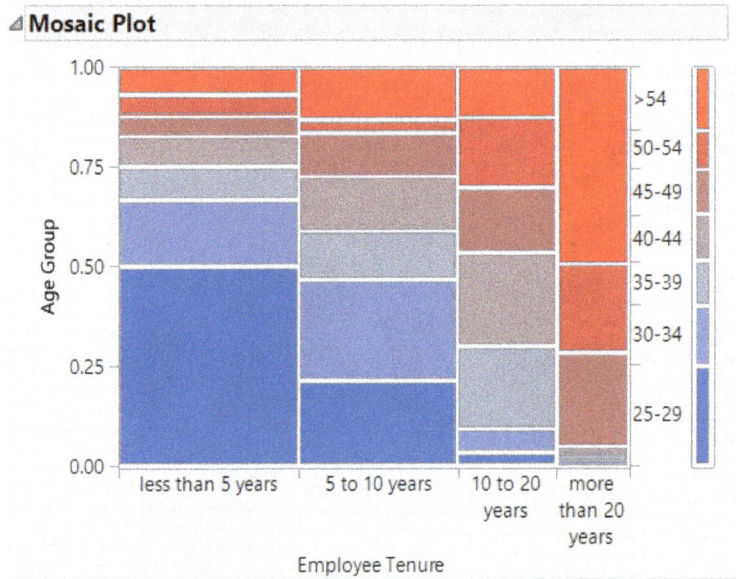

Data Access Table

To access the data table used for Figure 3.32, follow **Help ▶ Sample Data Library ▶ Consumer Preferences.jmp**.

Usage

Group-by-group counts are shown as proportional colored rectangles in a two-by-two arrangement. As shown, the graph displays a simple color chart of the proportion of seven age groups displayed by four employee tenure groups in a large company. Older employees are represented in red shades while younger employees are represented in blue shades. (See Figure 3.32.) From this graph, you can see half of the tenure group less than 5 years are between 25 and 29 years old. Half of the tenure group more than 20 years are greater than 54 years old.

Required

One nominal or ordinal column for the **Y, Response** column and at least one nominal or ordinal column for the **X, Factor** role.

Select **Analyze ▶ Fit Y by X**. Select a nominal or ordinal column for the **Y, Response** role and a nominal or ordinal column for the **X, Factor** role, and click **OK**.

Optional

Using the Graph Builder method, select **Graph ▶ Graph Builder** and drag a nominal or ordinal column to the **Y** drop zone and a nominal or ordinal column to the **X** drop zone.

Select the mosaic icon from the Elements palette.

3.6 Graphs Displaying Multiple Columns

Sometimes it is valuable to see a problem in more than two dimensions. This section uses JMP to visualize three or more columns at once. These graphs appear under the **Graph** menu and contain only a few built-in analytic procedures. (See Figure 3.33.) The Profiler, discussed at the end of this section, also appears in the results of the Fit Model platform in the Analyze menu.

Figure 3.33 Graph Menu

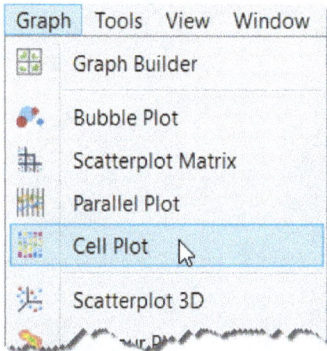

Like most JMP graphs, these multi-dimensional (or multi- column) graphs are interactive and enable you to select, rotate, and animate them. You can copy and paste these into other documents. At this book's release, many graphs can be saved as interactive HTML and retain their interactivity when opened in a web browser. Sharing these graphs is discussed in detail in Chapter 7.

Scatterplot 3D

Accessible from the **Graph** menu, this chart displays a three-dimensional scatterplot that can be rotated with your mouse. The Scatterplot 3D platform displays three columns at a time from the columns that you select.

Figure 3.34 Scatterplot 3D

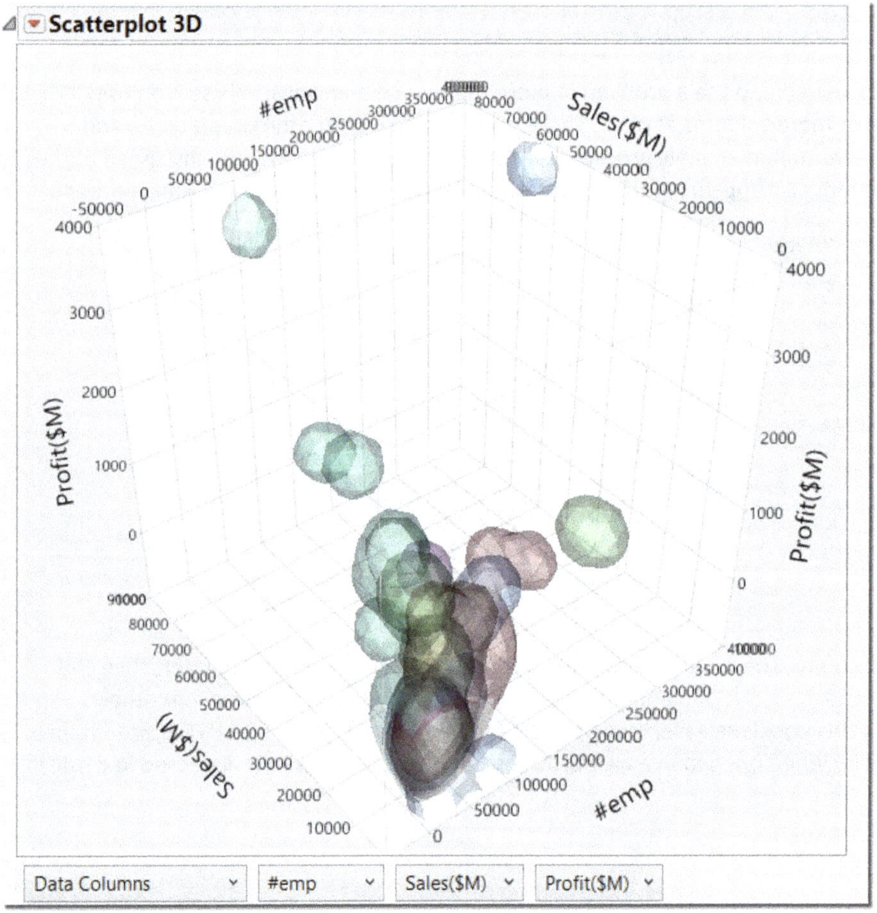

Data Access Table

To access the data table used for Figure 3.34, follow **Help ▶ Sample Data Library ▶ Financial.jmp**.

Usage

To view patterns among any two or three columns of data. This plot is very useful for exploring data in three dimensions. The chart displays sales (Y axis) by number of employees (X axis) by profits (Z axis) with colored 3D density contours by company type. (See Figure 3.34.) For example, note the two points with a very high employee count also have very high profits. The eye can detect possible differences among company types across the three columns. This is an interactive plot and can be rotated on any axis. To rotate the graph, click and hold the graph and move the mouse.

Required

Two or more columns of any modeling type (can be continuous, nominal, or ordinal). Three columns are required for a three-dimensional plot.

Select **Graph ▶ Scatterplot 3D**. Select at least two columns (three are recommended), place them in the **Y, Columns** role, and click **OK**. To include surfaces as displayed, from the red triangle, select **Nonpar Density Contour**, grouped by **Type**.

Treemap

A *treemap* is a graphical technique of observing patterns among groups that have many levels. Treemaps are especially useful in cases where histograms are ineffective because there are so many bars.

Figure 3.35 Treemap Airline Delays

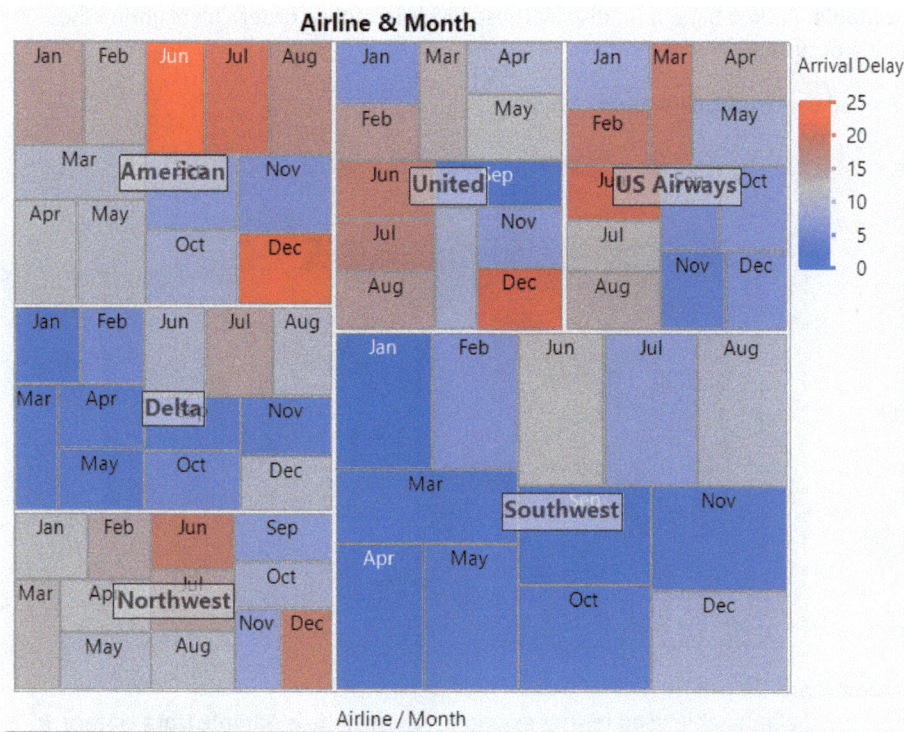

Data Table Access

To access the data table used for Figure 3.35, follow **Help ▶ Sample Data Library ▶ Airline Delays.jmp**.

Usage

For example, the chart displays airline arrival delays by month, where hot colors represent longer delays and cold colors represent shorter delays by airline. Larger squares represent longer delays as well. (See Figure 3.35.) These maps produce convenient visual rankings or groups within groups.

Required

At least one nominal, or ordinal column for the **X** role.

Select **Graph ▶ Graph Builder** and drag **Airline** and **Month** to the **X** drop zone. Select the treemap icon ⊞ from the Elements palette. Drag **Arrival Delay** to the **Color** drop zone. Click **Done**. Experiment!

Packed Bar Chart

A *packed bar chart* is the synthesis of an ordered bar chart and a treemap. Like treemaps, packed bars are especially useful in cases where there are dozens or even hundreds of levels of a categorical variable. Packed bar charts display these levels in ranked order, highlighting the most frequently occurring values.

Figure 3.36 Packed Bars from Billion Dollar Events

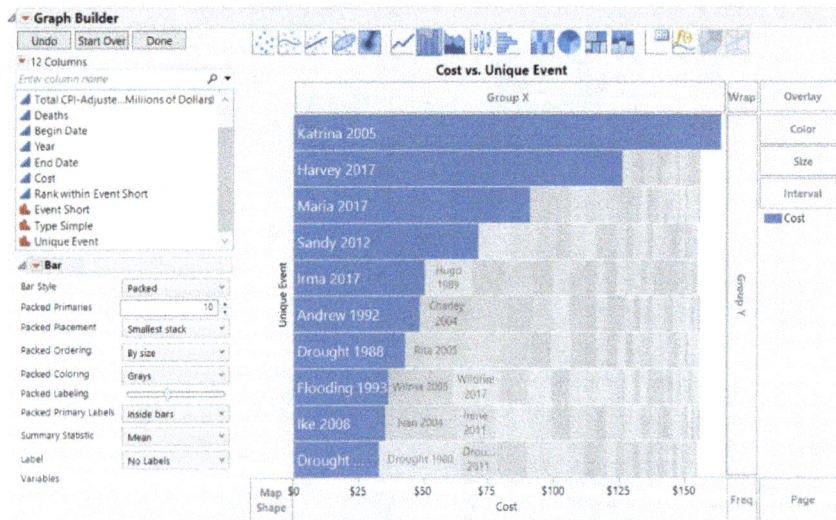

Data Table Access

To access the data table used for the example, follow **Help ▶ Sample Data Library ▶ Billion Dollar Events.jmp**.

Usage

For example, in Figure 3.36, the chart displays the top categories as a bar chart with blue bars. Secondary categories are labeled and in gray. In this example, you can clearly see that Katrina in 2005 had the largest impact, followed by Harvey and Maria in 2017.

Required

One continuous column for the X drop zone and one nominal for the Y drop zone.

Select **Graph** ▶ **Graph Builder**. Select **Unique Event** and drag it to the **Y** zone. Select

Cost and drag it to the **X** zone. Select the **Bar** element. In the Bar options panel:

- For the **Bar Style**, select **Packed**.
- Change **Packed Primaries** to **10**.
- Move the **Packed Labeling** slider down until it's about halfway.

Parallel Plot

A *parallel plot* is used when visualizing more than three dimensions in one graph. It is useful for seeing how the values of the many variables separate or stay together as you scan the plot from left to right. The parallel plot is scalable to any number of variables.

Figure 3.37 Parallel Plot of Passenger Information on the Titanic

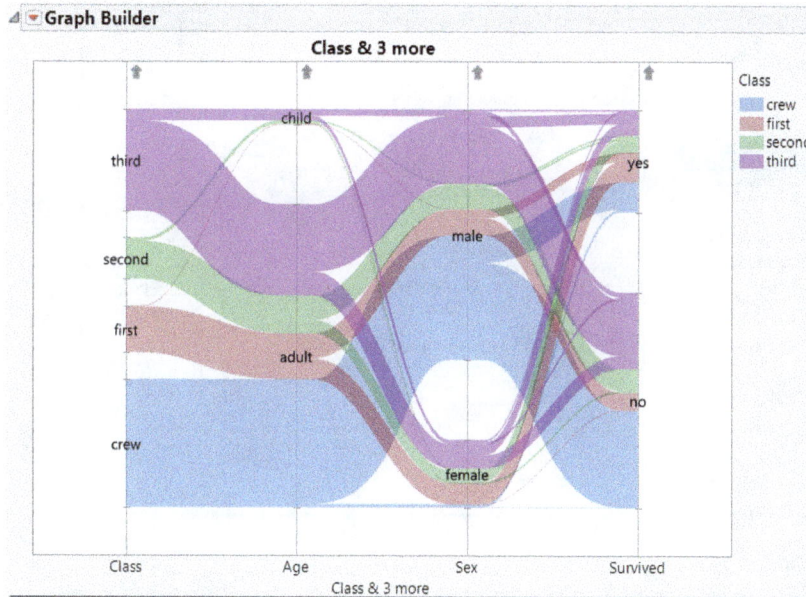

Data Table Access

To access the data table used for the example, follow **Help** ▶ **Sample Data Library** ▶ **Titanic.jmp**.

Usage

For example, in Figure 3.37, the chart displays the three top predictors of survival among the Titanic passengers. The thickness of the curved lines is reflective of the number of passengers in each of the groups. In this example, when clicking on the class you can see the relative passenger frequencies across the other category variables and whether this group survived or not. Try it!

Required

Continuous or nominal columns for the X drop zone.

Select **Graph ▶ Graph Builder**. Select **Class, Age, Sex and Survived** and drag the columns to the **X** zone. Select the **Parallel** ⧖ element. Select **Class** and drag it to the **Color** drop zone.

Bubble Plot

A *bubble plot* is an interactive scatterplot that represents its points as circles (bubbles). Optionally, the bubbles can be sized according to another column, colored by yet another column, and dynamically indexed by a time column. With the opportunity to see up to five dimensions at once (x position, y position, size, color, and time), bubble plots can produce dramatic animated visualizations and are effective at communicating complex relationships.

Figure 3.38 Bubble Plot Window with Assigned Variables

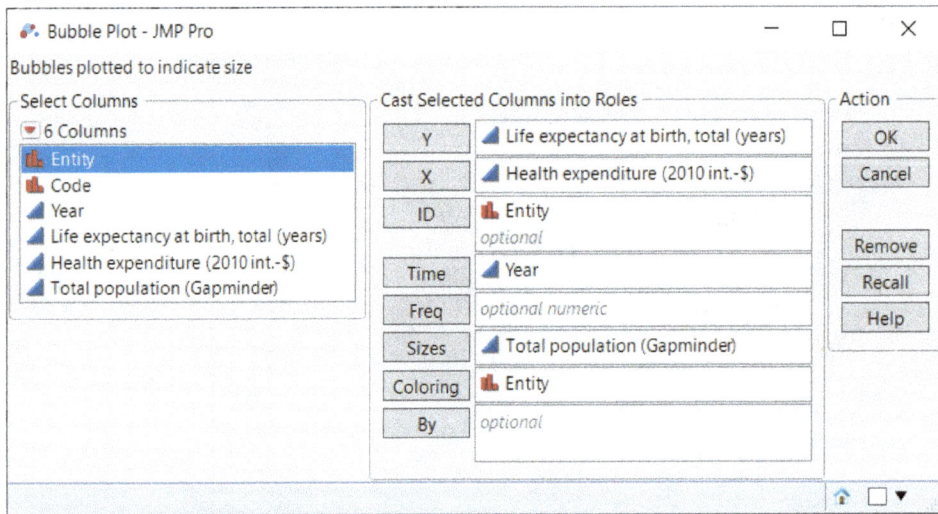

Data Table Access

To access the data table used for these chart examples, see the author page for this book. The original data comes from the World Bank: https://datacatalog.worldbank.org/dataset/world-development-indicators

Usage

Summarizing multi-column data in an interactive two-dimensional display. Frequently used where time is one of the columns. The bubble plot in this example displays the relationship between life expectancy at birth and health expenditure in 2010 U.S. dollars. (See Figure 3.39.)

Figure 3.39 Bubble Plot Life Expectancy Versus Health Expenditure

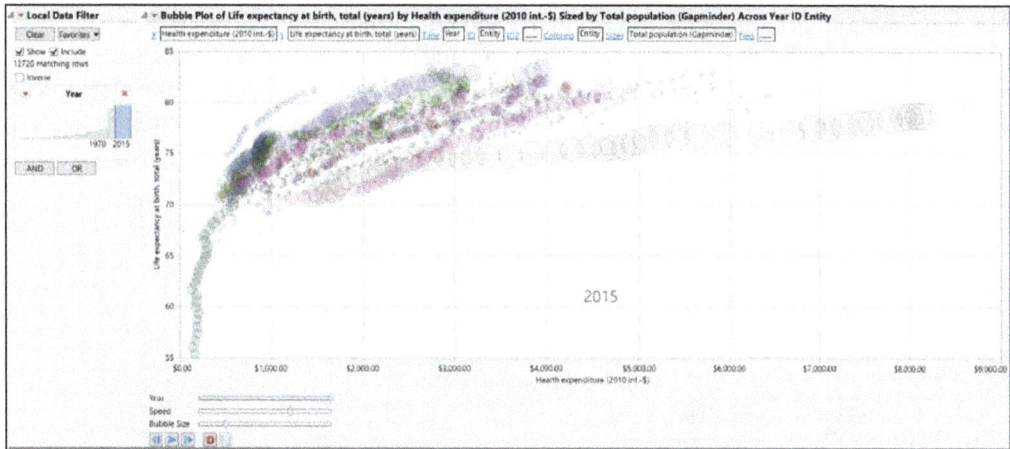

Required

One **X** column and one **Y** column of any type (continuous, nominal, or ordinal). For time animation, select a column for the **Time** role.

Select **Graph ▶ Bubble Plot**. Select **Life expectancy at birth, total (years)** and place it in the **Y** role. Select **Health expenditure (2010 int.-$)** and place it in the **X** role. For the animated time-based plot, select the **Time** column as **Year**. The **ID** column produces a label for each bubble. For this plot, select **Entity** as the **ID** role. The **Sizes** and **Coloring** columns can be specified to increase information density. For this plot, select **Total population (Gapminder)** as the **Sizes** role and select **Entity** for the **Coloring** role. (See Figure 3.38.)

To generate trail lines (as shown), select the **Trail Lines** and **Trail Bubbles** options from the red triangle. To save an interactive view of this plot for viewing in a web browser, select File ▶ **Save As** and choose **Interactive HTML with Data**.

Note

Bubble plots are particularly compelling when shared in a JMP Live environment. For more information about sharing results with JMP Live, see Section 7.6 "Publishing Reports in JMP Public (and JMP Live)."

To turn off the legend, select the red triangle, then select **Legend**.

Select the play ▶ button to start the animation. You can choose to filter the graph to include only data since 1970 as is done in the data filter shown. Use the **Year**, **Speed** and **Bubble Size** sliders to modify the data display and animation. Experiment!

Scatterplot Matrix

A *scatterplot matrix* is a chart that provides quick and orderly production of many bivariate graphs. These are assembled so that comparisons among many columns can be conducted visually and that correlation and data pattern can be easily detected. In addition, the plots can be customized and use advanced features (such as density ellipses) to provide for further analysis.

Figure 3.40a Scatterplot Matrix with Density Ellipses

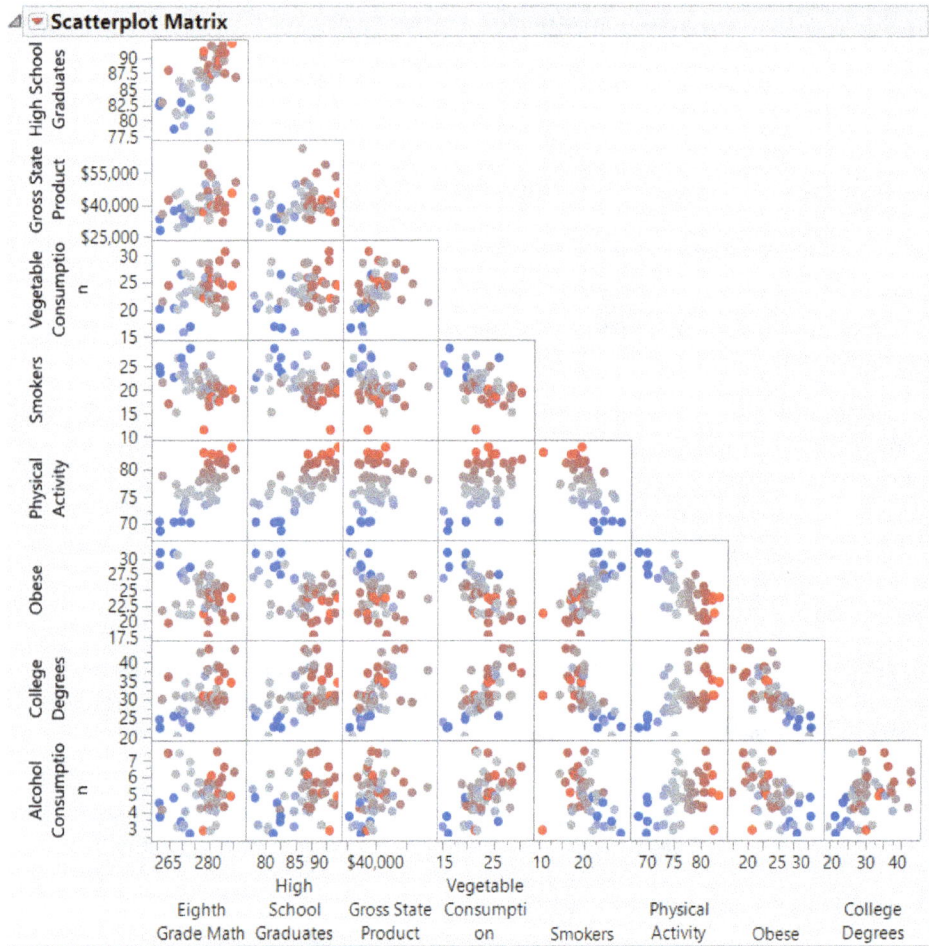

Figure 3.40b Scatterplot Matrix with Density Ellipses and Groups

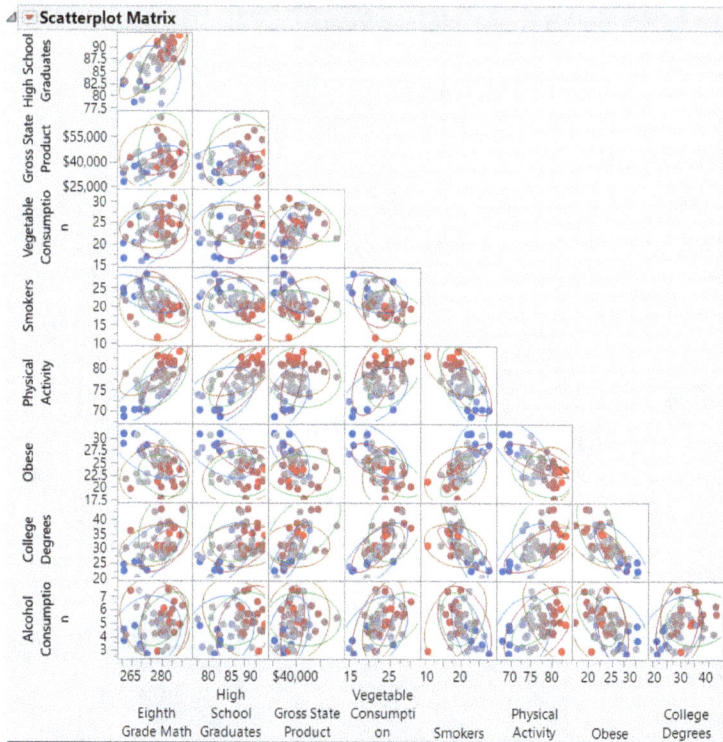

Data Access Table

To access the data table used for the above examples, follow **Help ▶ Sample Data Library ▶ US Demographics.jmp**.

Usage

In this example, the scatterplot matrix provides every bivariate combination of nine columns of US state demographic data. (See Figures 3.40a and 3.40b.) This chart quickly produces many correlation plots of all variables specified for easy identification of interesting groups and patterns.

Required

Two or more columns of any type (continuous, nominal, or ordinal) for the **Y, Columns** role. More than two columns are recommended.

Select **Graph ▶ Scatterplot Matrix**.

Select at least one column for the **Y, Columns** role. Select multiple Y and X columns for a matrix of graphs. Optionally, select a column and place it in the **X** role. Select a nominal or an ordinal column and place it in the **Group** role.

To include grouped ellipses as shown, include a column for the **Group** role in the window. Then for the plot, select **Density Ellipses** from the red triangle.

Profiler

The *profiler* is an interactive graph that provides a simple way to view complex relationships within a model. It lets you visualize what-if scenarios quickly and easily by enabling you to see the effect that changes in one column have on the remaining columns. This tool is especially useful when describing multiple variable models by demonstrating the sensitivity of changes in one or more X columns on the predicted Y.

The profiler displays profile traces for each X column. A profile trace is the predicted response as one column is changed (by dragging the vertical red dotted line in the graphs) while the others are held constant at the current values. The profiler recalculates the predicted responses (in real time) as you vary the value of an X column.

Figure 3.41 Prediction Profiler

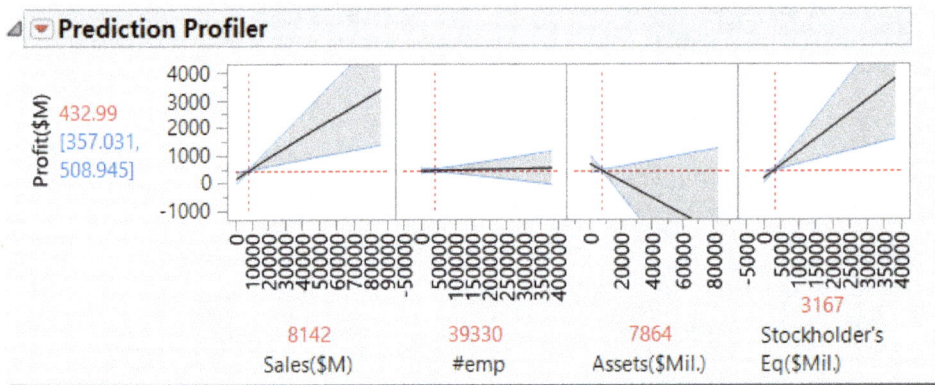

Data Table Access

To access the data table used for Figure 3.41, follow **Help ▶ Sample Data Library ▶ Financial.jmp**.

Usage

In this example, the chart displays the relationship between a continuous column Profits($M) and four predictors: sales [Sales($M)], number of employees (#emp), assets [Assets($Mil.)], and stockholder's equity [Stockholder's Eq($Mil.)]. (See Figure 3.41.)

The profiler shows that as sales increase, predicted profits increase, and as assets increase, profits decrease. To generate interactive, real-time predictions for profits, drag the vertical red trace lines in the profiler in JMP. To save the profiler as an interactive web page Choose **File ▶ Save As**, then choose **Interactive HTML with Data** as the file type.

Required

A formula column that can be created in one of two ways: from Fit Model (which generates a formula), or from a formula column either saved from Fit Model or entered in a data table by hand.

To create a formula in a column manually, see Section 2.4.

Create a Profiler from Fit Model

Select **Analyze ▶ Fit Model**, and select one or more columns for the **Y** role and one or more columns for the **Construct Model Effects**. Select the **Emphasis** pop-down menu, select **Effect Screening,** and click **Run Model**. The profiler appears at the bottom of the report.

Create a Profiler for Use Outside of JMP

From the Fit Model report window from the previous step, select the red triangle and select **Save Columns ▶ Prediction Formula**. This action saves a prediction formula to a JMP data table column as the last column in the table.

Once you have a prediction formula in a column, select **Graph ▶ Profiler** from the top menu. Select the prediction formula column that has appeared in the data table, place it in the **Y, Prediction Formula role**, and click **OK**. This implementation enables you to save the profiler as an Interactive HTML file with Data. Choose **File ▶ Save As** and choose **Interactive HTML with Data**.

3.7 Word Cloud

There are many times when data collected is free form text. For example, you might have comments on a product, reports about accidents or incidents, or open-ended questions on a questionnaire. The Text Explorer in JMP enables you to analyze this unstructured text. The process can include excluding specified words, combining similar terms by stemming or recoding, and analysis based on the terms counts. A *word cloud* is often an excellent way to visualize this analysis of text data. (See Figure 3.42.)

Figure 3.42 Word Cloud for Narrative Cause of Aircraft Incidents

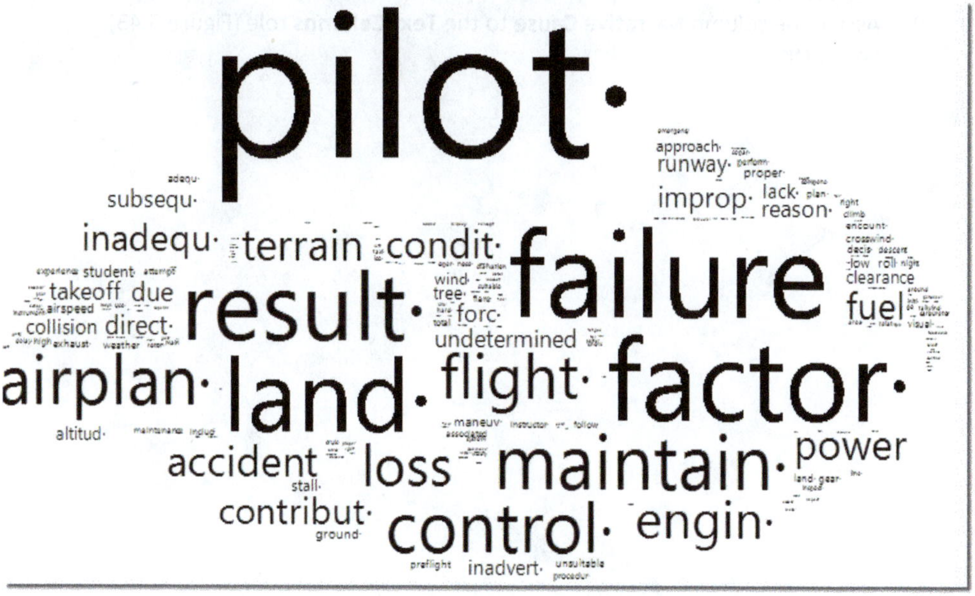

Data Table Access

To access the data table used for the above example, follow **Help ▶ Sample Data Library ▶ Aircraft Incidents.jmp**.

Usage

To illustrate, we will use the Aircraft Incidents sample data table. This has information about over 1900 aircraft incidents and accidents that occurred in 2001. Included in the data table is a narrative cause column containing comments on the cause of the incident. Examples of these entries include:

- Improper weather evaluation by both the pilot and pilot/passenger, and the pilot's inadvertent VFR flight into IMC resulting in his spatial disorientation. Factors were the pilot rated passenger's spatial disorientation, fog, and night conditions.

- The flight instructor's failure to ensure (supervision) the student had an adequate supply of fuel available, and the student's failure to refuel the aircraft sufficiently resulting in a fuel starvation/exhaustion condition and total power loss. Contributing factors were the trees and unsuitable terrain at the forced landing site.

- A loss of right engine power due to chaffed and frayed wiring on the right and left magnetos on the right engine resulting in a forced landing, an inadvertent stall, and subsequent impact with the ground.

Begin by opening the sample data table **Aircraft Incidents.jmp**.

1. Select **Analyze ▶ Text Explorer**.
2. Assign the column **Narrative Cause** to the **Text Columns** role (Figure 3.43).
3. Select **OK**.

Figure 3.43 Text Explorer for Narrative Cause of Aircraft Incidents

◢ ▼ **Text Explorer for Narrative Cause**

Number of Terms	Number of Cases	Total Tokens	Tokens per Case	Number of Non-Empty Cases	Portion of Non-Empty Cases
2573	1906	51330	26.9307	1902	0.9979

◢ **Term and Phrase Lists**

Term	Count		Phrase	Count	N
pilot's	1157		pilot's failure	484	2
failure	1045		failure to maintain	458	3
landing	846		pilot's failure to maintain	330	4
maintain	575		engine power	298	2
control	563		loss of engine	280	3
airplane	546		loss of engine power	273	4
factor	542		directional control	221	2
flight	537		forced landing	206	2
pilot	509		contributing factor	175	2
resulted	489		pilot's inadequate	174	2
loss	488		maintain directional control	163	3
engine	458		maintain directional	163	2
factors	457		undetermined reasons	158	2
terrain	391		failure to maintain directional	129	4
accident	370		pilot's improper	119	2
power	364		contributing factors	114	2
inadequate	335		airplane control	111	2
contributing	331		landing gear	103	2
fuel	323		control of the airplane	97	4
resulting	290		inadvertent stall	88	2
conditions	259		fuel exhaustion	79	2
improper	246		failure of the pilot	78	4
⋯ue	244		⋯iated with the accident	76	4

The results show there are 2574 terms found in the 1906 rows in the data table. The terms and phrases that appear most frequently are found at the top of their respective lists.

4. To visualize the terms and the frequency of their appearance, click on the red triangle and select **Display Options ▶ Show Word Cloud** (Figure 3.44).

Figure 3.44 Word Cloud for Narrative Cause of Aircraft Incidents

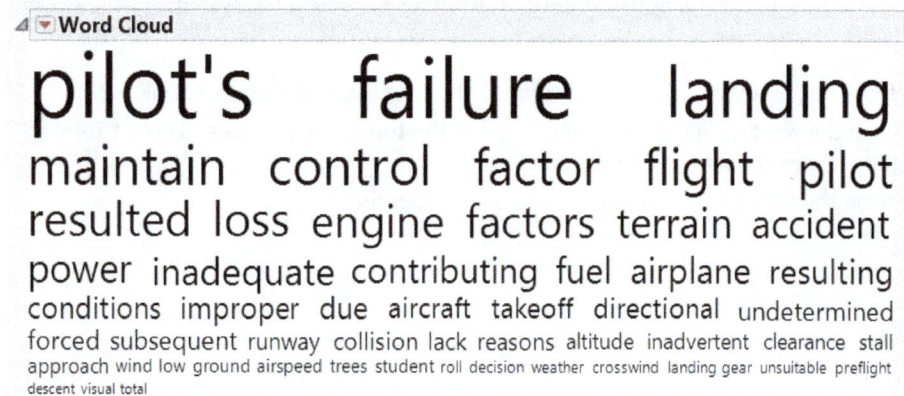

◢ ▼ **Word Cloud**

pilot's failure landing
maintain control factor flight pilot
resulted loss engine factors terrain accident
power inadequate contributing fuel airplane resulting
conditions improper due aircraft takeoff directional undetermined
forced subsequent runway collision lack reasons altitude inadvertent clearance stall
approach wind low ground airspeed trees student roll decision weather crosswind landing gear unsuitable preflight descent visual total

The words that appear largest in the word cloud are those that appear most often in the text being analyzed. A few things that you might note in these results:

- Pilot's and pilot are different words. Resulting and resulted also appear separately in the Word Cloud. Words like these could be combined by stemming the words. This is the process of reducing the terms to their roots so that all are treated as the same term in the analysis.

- Airplane and aircraft are both in the word cloud. These words might be substitutes for one another, so you might want to recode aircraft to read airplane (or the other way around if you prefer).

5. To recode aircraft to airplane, click on the red triangle at the top of the results and select **Term Options ▶ Manage Recodes...**

6. In the Manage Recodes window, under the User column, type in aircraft as the old value and airplane as the new value. (See Figure 3.45.)

Figure 3.45 Managing Recodes

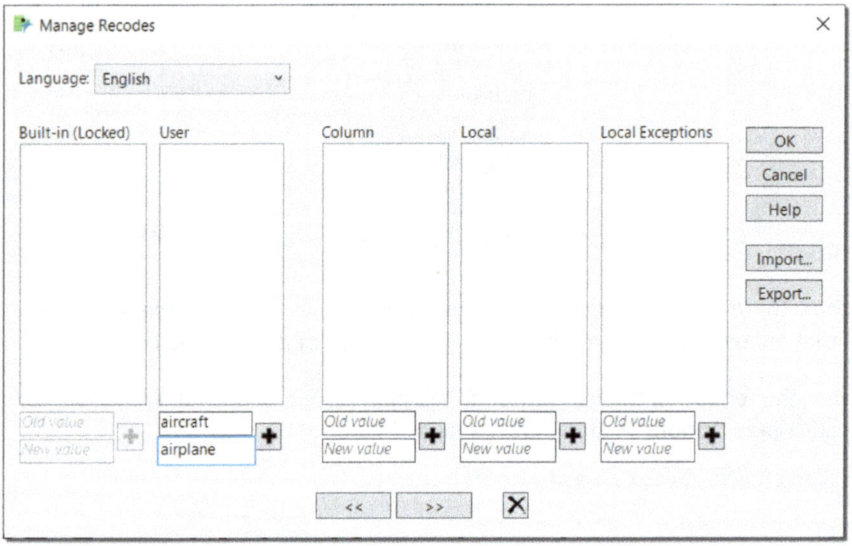

7. Click on the ✚ button to add this recode to the user list, and then select **OK**.

8. To stem the words, click on the red triangle at the top of the results and select **Term Options ▶ Stemming ▶ Stem for Combining** (Figure 3.46).

9. Examine the differences between the two word clouds.

Figure 3.46 Word Cloud After Recoding and Stemming

You will note that some of the words have moved (and changed size) in the new word cloud. You will also see dots at the end of some of the words that have been stemmed and then the stems combined.

The default organization of the word cloud is Ordered, with the most frequently occurring terms at the top left and diminishing in size as you move from left to right and top to bottom. You might prefer that the words be more randomly arranged in an oval (or cloud), with their size still representing the frequency of occurrence.

10. To change the layout of the Word Cloud, click the Word Cloud red triangle and select **Layout ▶ Centered** (Figure 3.47).

Figure 3.47 Word Cloud with Centered Layout

When exploring this unstructured text data with word clouds, it can be interesting to see whether there is a relationship between any other columns in the data table. For example, you could consider doing this analysis by the levels of a categorical column such as Fatal. To do so, redo this analysis with Fatal assigned as a By variable and consider the differences between the word clouds.

3.8 Summary

This chapter presented a series of the most frequently used graphs and their step-by-step recipes. They are presented in a cookbook style so that each graph can be recognized by its picture, name, or definition and easily replicated with simple steps. These charts represent the first tier of commonly used graphs. You can find additional graphs and instructions on how to create them in the JMP documentation, including *Essential Graphing*. Go to **Help ▶ JMP Documentation Library▶ Essential Graphing** for more information.

Chapter 4: Finding the Right Graph or Summary

In the last chapter, we presented an overview of the commonly used graphs in JMP with accompanying recipes. This works well when you know what you want your graph to look like. However, at other times, and particularly when you are exploring your data for the first time, it makes sense to take a different approach for several important reasons. First and foremost, you probably don't know what your data is going to tell you. Second, you can learn a great deal about your data when you explore it both visually and numerically. Reviewing your data in both ways can lead you to find the best way to display information.

In this chapter, we take advantage of a key feature of JMP: its ability to create graphs, maps, and summaries interactively and intuitively. While all platforms in JMP promote exploration, two features in particular – Graph Builder and Tabulate – are especially designed for this purpose. Because the former is geared toward interactively creating maps and graphs and the latter at numerical summaries, we cover them separately. Let us begin with finding the right picture of your data.

4.1 Using Graph Builder to Produce Graphs of Data

Graph Builder is a great tool to quickly see your data expressed in many different forms. In many ways, Graph Builder is like a blank canvas waiting for your artistic direction. With Graph Builder, it is helpful to begin thinking about your problem and the columns that are central to your questions at hand. Graph Builder is found under the **Graph** menu when you select **Graph** ▶ **Graph Builder**. (See Figure 4.1.)

As described in Chapter 3, Graph Builder works in a drag-and-drop manner. Simply click a column in the **Columns** box and drag it (without letting go) to one of the zones located around the canvas. You can hold the column in a zone (without letting go) to get a preview of the graph, then move it to another zone to see an alternative display. Once you release the mouse button, Graph Builder keeps that selection and enables you to customize the display or select another column to begin exploring relationships. You can repeat this process by adding more columns to view more complex displays.

Example 4.1: TechStock

You can find the data set at **Help ▶ Sample Data ▶ Business and Demographic ▶ TechStock.jmp**.

We will be working with one of the sample data files to illustrate the steps in this section. The file that we will use, **TechStock.jmp**, contains closing price data from selected technology stocks. The columns include:

- **Date:** the date expressed as dd/mm/yyyy
- **Open:** the opening price of the stock
- **High:** the highest price of the stock on that date
- **Low:** the lowest price of the stock on that date
- **Close:** the closing price of the stock on that date
- **Volume:** the number of shares traded on that date
- **Adj. Close*:** the adjusted price of the closing price
- **YearWeek:** the week of the year expressed as yyyy/week#

1. First, open the sample data table **TechStock.jmp**, as indicated previously.
2. Open Graph Builder. Select **Graph ▶ Graph Builder**. (See Figure 4.1.)

Figure 4.1 Open Graph Builder

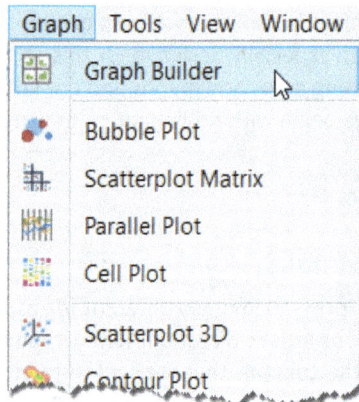

3. Notice the columns in your data table appear on the left in the **Columns** box of the Control Panel. (See Figure 4.2.)

Figure 4.2 Graph Builder Variables

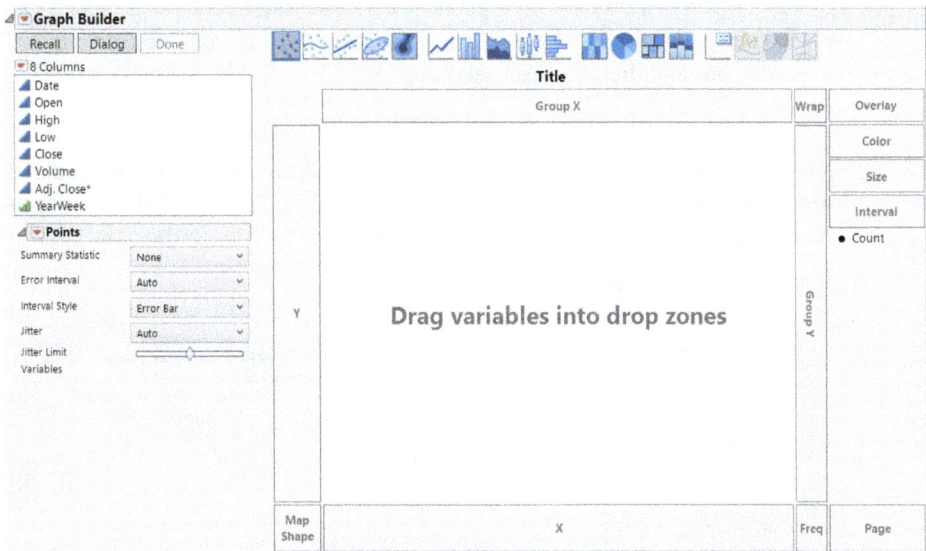

4. Drag the **Adj. Close*** column from the **Columns** box to one of the labeled zones located around the graph canvas. You will notice that as soon as the column is dragged into the zone, a graph of that data is immediately displayed. Keeping the mouse button depressed, experiment by dragging the column into different drop zones to see what happens. If you let go of the mouse button, you can always click the Undo or **Start Over** button to begin again.

5. Click, hold, and drag the **Adj. Close*** column to the **Y** drop zone and release the mouse button. You should see a point graph. (See Figure 4.3.) Your graph might appear different depending on the default size of the graph area.

Figure 4.3 Drag Adjusted Close to Y Drop Zone

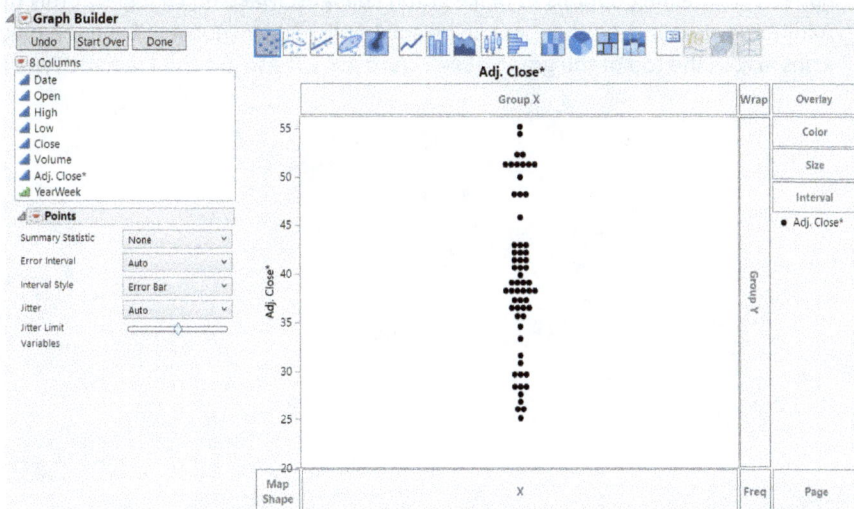

> **Note**
>
> If you change the **Jitter** setting on the graph from Auto to **Positive Grid**, you will see a familiar Dot Plot, but oriented vertically.

6. Now click and drag the **Date** column to the **X** drop zone. This action immediately produces a scatter plot with dates on the X axis. (See Figure 4.4.) Notice that the trend for the close prices is downward, and a smoothing line is fit to the data.

Figure 4.4 Drag Date to X Drop Zone

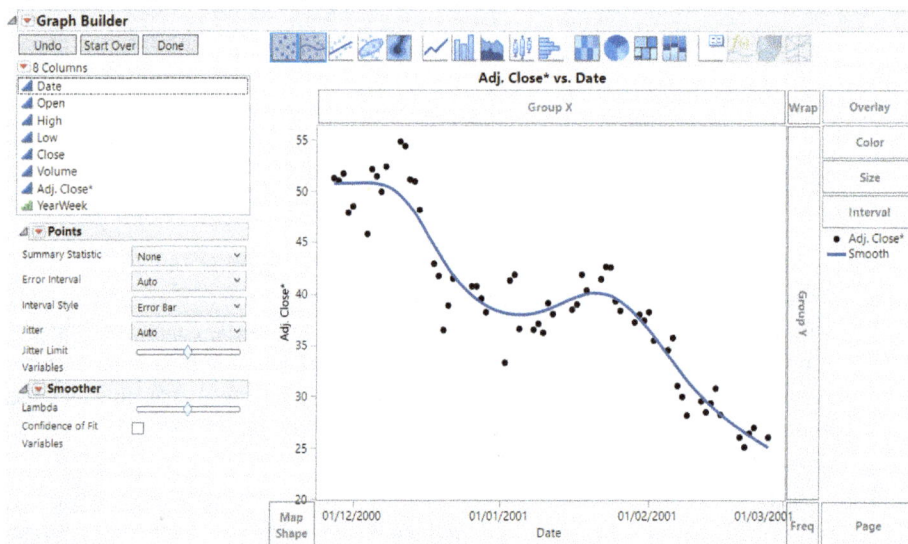

7. Right-click in the graph region to change the look of the graph. (See Figure 4.5.) A menu appears, containing options to make graph changes. Make no changes at this time. Additional visualization changes can be accomplished by choosing from the palette of icons across the top of the graph. (See Figure 4.6.) These are the Element Type Icons.

Figure 4.5 Right Mouse Click in the Graph Region

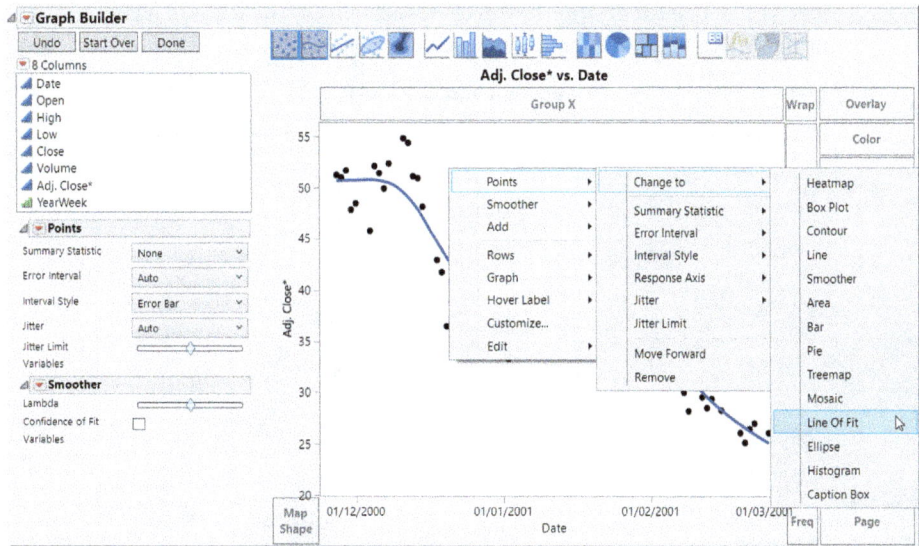

Figure 4.6 Element Type Icon Smoother

8. Select the icon that is second from the left on the palette of icons at the top (circled in Figure 4.6). This toggles the smooth curve on and off. Highlighted choices are enabled. Grayed out choices are disabled. Leave the smooth curve enabled.

9. Now double-click the **Date** axis. You see a window box of choices. (See Figure 4.7.) Change the **Tick Label Orientation** to **Vertical**. In **Grid Lines**, add a check for grid lines for the **Major** and **Minor** axes and click **OK**.

Figure 4.7 Change Tick Orientation and Gridlines

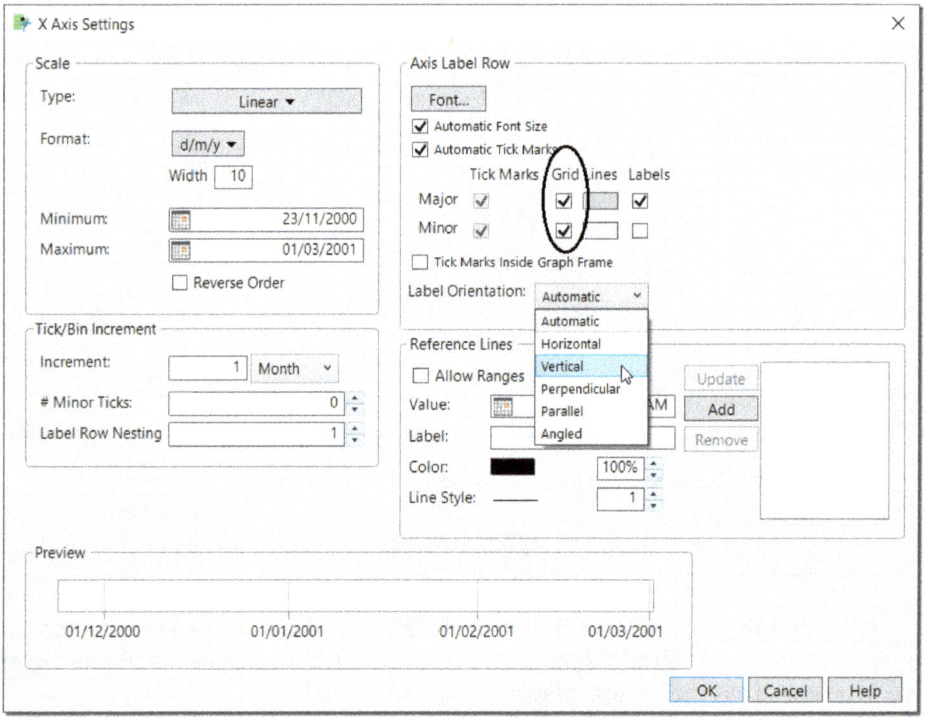

10. You should now see a graph with vertical dates and gridlines at the major tick date intervals. Now drag the **Volume** column to the area just above the Y axis label for **Adj. Close*** (Figure 4.8).

Figure 4.8 Drag Volume to Y Drop Zone

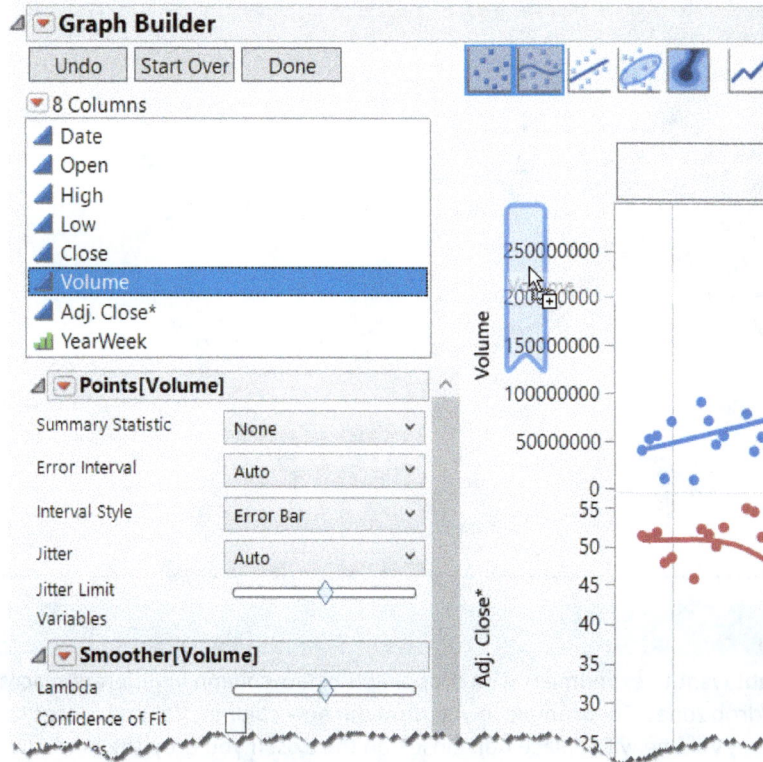

11. You will see a graph rendered (Figure 4.9).

Figure 4.9 Graph Rendered

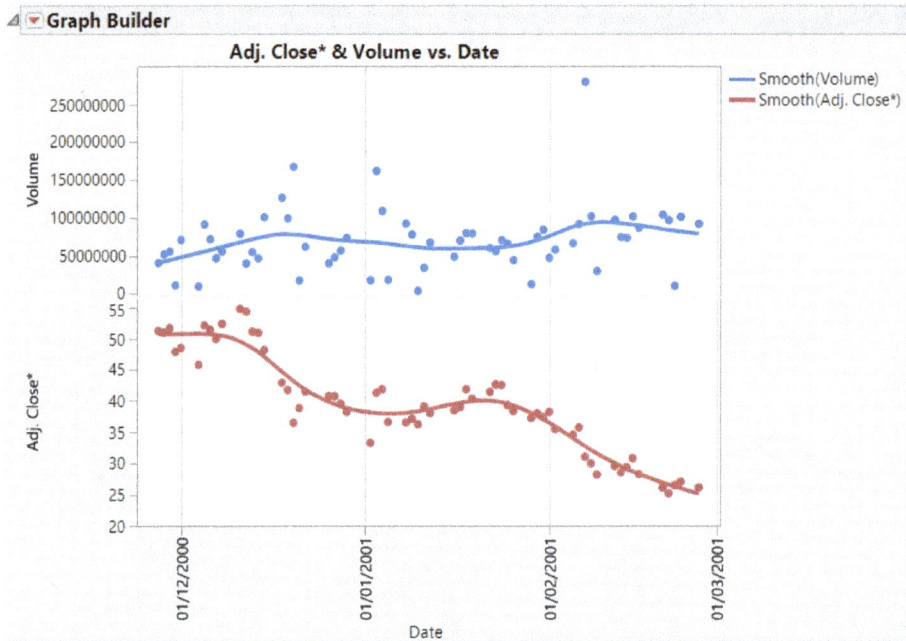

You might want to experiment with dropping the new column in different places in the Graph Builder drop zones. For example, if you drop the new column, Volume, closer to the center of the Y drop zone, Volume will replace Adj. Close* on the axis. If you drop the new column at the center of the drop zone just inside Adj. Close*, it will graph the values of the two columns on the same axis. This would only be appropriate if the magnitude of the two columns is similar.

12. Right-click the **Volume** graph points (see Figure 4.10) and select **Points ▶ Change to ▶ Bar**.

Figure 4.10 Change Points to Bars

The Element Type Icons at the top of the chart give an iconographic indication of what parts of the graph elements are currently enabled. (See Figure 4.11.)

Figure 4.11 Element Type Icons

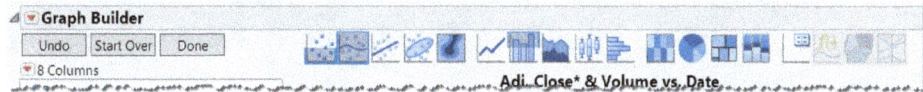

Elements of the graph's rendering can be controlled from this panel. Reading the icons for the current graph from left to right, points are enabled for the Adj. Close* Y axis only (indicated by the lower half of the icon being highlighted and corresponding to the lower half of the graph in Figure 4.12). Smoother is enabled for both Adj. Close* and for Volume. Bars are enabled for Volume only (the upper half of the graph).

13. This renders a graph showing trading volume and adjusted closing prices for the time period. (See Figure 4.12.)

Figure 4.12 Enable Smoother for Volume and Adjusted Close

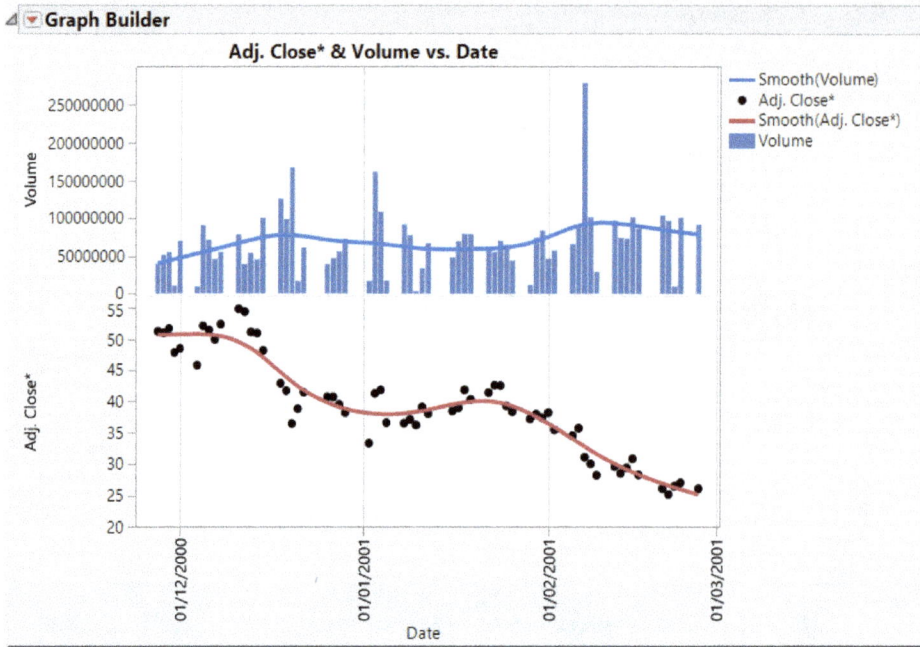

14. Now let's include the daily high and low prices in the graph. Drag the **High** column to the location just to the right of the Y axis for Adj. Close* (Figure 4.13).

Figure 4.13 Include Daily High and Low Prices

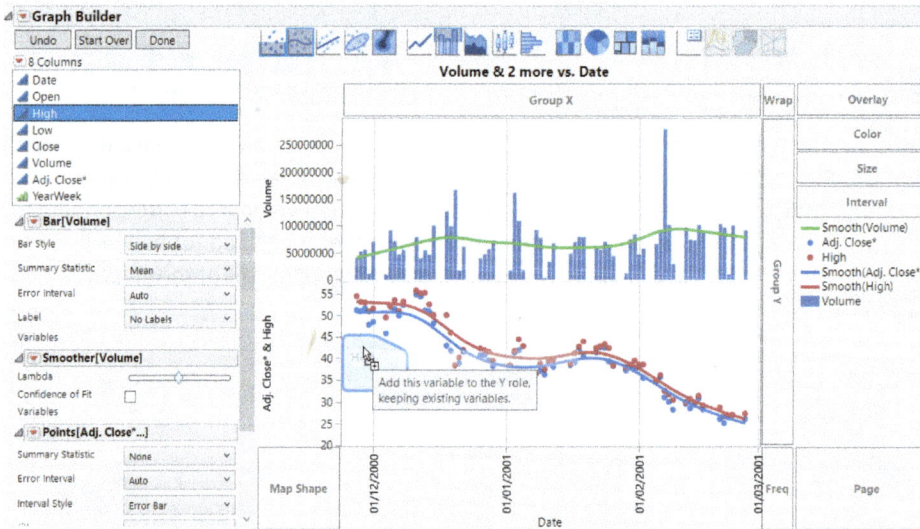

15. A red **High** line is now rendered that represents the maximum price of the stock portfolio by day. If you did not get the picture to match Figure 4.13 the first time, you can always click undo and try again. Now do the same with the Low column by clicking and dragging the **Low** column to the same location. A green Low line is now rendered that represents the minimum price for each day.

16. To complete the graph, edit the last few items. Double-click inside the legend. (See Figure 4.14.)

Figure 4.14 Edit the Legend

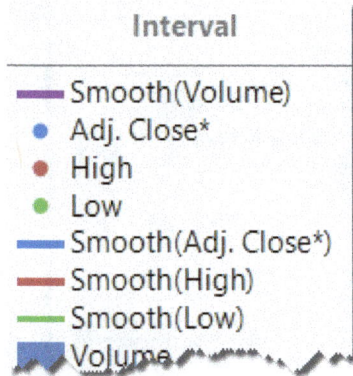

17. A **Legend Settings** window appears (Figure 4.15).

Figure 4.15 Legend Settings Window

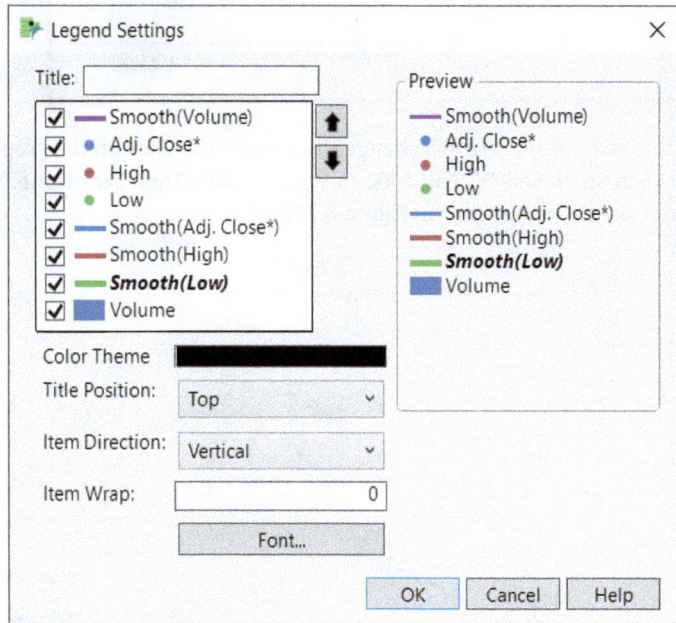

18. Click on the purple **Volume** line and remove the check mark for that item. Then look at in the **Preview** area of the dialog box. The dialog box should look like Figure 4.16. Click **OK**.

Figure 4.16 Volume Line and Preview

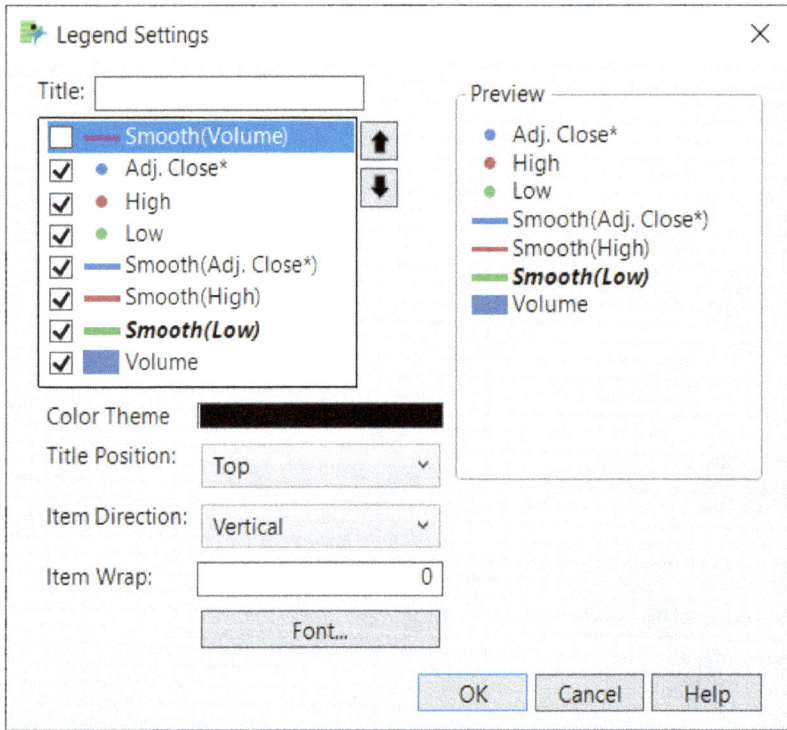

19. Now let's add a few last details. We will change the graph title. Click on the graph title and change it to **Tech Stock Sell-Off Fall 2000 to Winter 2001**. Then click **Done**. You should see a graph like the one shown in Figure 4.17.

Figure 4.17 Final Graph

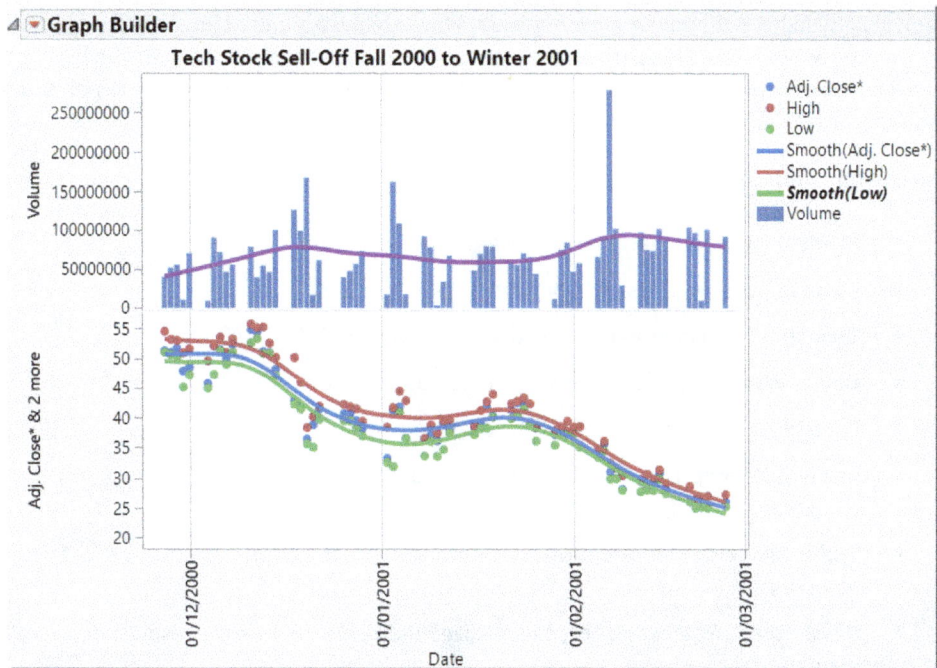

4.2 Using Graph Builder to Produce Maps

The Graph Builder platform also enables you to visualize data on a map, even at the street level. In JMP, maps can be created two ways: using shape files (which are data tables containing the boundaries and names of the shapes such as states or counties of the USA) or background maps (which are background images that recognize latitude and longitude and enable you to plot points accordingly).

Today, mapping data is abundant. JMP provides an easy and flexible environment in which to convey your data on maps. The example in the next section will illustrate how to create a map using the background map service called OpenStreetMaps. An Internet connection is required to use OpenStreetMaps, and you must be using JMP 11 or later.

To make maps in Graph Builder, simply use the same drag and drop methods that you practiced in the previous section. The key step is dragging a shape column to the Map Shape drop zone (to use a Shape file) or in using latitude and longitude data columns in the X and Y drop zones respectively (to use a background map). This action enables maps to be visualized with your data.

If you are unfamiliar with the special conditions of how map data is handled in JMP, see Section 2.7.

Example 4.2: San Francisco Crime

We will be working with one of the sample data files to illustrate the steps in this section. The file that we will use, **San Francisco Crime.jmp**, contains data on individual crime incidents in San Francisco from April 2012. You can find the data set at **Help ▶ Sample Data ▶ Graph Builder ▶ San Francisco Crime.jmp.**

The columns include:

- **Incident Number:** identifier represented as an integer
- **Date:** the date expressed as dd/mm/yyyy
- **Time:** hour and minute expressed as HH:MM: AM/PM
- **Day of Week:** day of the week expressed as a word
- **Category:** 32 offense categories, expressed as words
- **Incident Description:** 336 standardized sub categories of the Category column, expressed as words
- **Traffic Incident:** two categories standardized for indicators of traffic incident, expressed as words
- **Resolution:** 16 categories of standardized indicators of incident resolution, expressed as words
- **Police District:** 10 categories of geographic police district names in the city of San Francisco
- **Incident Address:** nearest cross-street
- **Longitude:** longitude expressed in degrees and directions east and west
- **Latitude:** latitude expressed in degrees and directions north and south
- **Reference:** a compact set of four hidden columns that automatically performs recoding
- **Police Department Color:** a row state column indicating a color marker indexed to each police district

1. First, open the sample data table by selecting **Help ▶ Sample Data ▶ Graph Builder ▶ San Francisco Crime.**
2. Open Graph Builder. Select **Graph ▶ Graph Builder**. (See Figure 4.18.)

Figure 4.18 Open Graph Builder

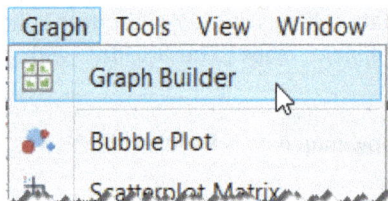

3. Notice the columns in your data table appear on the left in the **Columns** list in the Control Panel. (See Figure 4.19.)

Figure 4.19 San Francisco Crime Variables

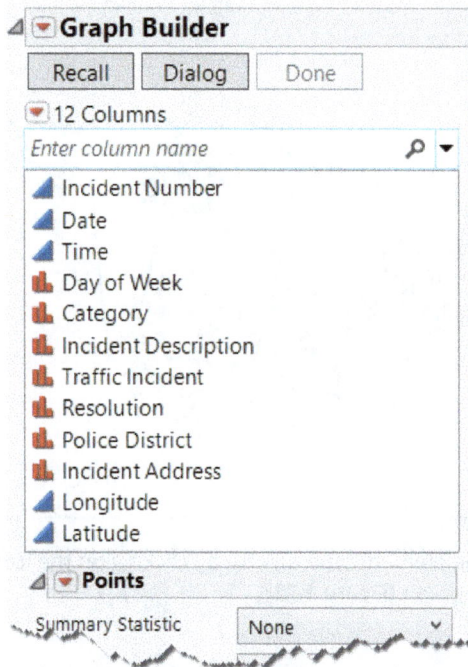

4. Now drag the **Latitude** column from the **Variables** list to the **Y** drop zone located on the left side of the graph canvas. Then drag the **Longitude** column from the **Variables** list to the **X** drop zone located at the bottom of the graph canvas. You should see a point graph with a smoother line. (See Figure 4.20.)

Figure 4.20 Latitude and Longitude Point Graph

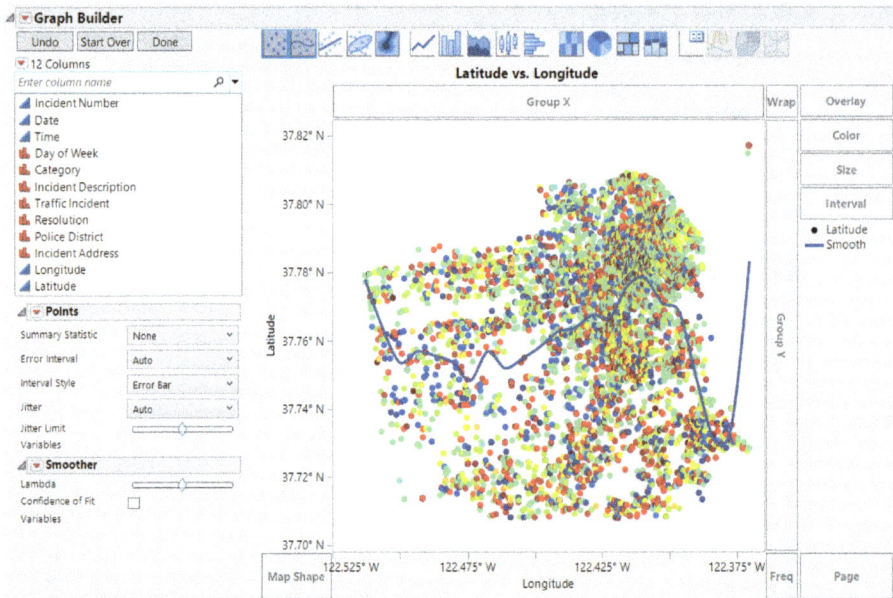

5. Let's deselect the second icon on the element type palette by clicking on the icon to remove the smoother line from the graph (Figure 4.21).

Figure 4.21 Remove Smoother Line

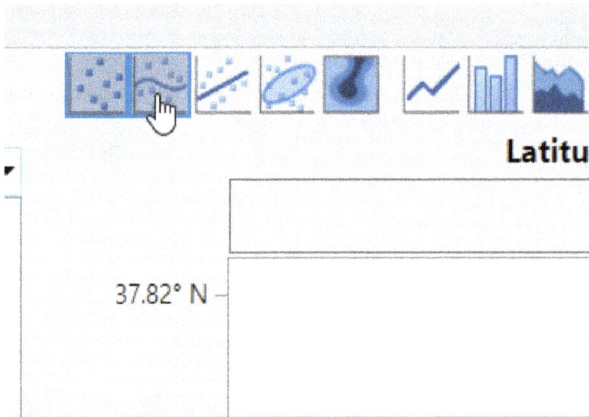

6. Now right-click in the middle of the graph area and choose **Graph ▶ Background Map...** from the submenu. (See Figure 4.22.)

Figure 4.22 Graph Builder Background Map

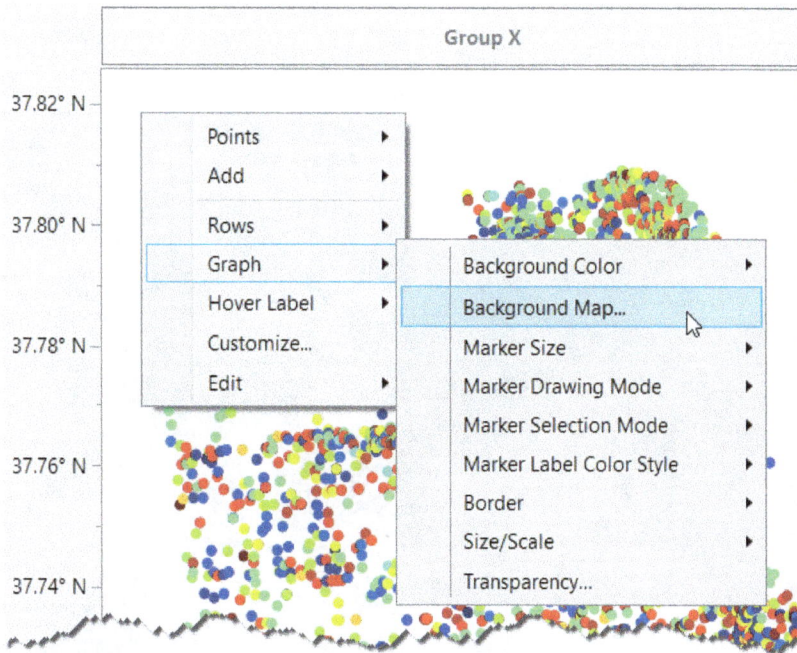

7. Select **Street Map Service** and click **OK**. (See Figure 4.23.)

Figure 4.23 Graph Builder Street Map Service

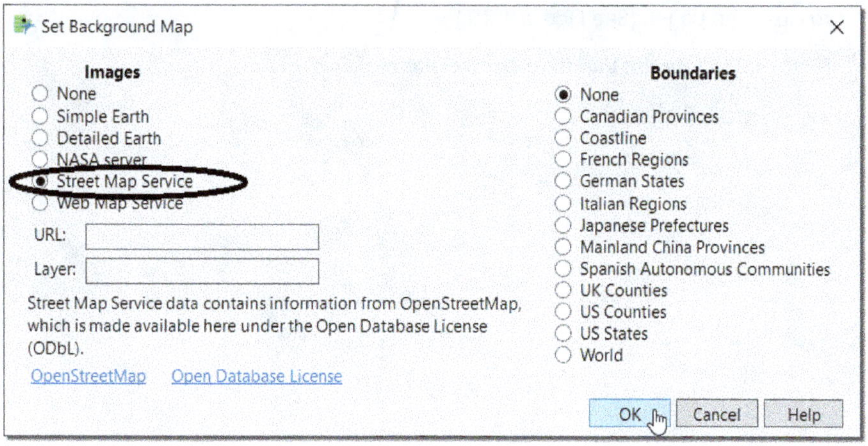

8. The addition of San Francisco streets now appears as a background to the color-coded crime locations. (See Figure 4.24.)

Figure 4.24 Background Street Map Overlay – Color-Coded Crimes

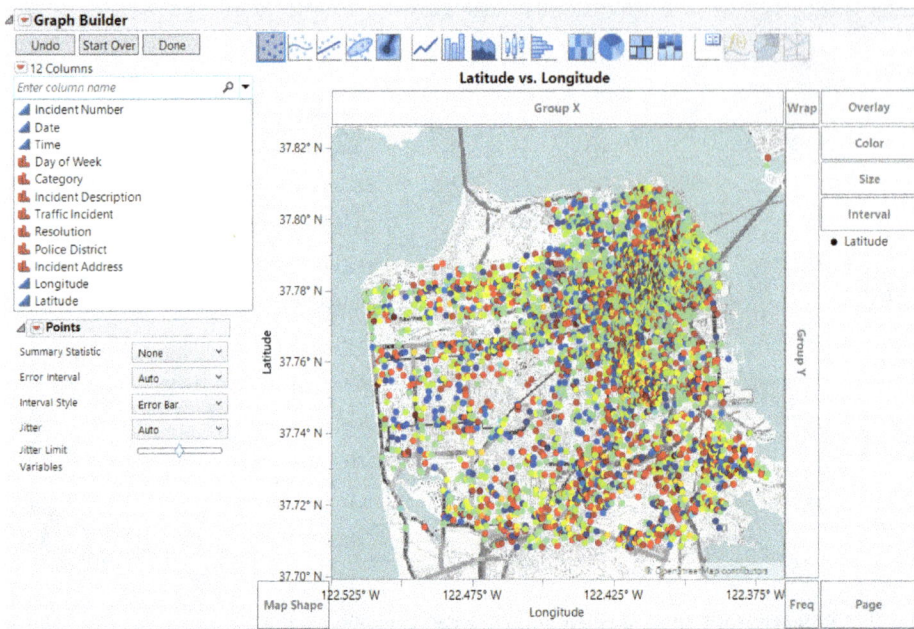

9. You are now finished with the Control Panel. Click the **Done** button in the control panel to close the panel. (See Figure 4.25.)

Figure 4.25 Close the Variables Control Panel

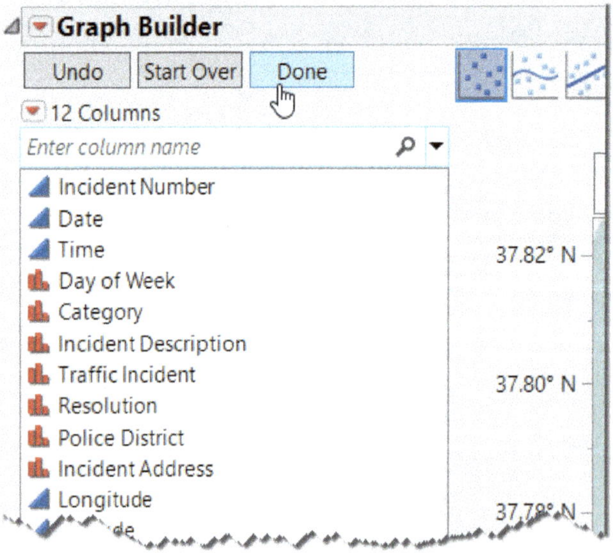

10. Now let's focus on assaults in one particular police district. To do this, we will use a powerful feature called the **Local Data Filter**. From the red triangle for Graph Builder, choose **Local Data Filter**. (See Figure 4.26.)

Figure 4.26 Select Local Data Filter

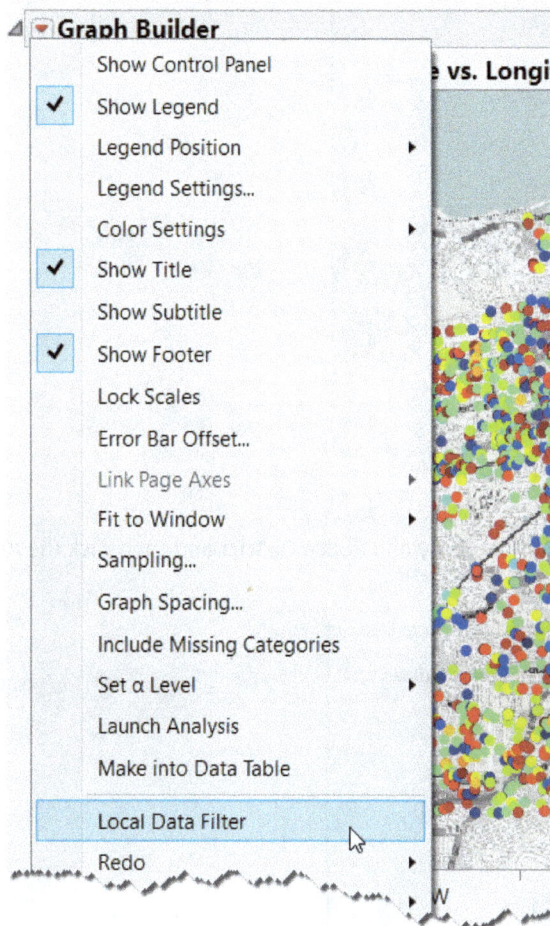

11. A **Local Data Filter** panel appears to the left and inside the Graph Builder window. (See Figure 4.27.)

Figure 4.27 Local Data Filter Panel

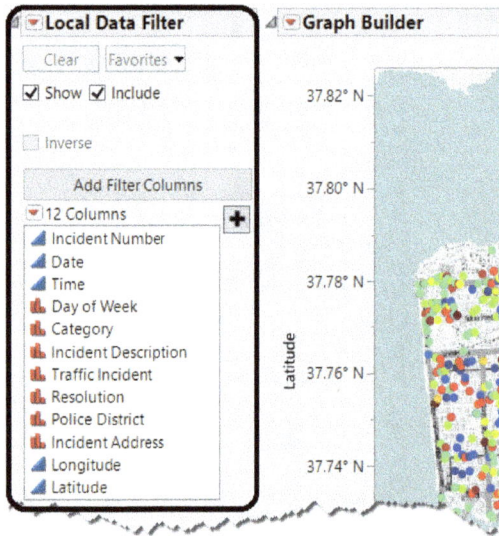

12. Hold down the **Ctrl** key, select **Category** and **Police District** and then click the **Add** button. (See Figure 4.28.)

Figure 4.28 Data Filter Category and Police District

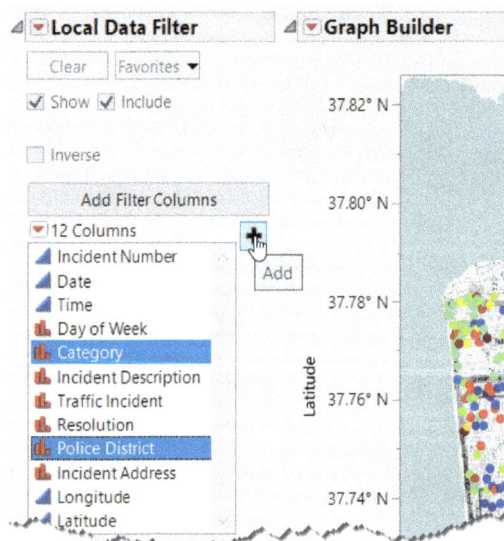

13. A list of crimes and police districts appears in the local data filter. Select **ASSAULT** from the **Category** list, hold down the **Ctrl** key and select **MISSION** from the **Police District** list. The map automatically zooms to the Mission District and maps assault crimes as blue dots. (See Figure 4.29.)

Figure 4.29 Data Filter Assault and Mission

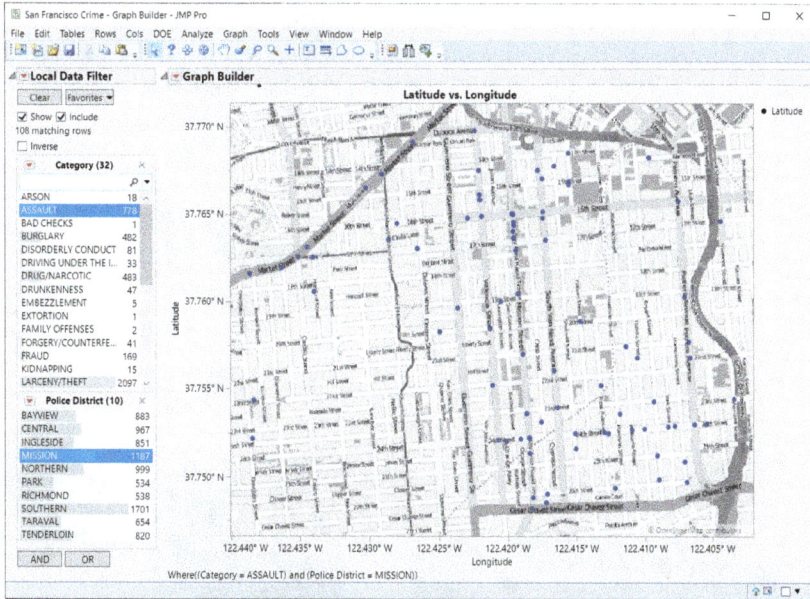

14. Move your mouse pointer to hover over any of the blue dots. Labeling information for the crime is automatically displayed (Figure 4.30).

Figure 4.30 Display Label Information

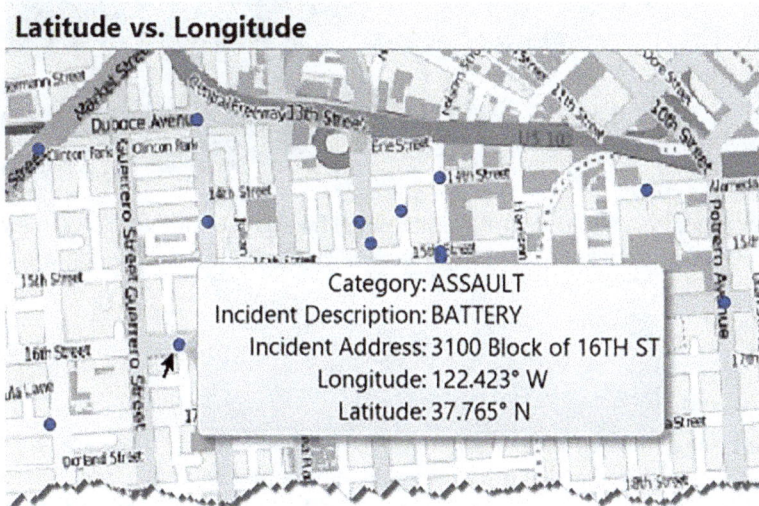

Category: ASSAULT
Incident Description: BATTERY
Incident Address: 3100 Block of 16TH ST
Longitude: 122.423° W
Latitude: 37.765° N

> **Note**
>
> Column information can be used as labels by turning on the label property in the columns list in the data table. If labeling columns is unfamiliar to you, see "Adding Labels to Data" in Chapter 2.

15. Let's add contours to this map. Contours will help plot the density of assaults for the Mission District in San Francisco during this time period. From the red triangle for Graph Builder, select **Graph Builder ▶ Show Control Panel**. Hold the **Shift** key and select the contour icon (circled in Figure 4.31).

Figure 4.31 Add Contours for Density of Assaults

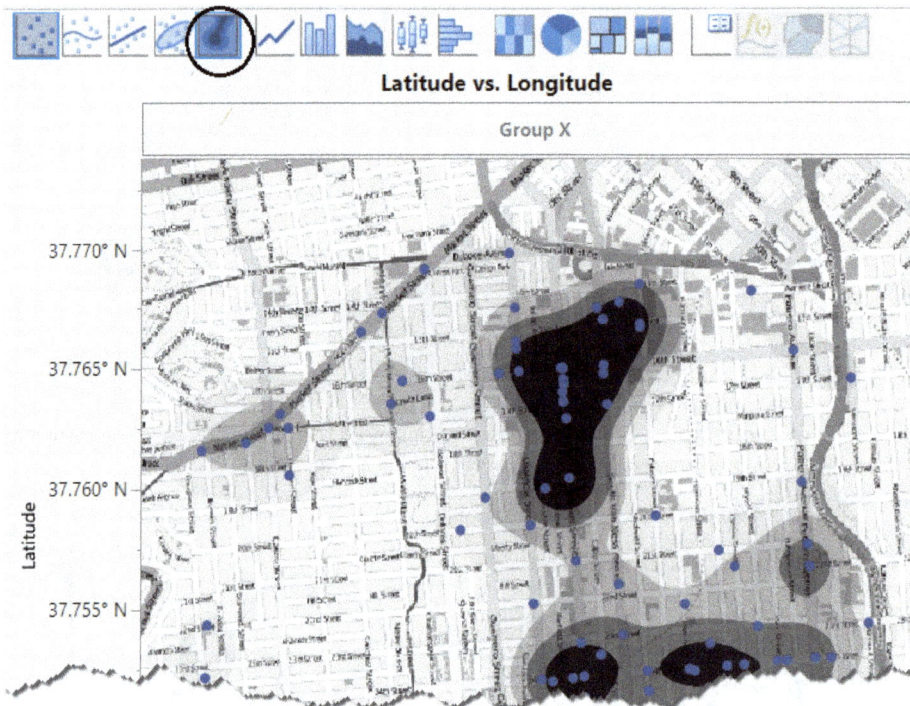

16. Many additional visualizations are possible. Feel free to experiment. You can always click the **Undo** button if you want to go back. Click the **Done** button when you are finished.

4.3 Using Control Chart Builder

As you are aware by now, the approach taken in JMP is one of discovery. There are two graph platforms that are specifically aimed at discovering the nature of your data as you create the graph. Graph Builder was introduced earlier. Control Chart Builder is more focused, as it is used to examine data from a process to assist in identifying unexpected variations or changes in the process.

To open Control Chart Builder select **Analyze ▶ Quality and Process ▶ Control Chart Builder**. As with other platforms, you see a list of your data table columns on the left side. Similar to Graph Builder, there are drop zones in which to place columns, but there is no dialog box to assign columns to roles in the graph.

Examine the drop zones in Control Chart Builder:

- The **Y drop zone** on the left of the graph area is the only required element for a control chart. This is for the process variable to be examined.
- The **Subgroup drop zone** at the bottom of the graph area is for a variable that naturally divides the data into groups.
- The **Phase drop zone** is at the top of the graph. A phase is most often used to identify known changes in the process during the period graphed.

Figure 4.32 Control Chart Builder showing Drop Zones

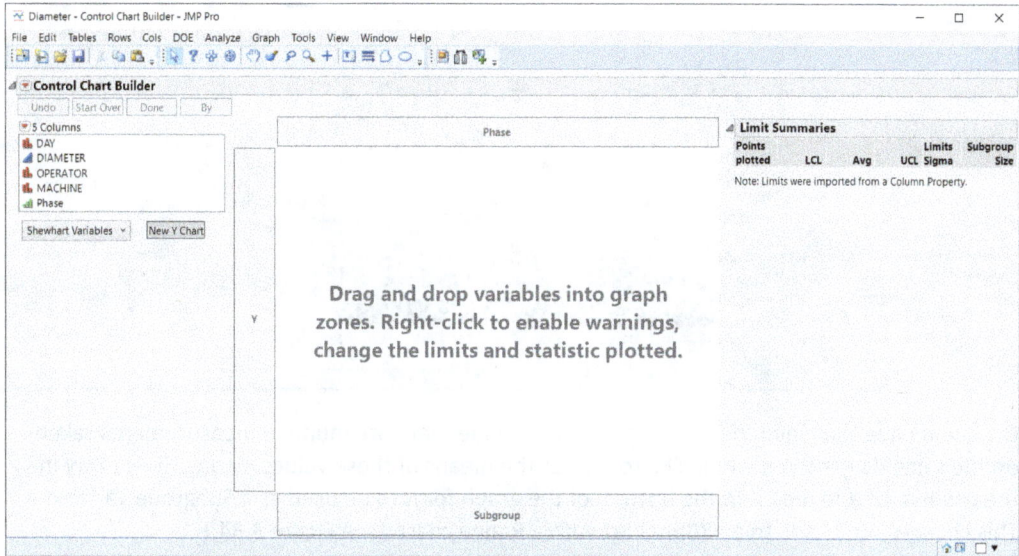

In addition to the drop zones, two buttons can optionally be used to create multiple charts:
- The **By** button is used for data with a single process column using multiple levels of another column. For example, to create separate charts for each location where data is collected.
- The **New Y Chart** button is to select another column to graph in a different chart using the same properties of the first chart.

Example 4.3: Diameters

We will use **Diameters.jmp** to demonstrate using Control Chart Builder with the drop zones. The data table contains measurements of the diameter of tubing used for medical applications. For each diameter measured, the date and operator were recorded. Initial measurements were taken from May 1 to May 20, after which some adjustments were made to the manufacturing process. Additional measurements were then taken from May 21 to June 9. These two time periods are indicated as Phase 1 and 2.

You can find the data set at **Help ▶ Sample Data ▶ Control Charts ▶ Diameter.jmp**.

To begin, create a chart by dragging a continuous process variable into the Y drop zone. For this data table, click on the column **Diameter** and drag it to the Y drop zone. By default, this creates an Individual Measurements chart to examine the data values as well as a Moving Range chart to examine the variability in the data. (See Figure 4.33.)

Figure 4.33 Individual Measurements and Moving Range Chart

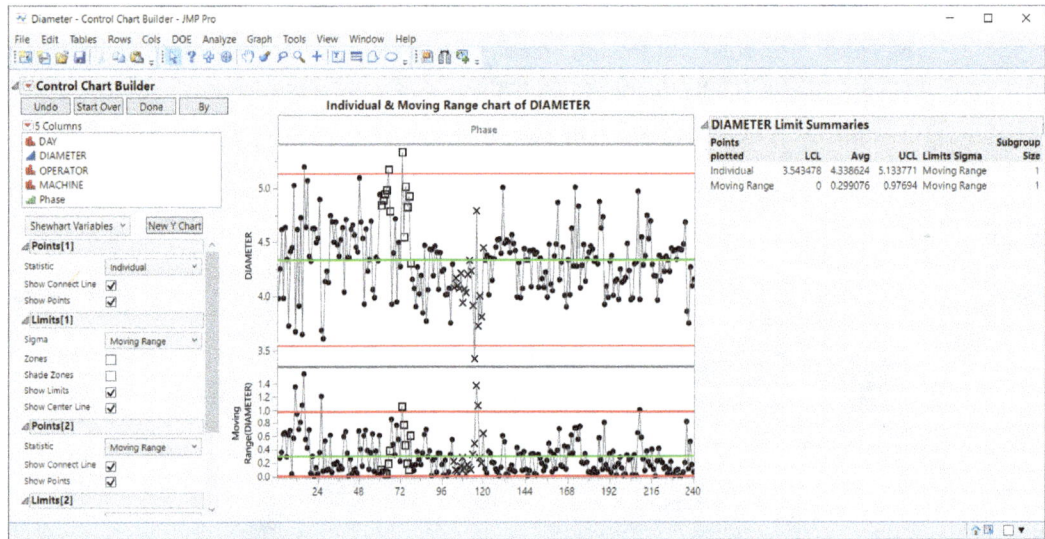

Once you have examined these charts, you determine there are multiple measurements taken on the same day and you would like to look at the means of these values by day. Select **Day** in the column list and drag it to the bottom of the graph for it to be used as a Subgroup variable. This changes the charts to an **XBar** chart with a **Range** chart. (See Figure 3.34.)

Figure 4.34 XBar and Range Charts

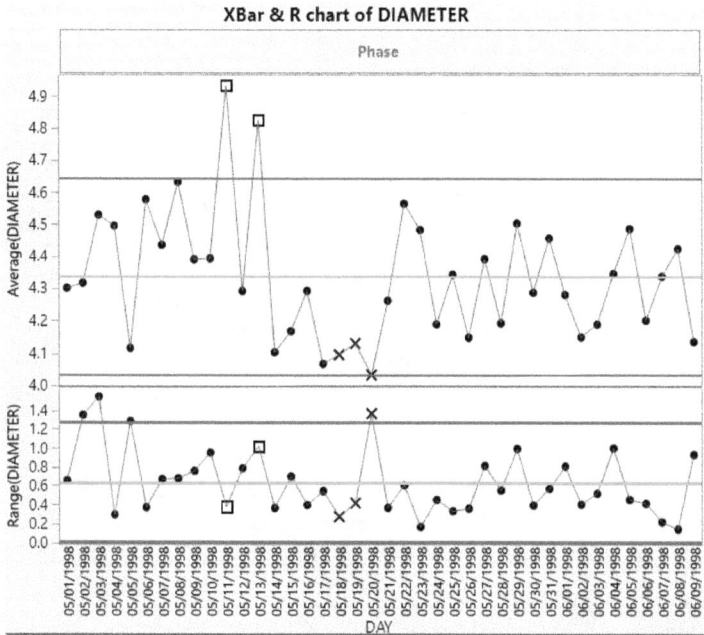

Finally, if you want to examine each of the operators across the days, you can select **Operator** in the columns list and drag it to the outside of the horizontal axis to get Day in Operator as your subgroups. (See Figure 4.35.)

Figure 4.35 XBar and Range Charts with Nested Subgroups

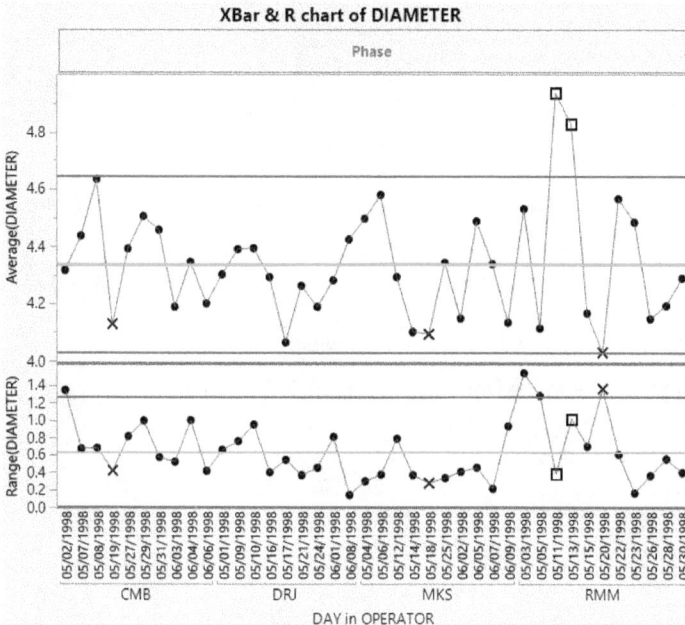

> **Note**
>
> Instead of using Operator as a second subgroup variable, you could choose to use it as a By variable to create a separate control chart for each operator. To do so, instead of dragging Operator to the X drop zone, select **Operator** in the Columns list, then click the **By** button.

4.4 Using Tabulate

Now that you are familiar with using drop zones to generate a graph, we introduce the Tabulate platform. Tabulate is used in a similar fashion to Graph Builder, but instead of creating graphs, your objective is to create numerical summaries for their own sake or for the development of new data tables based on summarized data for further analysis.

Example 4.4: Movies

We will use **Movies.jmp** to demonstrate Tabulate. This data table is a listing of popular US movies that were released from 1937 through 2003. The movies are categorized by type, rating, and director and contain earned gross dollar amounts for both the US domestic market and the worldwide market.

> **Note**
>
> Dollar amounts are expressed in millions of dollars for both domestic and worldwide markets.

You can find the data set at **Help ▶ Sample Data ▶ Business and Demographic ▶ Movies**.

The columns include:

- **Movie:** name of movie
- **Director:** director of movie
- **Type:** genre/category of movie (for example: comedy, family)
- **Rating:** US movie rating system (for example: general audience [G], adult [R])
- **Year:** year of movie release (for example: 1937)
- **Domestic $:** US domestic revenue in $ earned by the movie in that year
- **Worldwide $:** Worldwide revenue in $ earned by the movie in that year

1. After opening the data table, open **Tabulate** (select **Analyze ▶ Tabulate**). (See Figure 4.36.)

    Figure 4.36 Open Tabulate Platform

    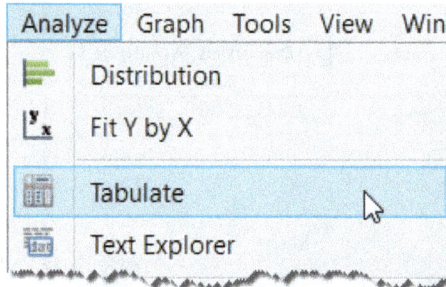

2. The Tabulate platform presents a control panel with your columns, a statistics list, and drop zones for rows and columns (see Figure 4.37)

    Figure 4.37 Tabulate with Domestic and Worldwide Columns

    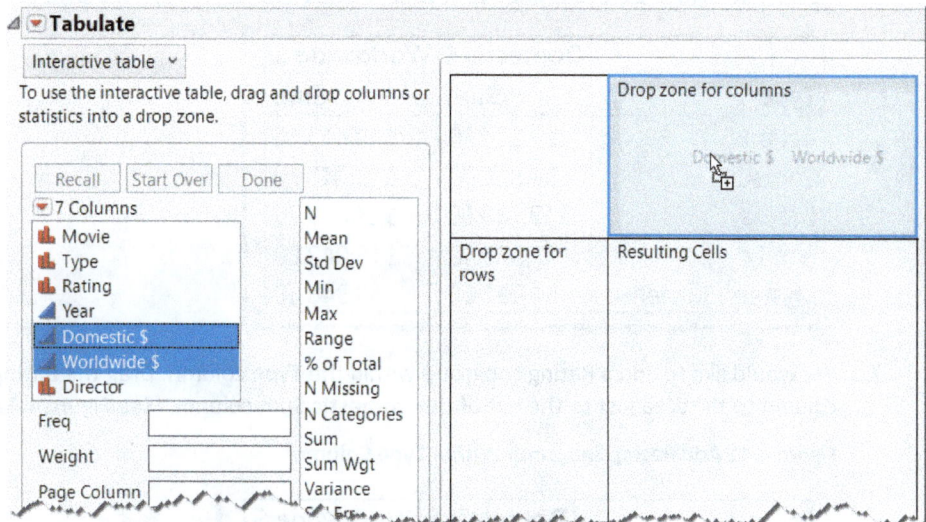

3. Click on both the **Domestic $** and **Worldwide $** columns and drag them to the part of the table labeled **Drop zone for columns**. (See Figure 4.37.)

4. A sum of movie revenue by domestic $ and worldwide $ appears. (See Figure 4.38.)

    Figure 4.38 Sum of Domestic and Worldwide Movie Revenue

	Domestic $	Worldwide $
	Sum	Sum
	$43,622.40	$84,048.60

5. Now drag the **Type** column to the row area. (See Figure 4.39.)

Figure 4.39 Drag Type Column to Row Area

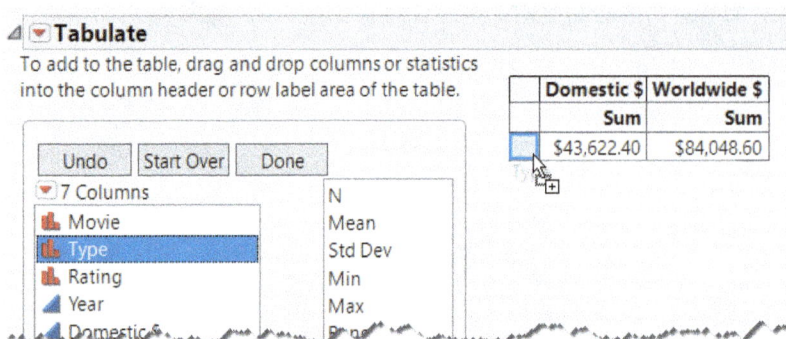

6. Release the mouse button. The table now sums the revenue by type for both domestic and worldwide (Figure 4.40).

Figure 4.40 Sum of Domestic and Worldwide Movie Revenue by type of movie

Type	Domestic $ Sum	Worldwide $ Sum
Action	$9,224.20	$19,610.10
Comedy	$9,818.00	$15,746.10
Drama	$12,789.80	$24,856.70
Family	$7,575.40	$15,042.80
Mystery-Suspense	$4,031.60	$8,348.60

7. We would like to add a Rating subgroup within the **Type** column. Drag the **Rating** column to the area just to the left of the **Domestic Sum** column. (See Figure 4.41.)

Figure 4.41 Add Rating Subgroup Within Type Column

Type	Domestic $ Sum	Worldwide $ Sum
Action	$9,224.20	$19,610.10
Comedy	$9,818.00	$15,746.10
Drama	$12,789.80	$24,856.70
Family	$7,575.40	$15,042.80
Mystery-Suspense	$4,031.60	$8,348.60

8. Release the mouse button and a **Rating** column appears. (See Figure 4.42.)

Figure 4.42 Table with Rating Column Added

Type	Rating	Domestic $ Sum	Worldwide $ Sum
Action	G	$100.50	$100.50
	PG	$899.00	$1,482.40
	PG-13	$5,732.60	$12,269.00
	R	$2,492.10	$5,758.20
Comedy	PG	$2,825.70	$4,409.70
	PG-13	$4,634.50	$7,498.60
	R	$2,357.80	$3,837.80
Drama	G	$198.70	$390.50
	PG	$5,662.10	$9,579.10
	PG-13	$3,719.90	$8,897.50
	R	$3,209.10	$5,989.60
Family	G	$3,640.10	$7,091.10
	PG	$3,791.10	$7,658.30
	PG-13	$144.20	$293.40
Mystery-Suspense	PG-13	$1,965.20	$4,342.80
	R	$2,066.40	$4,005.80

Note

You can click the **Undo** button at any time to undo the last action, or you can click **Start Over** to start over again.

9. Drag the Mean from the **Statistics** column and drop it onto one of the **Sum** column headings. (See Figure 4.43.)

Figure 4.43 Drag Mean Column to Sum Column

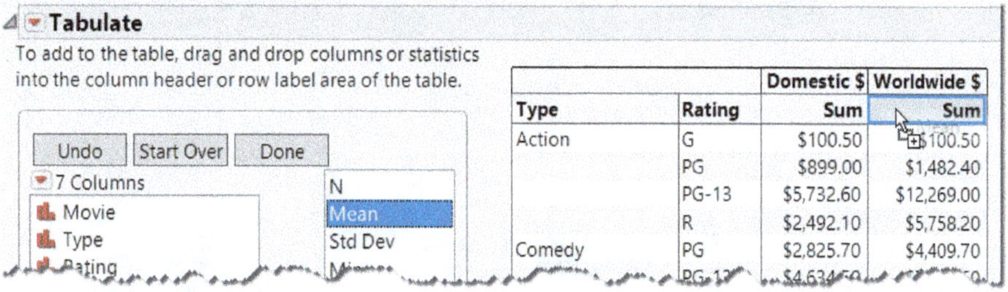

10. Release the mouse button. The **Sum** columns have been transformed to **Mean** columns. (See Figure 4.44.)

Figure 4.44 Sum Columns Changed to Mean Columns

		Domestic $	Worldwide $
Type	**Rating**	**Mean**	**Mean**
Action	G	$100.50	$100.50
	PG	$149.83	$247.07
	PG-13	$184.92	$395.77
	R	$138.45	$319.90
Comedy	PG	$141.29	$220.49
	PG-13	$144.83	$234.33
	R	$138.69	$225.75
Drama	G	$198.70	$390.50
	PG	$161.77	$273.69
	PG-13	$232.49	$556.09
	R	$128.36	$239.58
Family	G	$158.27	$308.31
	PG	$180.53	$364.68
	PG-13	$144.20	$293.40
Mystery-Suspense	PG-13	$151.17	$334.06
	R	$129.15	$250.36

11. Now drag the **Sum** statistic to the right of the **Domestic $**, **Mean** column. (See Figure 4.45.) Release the mouse button. **Sum** columns then reappear in the table (Figure 4.46).

Figure 4.45 Drag Sum Column to the Right of Domestic $ Column

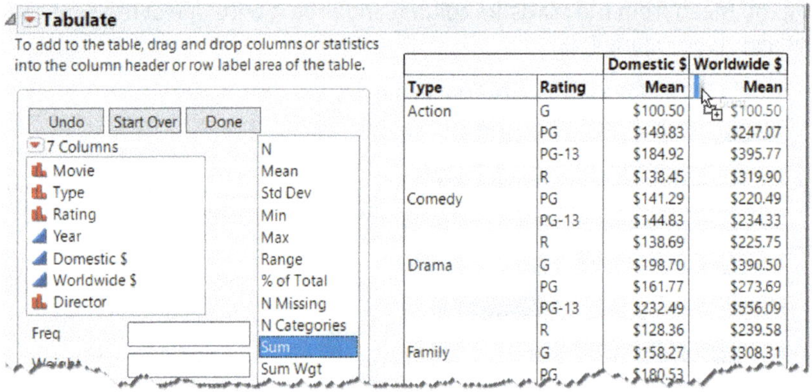

Figure 4.46 Sum Column Added

Type	Rating	Domestic $		Worldwide $	
		Sum	Mean	Sum	Mean
Action	G	$100.50	$100.50	$100.50	$100.50
	PG	$899.00	$149.83	$1,482.40	$247.07
	PG-13	$5,732.60	$184.92	$12,269.00	$395.77
	R	$2,492.10	$138.45	$5,758.20	$319.90
Comedy	PG	$2,825.70	$141.29	$4,409.70	$220.49
	PG-13	$4,634.50	$144.83	$7,498.60	$234.33
	R	$2,357.80	$138.69	$3,837.80	$225.75
Drama	G	$198.70	$198.70	$390.50	$390.50
	PG	$5,662.10	$161.77	$9,579.10	$273.69
	PG-13	$3,719.90	$232.49	$8,897.50	$556.09
	R	$3,209.10	$128.36	$5,989.60	$239.58
Family	G	$3,640.10	$158.27	$7,091.10	$308.31
	PG	$3,791.10	$180.53	$7,658.30	$364.68
	PG-13	$144.20	$144.20	$293.40	$293.40
Mystery-Suspense	PG-13	$1,965.20	$151.17	$4,342.80	$334.06
	R	$2,066.40	$129.15	$4,005.80	$250.36

12. The decimal and display formats can be changed. In the previous example, there are values displayed to the right of the decimal point for the **Mean** column. We would like to eliminate these values in the table. To adjust the display of decimals, click **Change Format**. (See Figure 4.47a.)

Figure 4.47a Change Format

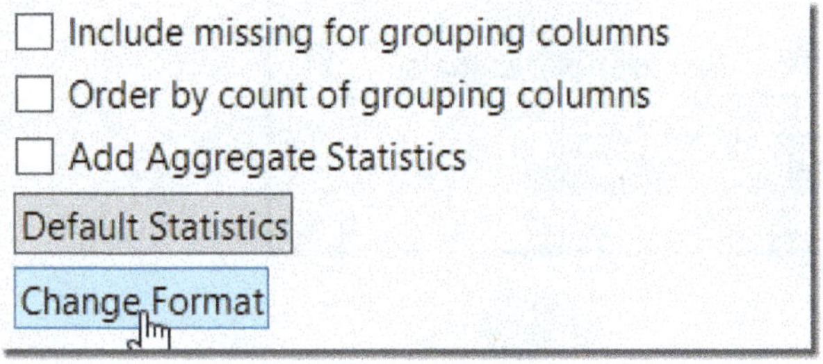

☐ Include missing for grouping columns
☐ Order by count of grouping columns
☐ Add Aggregate Statistics

Default Statistics

Change Format

13. An additional format panel appears. In the **Format** pane, select **Use the same decimal format**. (See Figure 4.47b.)

Figure 4.47b Change Decimal Display

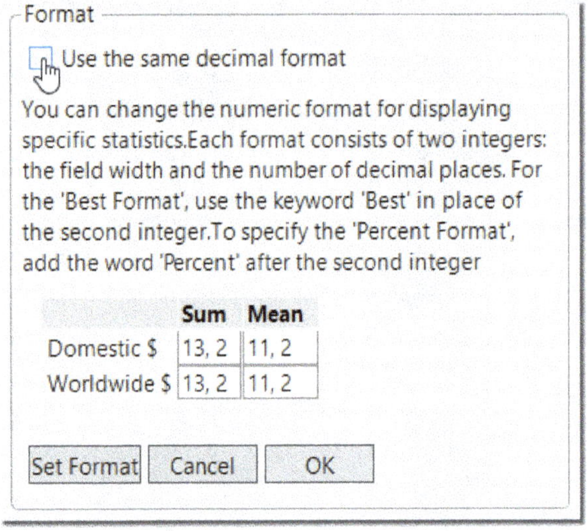

Format

☐ Use the same decimal format

You can change the numeric format for displaying specific statistics.Each format consists of two integers: the field width and the number of decimal places. For the 'Best Format', use the keyword 'Best' in place of the second integer.To specify the 'Percent Format', add the word 'Percent' after the second integer

	Sum	Mean
Domestic $	13, 2	11, 2
Worldwide $	13, 2	11, 2

[Set Format] [Cancel] [OK]

14. In the revised Format panel, click **Fixed Dec** and enter a 0 for **Number of Decimals**. (See Figure 4.47c.) Then, click **Set Format** and **OK**.

Figure 4.47c Specify Fixed Decimal Width

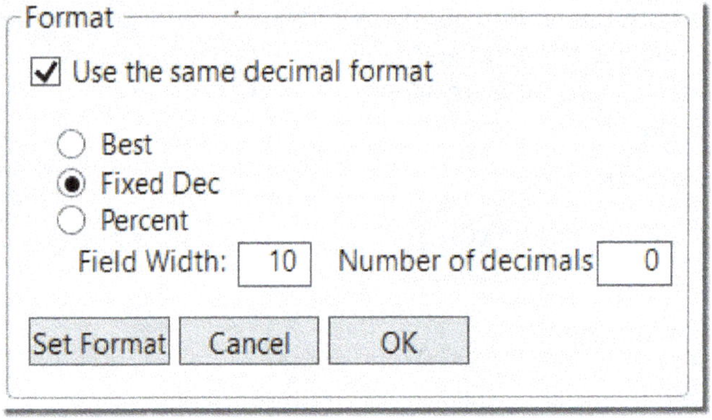

Format

☑ Use the same decimal format

○ Best
◉ Fixed Dec
○ Percent
Field Width: [10]   Number of decimals [0]

[Set Format] [Cancel] [OK]

15. The table is now complete. Click **Done**. (See Figure 4.48.)

Figure 4.48 Finalize Changes

Recall that additional options are available on the red triangle. Click the red triangle next to Tabulate and select **Make Into Data Table** (see Figure 4.49), which will produce a JMP data table of the Tabulate results (Figure 4.50). **Make Into Data Table** will accomplish two tasks simultaneously; you will have a summary table for reporting and a data table of summarized data ready for further analysis or visualization.

Tabulate, therefore, is a useful tool for reshaping and reorganizing data for further analysis or graphing in addition to producing tabular reports.

Figure 4.49 Make Data Table from Tabulate Results

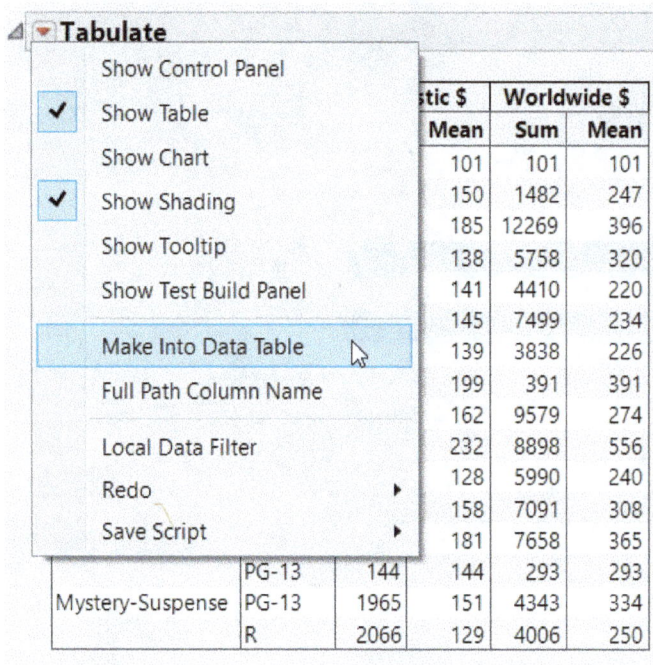

Figure 4.50 Newly Created Data Table

	Type	Rating	Sum(Domestic $)	Mean(Domestic $)	Sum(Worldwide $)	Mean(Worldwide $)
1	Action	G	101	101	101	101
2	Action	PG	899	150	1482	247
3	Action	PG-13	5733	185	12269	396
4	Action	R	2492	138	5758	320
5	Comedy	PG	2826	141	4410	220
6	Comedy	PG-13	4635	145	7499	234
7	Comedy	R	2358	139	3838	226
8	Drama	G	199	199	391	391
9	Drama	PG	5662	162	9579	274
10	Drama	PG-13	3720	232	8898	556
11	Drama	R	3209	128	5990	240
12	Family	G	3640	158	7091	308
13	Family	PG	3791	181	7658	365
14	Family	PG-13	144	144	293	293
15	Mystery-Suspense	PG-13	1965	151	4343	334
16	Mystery-Suspense	R	2066	129	4006	250

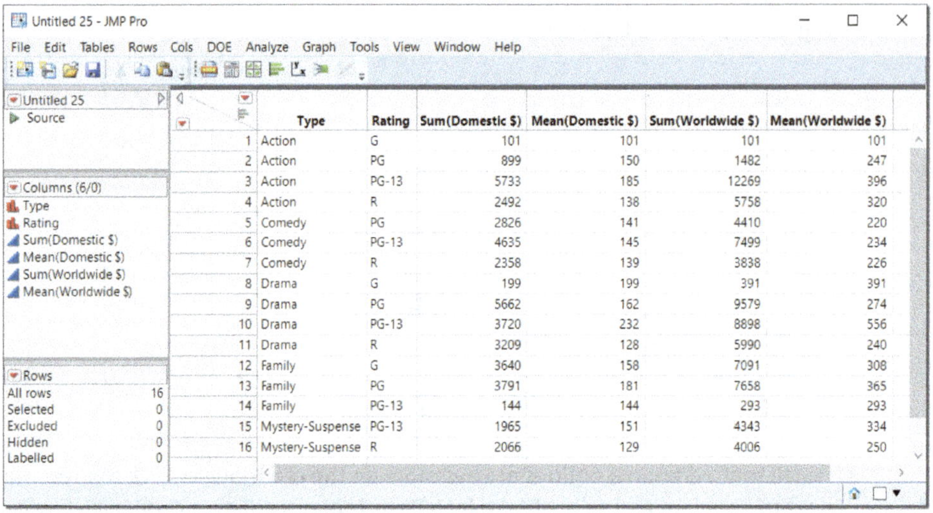

What do you think you can do with the new table in Figure 4.50? Try putting your new skills to work.

16. Use Graph Builder to make a graph out of the data table you just produced (select **Graph ▶ Graph Builder**).
17. Now, press **Ctrl** and select the columns **Sum (Domestic $)** and **Sum (Worldwide $)** so that they are both highlighted (see Figure 4.51).

Figure 4.51 Select Sum Columns

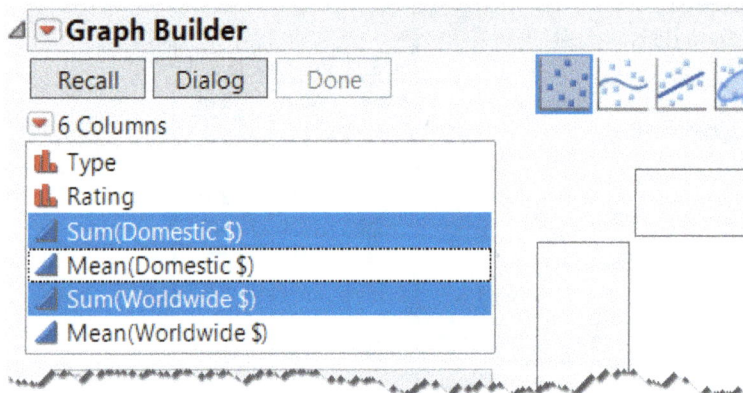

18. Drag the two selected columns to the **Y** drop zone. (See Figure 4.52.)

   Figure 4.52 Drag Sum Columns to Y Drop Zone

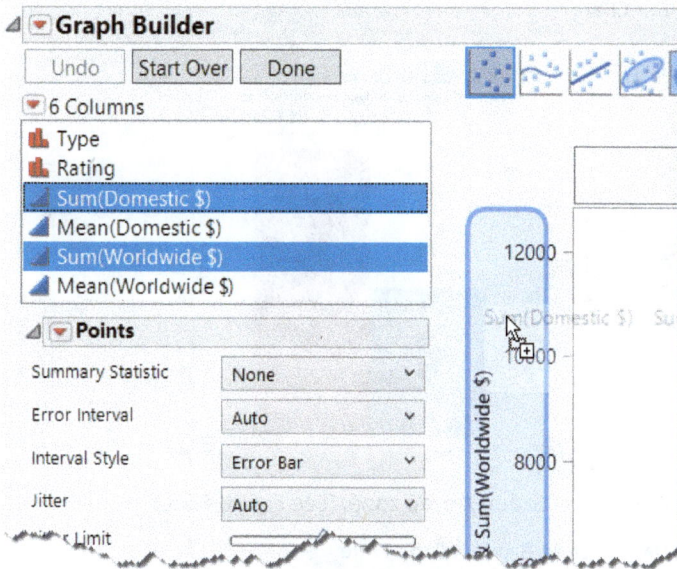

A point chart initially appears with blue and red dots.

19. Select the Bar element type icon. (See Figure 4.53.)

Note

Of course, you could right-click in the graph as you learned earlier in the chapter.

Figure 4.53 Convert Point Chart to Bar Chart

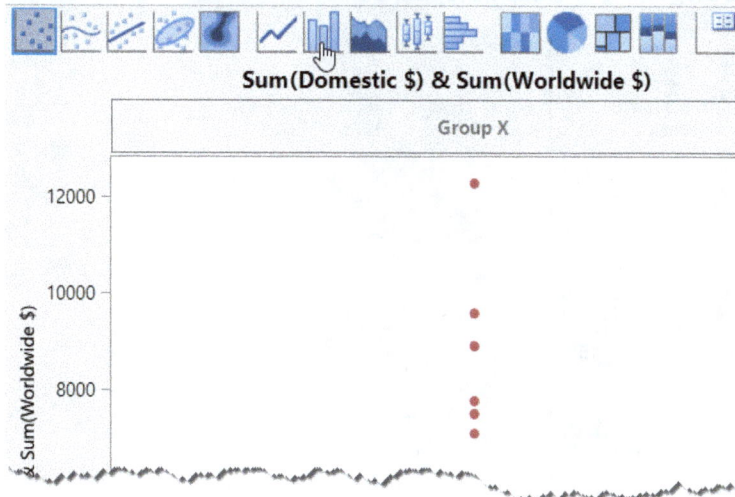

A bar chart appears, showing the worldwide and domestic revenue comparisons. (See Figure 4.54.)

Figure 4.54 Bar Chart

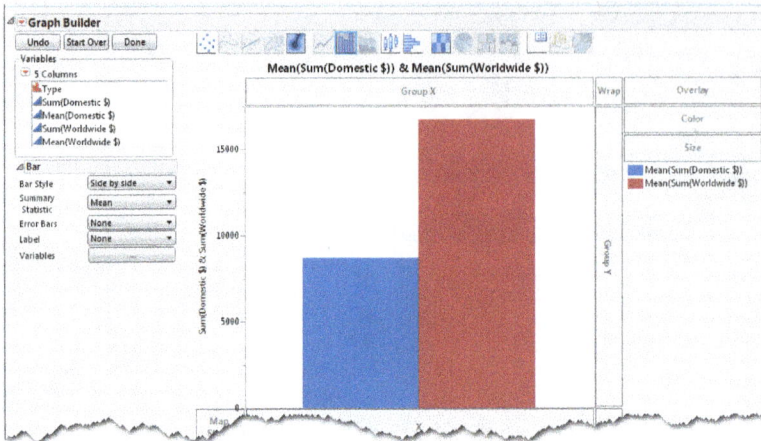

20. Now, drag **Type** to the **Group X** drop zone. (See Figure 4.55.)

Figure 4.55 Drag Type to Group X Drop Zone

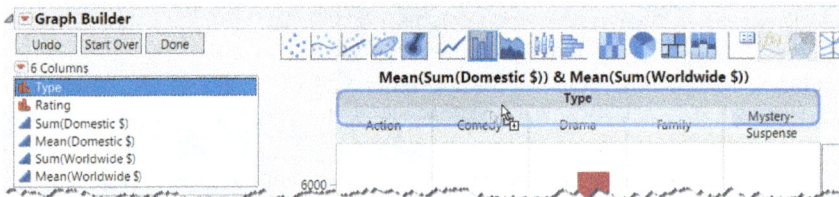

21. Now, drag **Rating** to the **Group Y** drop zone. (See Figure 4.56.)

Figure 4.56 Drag Rating to Group Y Drop Zone

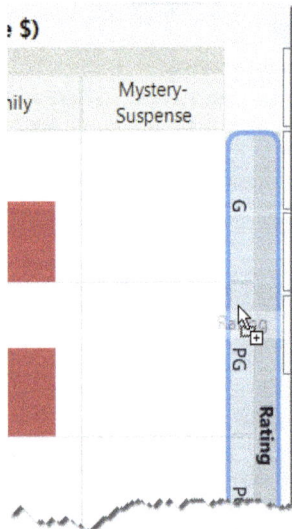

The nearly finished graph appears (Figure 4.57).

Figure 4.57 Nearly Finished Graph

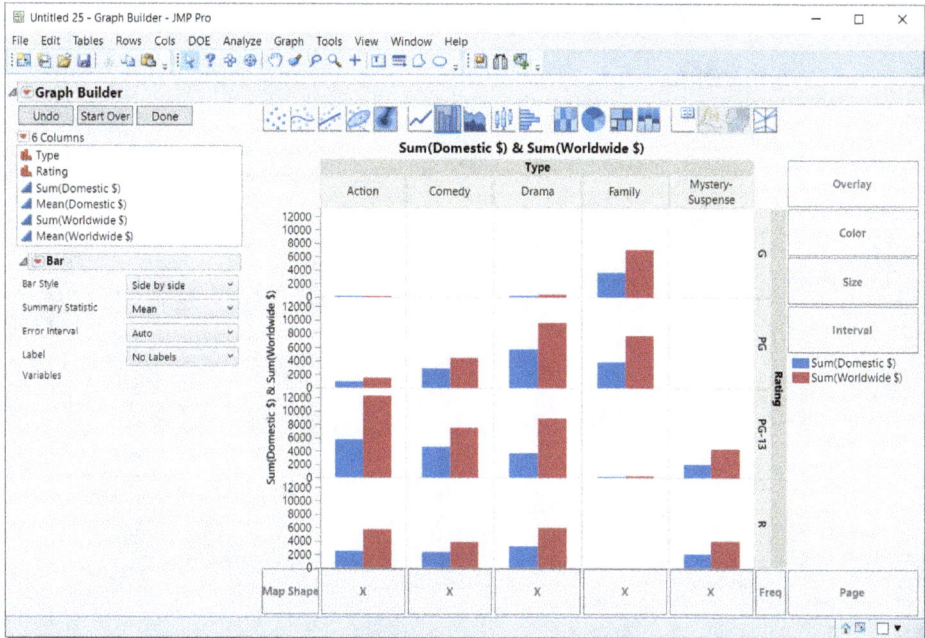

22. Click **Done,** and the finished graph appears. (See Figures 4.58 and 4.59.)

Figure 4.58 Click Done to Finalize

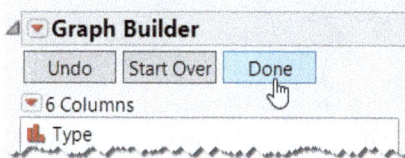

Figure 4.59 Final Graph

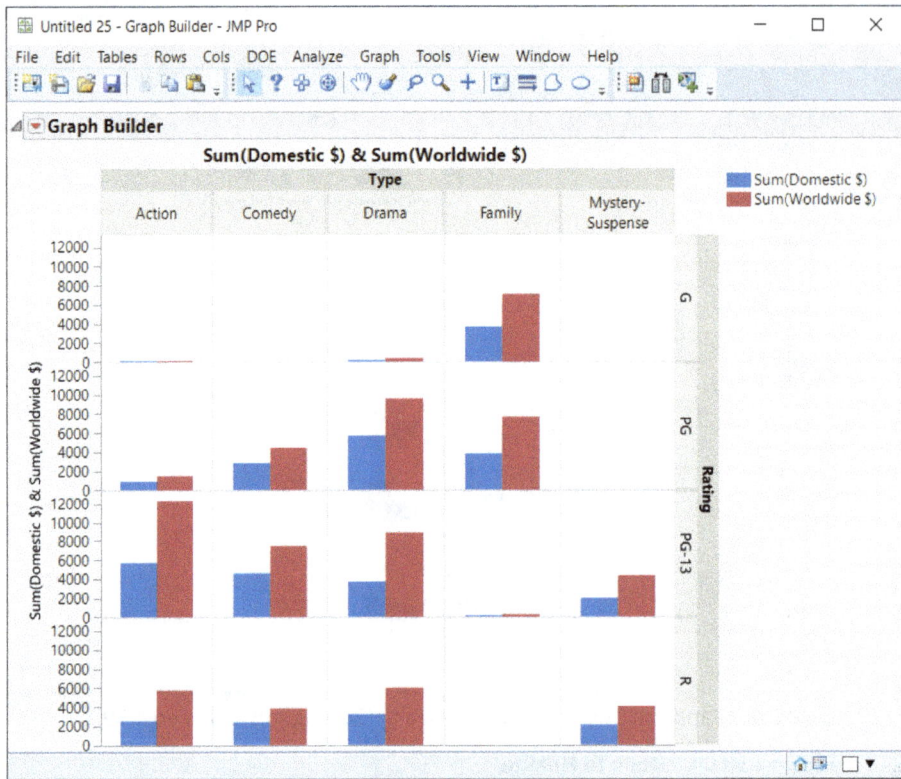

4.4 Summary

In this chapter, we examined a key feature of JMP, which is its ability to create graphs and summaries interactively and intuitively in an environment that enables quick exploration and discovery. This chapter introduced you to two tools, Graph Builder and Tabulate.

Graph Builder shows you quick previews of graphs and maps as you drag columns to drop zones, enabling you to flip through choices until you find just the right one to tell the graphical story in your data. With a local data filter, Graph Builder enables you to drill down to the right level in your data. In contrast, the methods used in Chapter 3 assume that you have a specific graph in mind and need directions to reproduce it.

Tabulate shows you quick previews of numerical table summaries as you drag columns to drop zones. Tabulate enables you to flip through choices until you find the right numerical table summary of your data.

Chapter 5: Problem Solving with One and Two Columns

In contrast to earlier chapters that have focused on describing data and producing graphs and maps, this chapter is about getting answers to your questions and making sense of your data in a condensed and rapid way. This problem-solving activity is often a process of trial and error, and it does not lend itself to brief descriptive steps. It takes thought and practice to do this well, but JMP is the perfect companion on this journey. This chapter helps you develop some appreciation and basic JMP skills in this problem-solving process.

Just as we discussed in the first chapter, JMP provides a navigation framework that is designed around the workflow of the problem solver. So, what do we mean by, "the workflow of the problem solver?" First, we are talking about the class of problems that are measurable or countable or that already have data that is written down. If you need assistance importing or accessing your data, see Chapter 2. By workflow, we are referring to the process by which you analyze data to arrive at some understanding or insight. This involves thinking about the questions that you may have of the data and recognizing how the questions translate to the menu items. Users who learn this find JMP to be a very intuitive partner in that problem-solving process.

In the problem-solving process, the answer to one question often prompts other questions. JMP is designed to help you answer these follow-up questions quickly, resulting in more rapid learning. Throughout this chapter, we use simple examples with scenarios that prompt questions you might want to answer. We show you how JMP's menus translate to these questions and how the results help you answer them. Just as many real-world problems start with basic questions and understanding and evolve into more complex ones, we start off with the basics here as well.

5.1 Introduction

The transformation of data into information is a process that involves a few basic JMP platforms that we introduce in this chapter, including **Distribution** and **Fit Y by X**. Distribution and Fit X by Y correspond to analyzing the characteristics of one and two columns, respectively, which is the scope of this chapter. The tools are found in the **Analyze** menu. Table 5.1 outlines the tool name, how many columns it supports, and its statistical terms. Don't worry if you do not recognize the statistical terms and acronyms; we will present the basic ideas as we go. Appendix B describes the statistical terms and concepts in this chapter.

The organization of the items on the Analyze menu is the same framework discussed in Section 1.3. Within this framework, we cover just a few menu items, but in the process, you gain access to over 100 statistical methods.

Table 5.1 Tool Name, Columns Supported, and Statistical Terms

Analyze Menu	How Many Columns	Statistical Terminology
Distribution	Single Columns	One sample univariate methods, histograms, box plots, quantiles, summary statistics, distribution and more
Fit Y by X	Two Columns	Bivariate, scatterplot, contingency, logistic, oneway ANOVA, nonparametrics and more

Note

If there is magic to JMP, the Analyze menu is it. The arrows in Table 5.1 indicate how the menu works. First and simple questions are answered with the Distribution platform and then further questions are often addressed with items farther down the menu like Fit Y by X until you get the answers that you need to solve your problem. Thus, you proceed from the simple to just the level of complexity that you need to answer your questions as you work your way down the Analyze menu. Sometimes, a discovery is made with a more complex method farther down the Analyze menu that leads you to confirm it with a simpler method back up the Analyze menu.

5.2 Analyzing One Column

JMP's goal in using this menu framework is to expose you to powerful methods in a logical order that enables you to learn about your data progressively and rapidly. The order is built into the menu structure.

Figure 5.1 displays the first item on the Analyze menu. It is the Distribution platform. It's first on the menu because this platform answers many basic questions that you should ask at the beginning of analysis of data.

Figure 5.1 Analyze Distribution

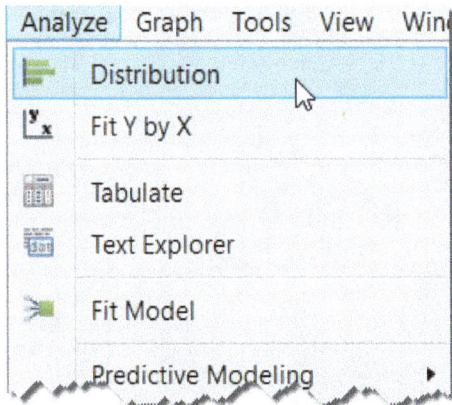

Another way to remember Distribution as a starting point is that the platform enables you to look at one column at a time and produces results for those individual columns. The technical terminology for statistical calculations for one column (or variable) of data is *univariate one-sample statistics*, meaning analysis of the properties of one variable represented as a sample. There are dozens of one-sample univariate statistics, and most are conveniently arranged in the Distribution platform.

Example 5.1: Financial

We will be using the financial performance data in the **Financial.jmp** data file to illustrate the analyses explored in this chapter. The data are from financial performance data for Fortune 500 companies selected from the April 23, 1990 *Fortune* magazine issue. You can find the data set at **Help ▶ Sample Data Library ▶ Financial.jmp**.

This data file includes columns for:

- **Type:** type of company
- **Sales($M):** yearly sales in millions of dollars
- **Profit($M):** yearly profits in millions of dollars
- **#emp:** number of employees at time of measurement
- **Profits/emp:** profits per employee in thousands of dollars
- **Assets($Mil.):** assets in millions of dollars
- **Sales/emp:** sales per employee in thousands of dollars
- **Stockholder's Eq($Mil.):** stockholder's equity in millions of dollars

Let's start working our way through an analysis. We will use this exercise to practice asking the early questions of unknown data using the Distribution platform in JMP. Let's open a data table on which to explore some analyses:

1. Select **Help ▶ Sample Data Library** (Figure 5.2).

Figure 5.2 Help Sample Data

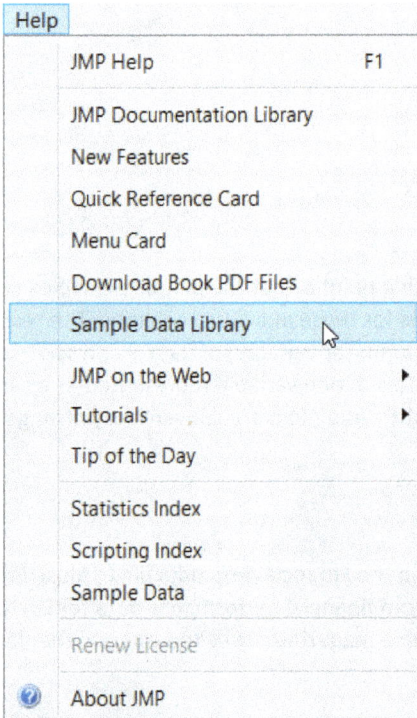

Help		
JMP Help		F1
JMP Documentation Library		
New Features		
Quick Reference Card		
Menu Card		
Download Book PDF Files		
Sample Data Library		
JMP on the Web		▶
Tutorials		▶
Tip of the Day		
Statistics Index		
Scripting Index		
Sample Data		
Renew License		
About JMP		

2. Then select **Financial.jmp**. (See Figure 5.3.) Some financial performance data are displayed. (See Figure 5.4.)

Figure 5.3 Select Financial.jmp

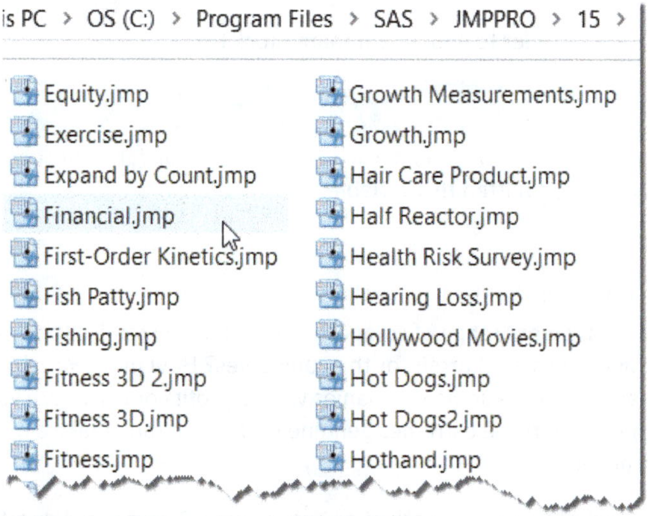
is PC > OS (C:) > Program Files > SAS > JMPPRO > 15 >
Equity.jmp
Exercise.jmp
Expand by Count.jmp
Financial.jmp
First-Order Kinetics.jmp
Fish Patty.jmp
Fishing.jmp
Fitness 3D 2.jmp
Fitness 3D.jmp
Fitness.jmp
Growth Measurements.jmp
Growth.jmp
Hair Care Product.jmp
Half Reactor.jmp
Health Risk Survey.jmp
Hearing Loss.jmp
Hollywood Movies.jmp
Hot Dogs.jmp
Hot Dogs2.jmp
Hothand.jmp

Figure 5.4 Financial Performance Data

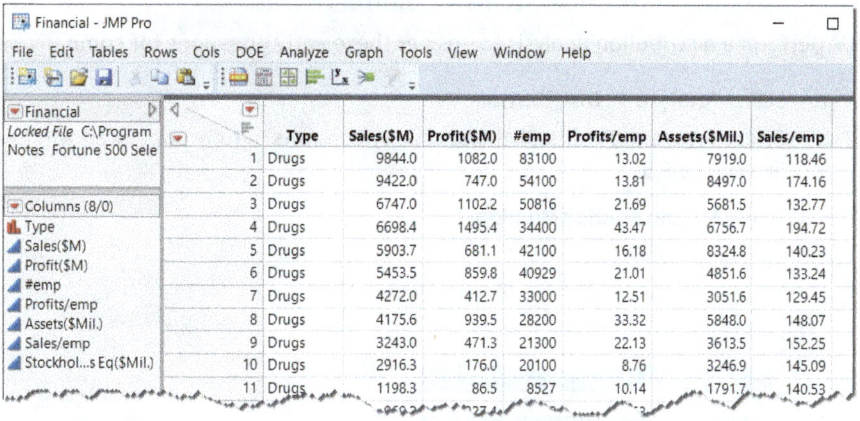
Financial - JMP Pro
File Edit Tables Rows Cols DOE Analyze Graph Tools View Window Help
Financial
Locked File C:\Program
Notes Fortune 500 Sele
Columns (8/0)
Type
Sales($M)
Profit($M)
#emp
Profits/emp
Assets($Mil.)
Sales/emp
Stockhol...s Eq($Mil.)

	Type	Sales($M)	Profit($M)	#emp	Profits/emp	Assets($Mil.)	Sales/emp
1	Drugs	9844.0	1082.0	83100	13.02	7919.0	118.46
2	Drugs	9422.0	747.0	54100	13.81	8497.0	174.16
3	Drugs	6747.0	1102.2	50816	21.69	5681.5	132.77
4	Drugs	6698.4	1495.4	34400	43.47	6756.7	194.72
5	Drugs	5903.7	681.1	42100	16.18	8324.8	140.23
6	Drugs	5453.5	859.8	40929	21.01	4851.6	133.24
7	Drugs	4272.0	412.7	33000	12.51	3051.6	129.45
8	Drugs	4175.6	939.5	28200	33.32	5848.0	148.07
9	Drugs	3243.0	471.3	21300	22.13	3613.5	152.25
10	Drugs	2916.3	176.0	20100	8.76	3246.9	145.09
11	Drugs	1198.3	86.5	8527	10.14	1791.7	140.53

Let's assume that your objective is to use this financial data to help you select company stock types to add to your portfolio. Your goal is to select stock types that will maximize the likelihood of positive returns and to use those returns to fund your favorite charity. We assume that:

- The most profitable companies will also tend to have the highest positive returns.
- You have enough data to make a reasonable prediction.

- Each row represents a type of company stock. You will need to select about 10 types of company stocks to sufficiently diversify your portfolio.

- Market conditions for the next six months will remain mostly the same for all sectors.

What questions will you ask in order to choose company stock types?

Note

Time is an important variable to consider in particular with financial data. For the purposes of illustration, however, we have omitted this variable.

Questions That Involve One Column

In this example, you might ask what range of company profits has existed for these stocks. Or, what has been the average (the mean) profit for the companies? How much variability (standard deviation) has there been? Are there some companies whose profits or losses are very extreme (outliers) relative to others? Are these extremes genuine or did someone make a mistake when entering the data (data quality)?

These are the types of questions you might ask of any set of data. These initial questions and many more are all answered with the Distribution platform from the Analyze Menu.

Using Distribution to Understand a Column of Data

Let's perform a distribution analysis to answer these early questions for company performance.

1. Select **Analyze ▶ Distribution**.
2. Select **Type** and **Profits($M)** and click **Y, Columns**. A fully populated window should look like Figure 5.5.

Figure 5.5 Distribution Window

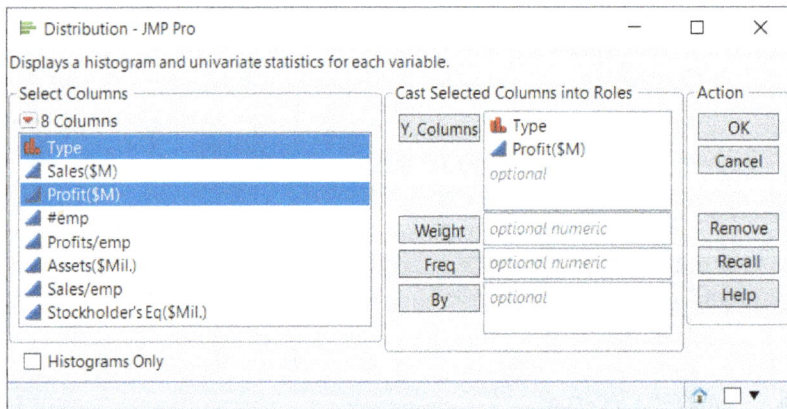

3. Click **OK**.

You are presented with a result (Figure 5.6). In this example, you can see several things that might capture your attention. There are six black dots at the high end of the range. There are the most profitable companies.

Figure 5.6 Distribution Report

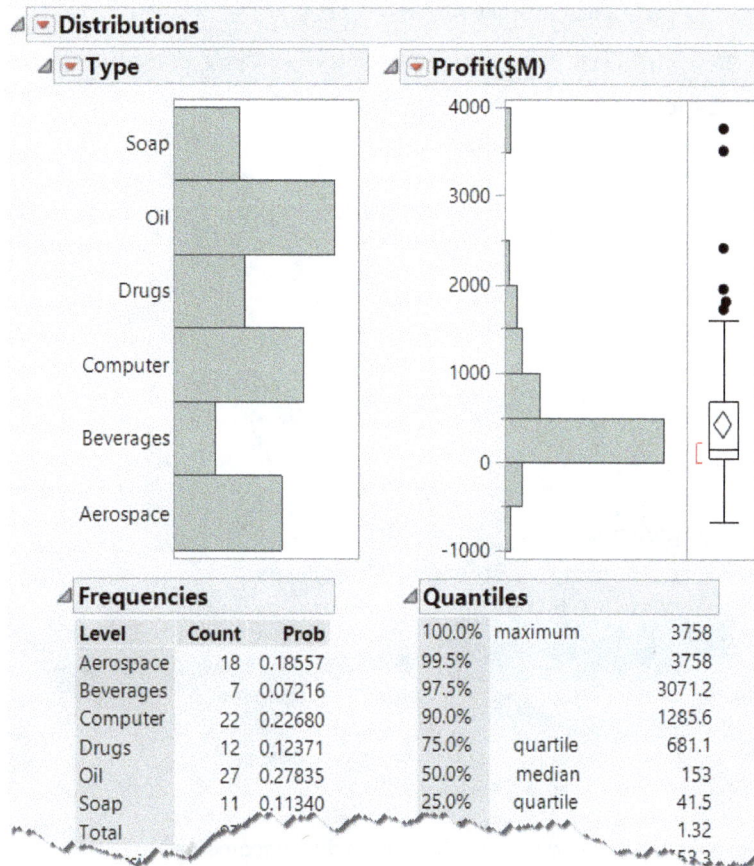

You might also notice that some company types in the portfolio are more heavily represented than others. The size of the bars in the Type distribution graph indicates that Oil has the most companies and Beverages the least. The frequencies are shown in the table below the graph. Oil has 27 companies in this data while Beverages has only seven. The average profit for all of the companies is slightly more than $426 million.

Note

Did you know that graphs and tables of numerical results appear together in report windows by design? People can learn faster when both graphs and numerical results appear together in the same context. This is a guiding principle of all JMP reports.

Now, let's try something new to help us answer some important questions.

4. Draw a box with your pointer around some of the highest profit entries. (See Figure 5.7.) Just take the pointer, left-click and drag down over the dots in the profit graph, and a rectangle appears.

Figure 5.7 Draw a Box to Show Highest Profits

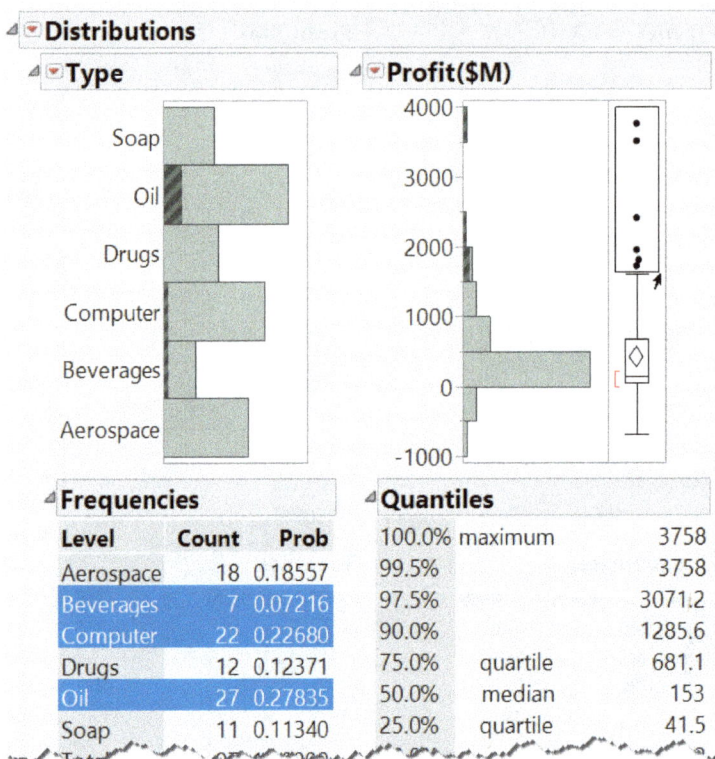

5. Notice that as you draw the box, the round dots become dark.
6. Notice also the graph bars on the left for company type. Certain types of company bars have turned dark green, including Oil, Computer, and Beverages. This means that companies with the greatest profits are also those mostly associated with the oil sector. Notice the tiny slivers of Computer and Beverages that turned dark green, which indicates that a small number of highly profitable companies are in these two business sectors.
7. Now select the **Window** menu from the top and select **Financial**.

    This brings the data table to the foreground. You can also select the data table icon from the lower right corner in the associated report window to bring the data table to the foreground. (See Figure 5.8.)

Figure 5.8 Select the Report Icon to Bring Data Table to the Foreground

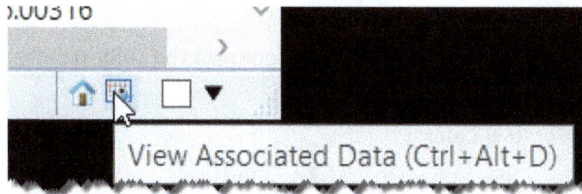

Note

Nearly all graphs that appear in report windows in JMP are tied directly to any other displayed graphs AND to rows in the corresponding data table. By selecting graphical attributes, you see those corresponding values represented in other graphs AND highlighted in the data table.

8. Scroll down to row 35. (See Figure 5.9.)

Figure 5.9 Rows Selected

		Type	Sales($M)	Profit($M)	#emp	Profits/emp	Assets($Mil.)	Sales/emp
	32	Computer	855.1	31.0	7523	4.12	615.2	113.66
	33	Computer	784.7	89.0	4708	18.90	955.8	166.67
	34	Computer	709.3	41.4	5000	8.28	468.1	141.86
	35	Oil	86656.0	3510.0	104000	33.75	83219.0	833.23
	36	Oil	50976.0	1809.0	67900	26.64	39080.0	750.75
	37	Oil	32416.0	2413.0	37067	65.10	25636.0	874.52
	38	Oil	29443.0	251.0	54826	4.58	33884.0	537.03
	39	Oil	24214.0	1610.0	53648	30.01	30430.0	451.35
	40	Oil	21703.0	1405.0	31338	44.83	27599.0	692.55
	41	Oil	17755.0	965.0	53610	18.00	17500.0	331.19
	42	Oil	15905.0	1953.0	26600	73.42	22261.0	597.93
	43	Oil	12492.0	219.0	21800	10.05	11256.0	573.03
	44	Oil					9257.0	

Do you see how the highlighted rows match those same points that you highlighted by drawing the box in the Distribution window?

Note

The highlighting is what we refer to as JMP's dynamic linking, which automatically links graphs to data, data to graphs, and graphs to other graphs.

Because these highlighted rows show high profits, let's mark them so that we can always find them.

9. Select **Rows ▶ Colors**, and then select the color red from the color palette. (See Figure 5.10.)

Figure 5.10 Select Red for Row Color

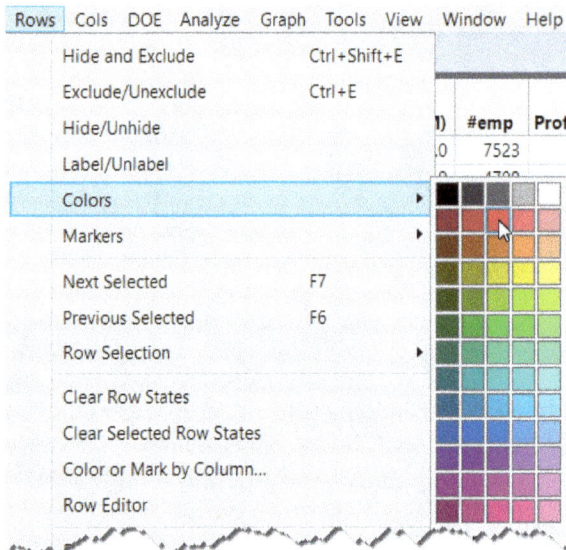

10. Select **Rows ▶ Markers**, and select the marker type **X**. (See Figure 5.11.)

Figure 5.11 Select Marker Type X for Rows

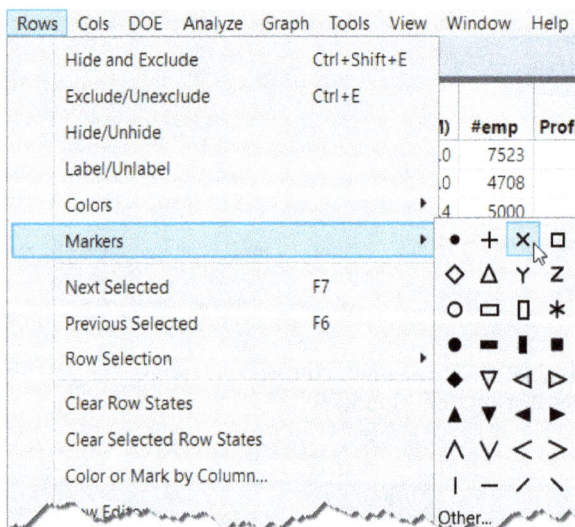

These rows in the data table are now marked red and appear with a marker type of **X**, as shown in the row number column of the data table. (See Figure 5.12.)

Figure 5.12 Rows Colored Red and Marked with X

11. Select **Window** and bring the **Financial – Distributions of Profit($M), Type** result to the foreground. Focus on the Distributions result for **Profit($M)**. (See Figure 5.13.)

Figure 5.13 Top Profit Marked with Red Xs

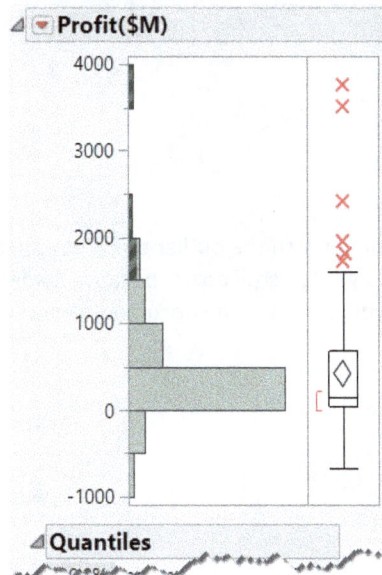

The distribution graph for profit also shows the red X markers as well as a box plot. The top X represents a company with over $3.7 billion in profits. (See Figure 5.13.) On the bottom of the range is a very unprofitable company with approximately $1 billion loss (Figure 5.13).

If a box plot or any graphical or numerical result is unfamiliar to you, use the question mark tool (?) from the **Tools** menu to identify the item in question and locate the corresponding documentation about the item.

12. Select the question mark (**?**) from the toolbar. The pointer changes to a question mark. (See Figure 5.14.)

Figure 5.14 Question Mark Tool from Toolbar

13. Then, move the question mark on top of the item that you are unfamiliar with and click on that item. (See Figure 5.15.) In this example, it is the item next to the distribution graph.

Figure 5.15 Question Mark on Distribution

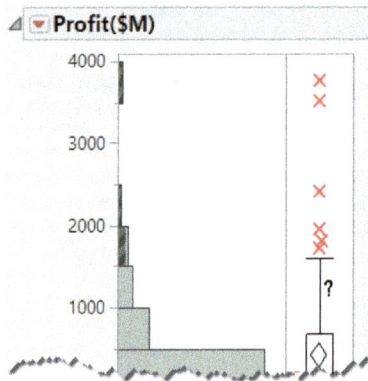

The section of the documentation associated with the outlier box plot appears automatically. (See Figure 5.16.) Don't forget to scroll down, because sometimes the topic of interest is a little below where you landed in the documentation.

Figure 5.16 Help for Box Plots

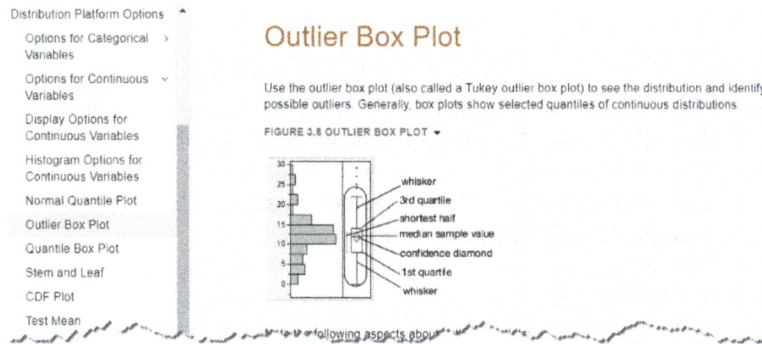

> **Note**
>
> Unfamiliar graphics, geometric shapes, and results are explained using the context-sensitive question mark tool. Just select the question mark tool (**?**) from the **Tools** menu and click on the part of the result that you want to learn about. You can learn statistics while you explore your data!

Summary of 5.2: What You Learned from the Distribution Platform

By observing the distribution, you found that the range of profits for the companies extended from a very unprofitable company at approximately -$1 billion to the most highly profitable company at over $3.7 billion (Figure 5.13). Companies were mostly on the profitable side, however, and most companies had moderate profits.

You might have noticed that the average, or mean, profits for all listed companies was $426 million.

The distribution results indicated that the variability, or standard deviation, of profits was about $717 million.

You found that all the extreme values were on the high side. By identifying them, you determined that most of these companies came from the oil sector and a few came from other industry types. You also identified an unfortunately unprofitable company.

5.3 Comparing a Continuous Column to a Categorical Column

Building upon the Distribution platform, the answers to questions you asked about one column have motivated you to ask further questions about the relationship between two columns. For example, what might be the relationship of profits to other columns in the data table? In fact, you have already identified an interesting relationship between the company type and its profits using dynamic linking, as some types of companies are associated with higher profits. This visual relationship suggests something is happening with profits that includes more than just one column. There may be a relationship between at least two columns that we will explore in this section.

The second item under the Analyze menu is **Fit Y by X**. It is designed to explore relationships between one column and another column. These are sometimes called bivariate relationships (meaning relationships between two variables). You might have noticed a pattern developing for items under the Analyze menu. The first item is Distribution, which is useful for looking at one column at a time. The second item is Fit Y by X for looking at the relationship between two columns. Figure 5.17 provides a description of each menu item.

Figure 5.17 Analyze Menu Item Descriptions

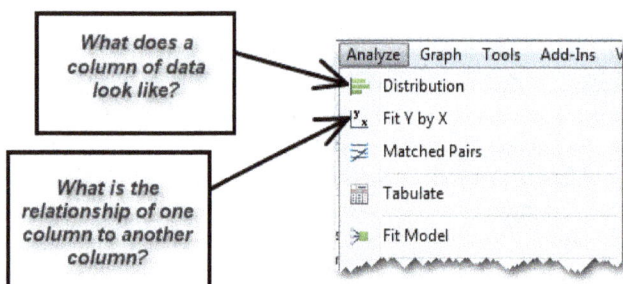

Let's continue with our example and perform a Fit Y by X analysis. We already suspect that there are interesting profit differences among the different company types, and we have discovered a few by just looking at one column at a time using the Distribution platform and dynamic linking. We will now formalize that inference.

The relationship that we want to explore includes the **Profit($M)** column and the **Type** column.
1. Select **Analyze ▶ Fit Y by X**. (See Figure 5.18.)

Figure 5.18 Select Fit Y by X

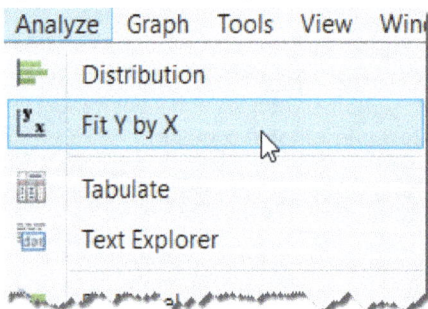

2. In the window (see Figure 5.19), select **Profit($M)** and click **Y, Response.**

Figure 5.19 Fit Y by X Profit by Type Oneway Analysis

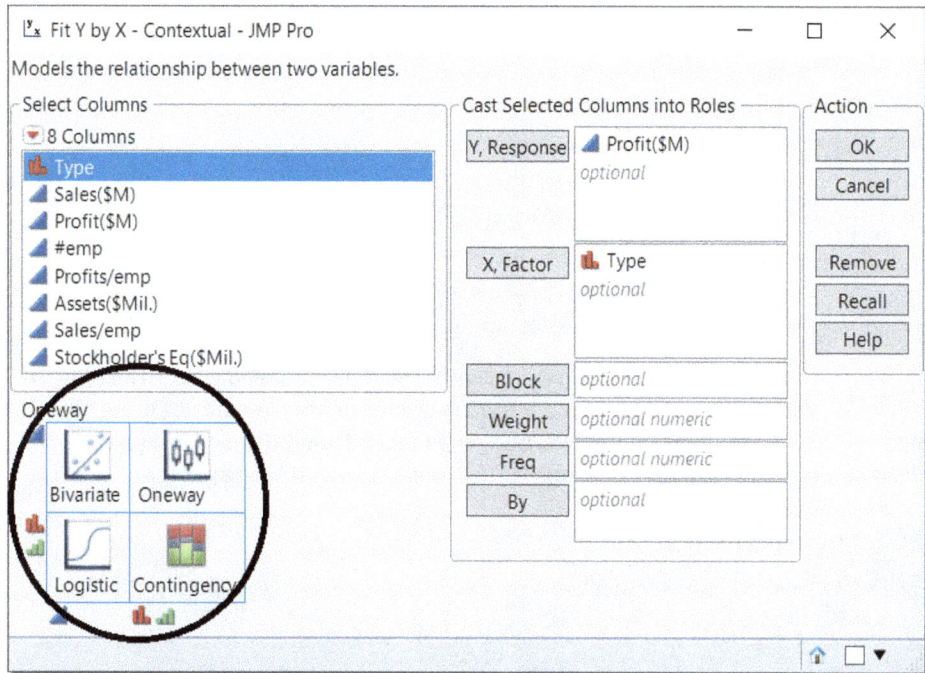

3. Select **Type** and click **X, Factor.**
4. Click **OK**.

Before we continue with the example, let's examine that last Fit Y by X window. We will focus on the preview or circled area. (See Figure 5.19.)

Notice that each modeling type has its corresponding icon (continuous, nominal, or ordinal, which are described in section 2.3). The modeling type of the column that you cast into a role determines the type of analysis that is produced. (See Figure 5.20.) In this case, we have selected **Profit($M)** for the **Y, Response** (the vertical axis), which is continuous, and **Type** for the **X, Factor** (the horizontal axis), which is nominal. The arrows in Figure 5.20 are a close-up view of the Fit Y by X window shown in Figure 5.19, which shows these selections. Where they intersect on the matrix indicates the type of analysis that will be produced; in this case, a oneway.

Figure 5.20 Oneway Analysis Illustration

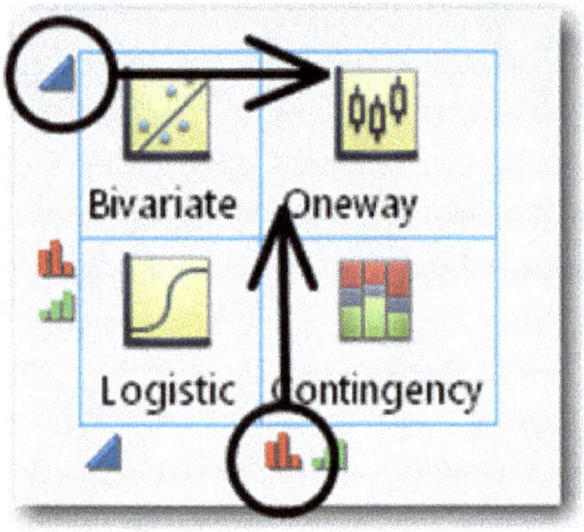

For now, don't worry about terms like *oneway* in the preview. If you want to learn about the result, you can place the question mark tool on the result after generating it. Just note that the picture previews are there so that you can get an idea of what types of analyses will be produced when you cast certain types of columns into Y and X roles within the platform.

When you click **OK**, a oneway analysis graph appears. (See Figure 5.21.)

Figure 5.21 Oneway Analysis Graph

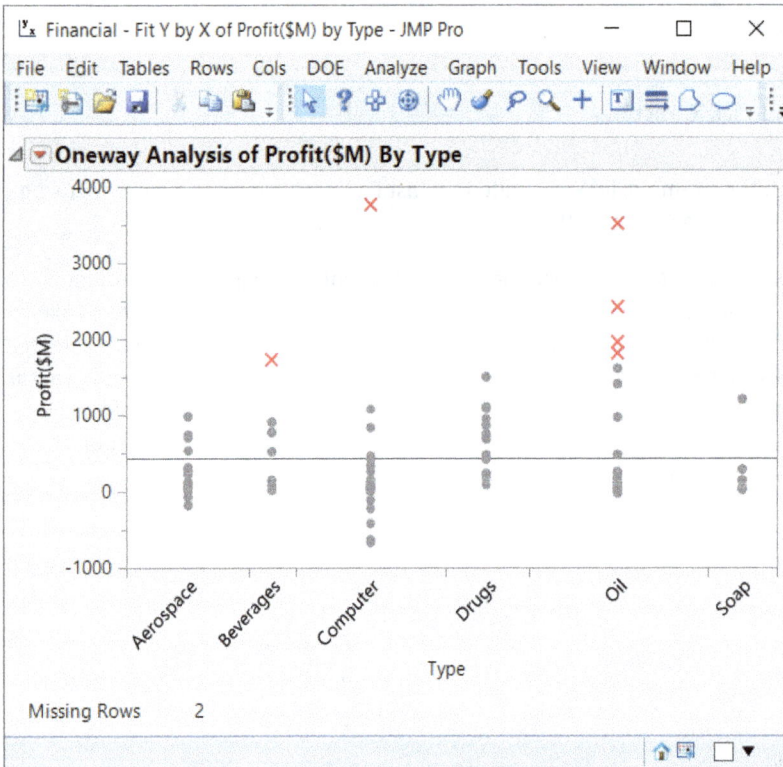

The Fit Y by X (two-column) analysis confirms what we started to observe earlier. We see that the highest profits have come from a mix of mostly oil companies, one computer company, and one beverage company. These are conveniently marked by Xs from a previous step.

Based on the observed best performers among the company types, we might start asking more complex questions like these:

- Are there differences in profits among the company types?

- How big and in what direction (negative or positive) are these differences?

- Should I choose more companies from one particular company type than from another type?

These questions and others can be answered using Fit Y by X because they involve two columns. How? As you make selections from the choices in the Analyze menu, the types of questions you start asking at each step are anticipated for you.

You might notice the red triangle associated with the results that you have generated. (See Figure 5.22.)

Figure 5.22 Red Triangle

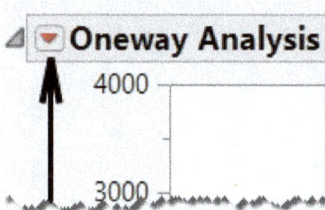

As we introduced earlier, red triangles anticipate questions that you might have at any stage of analysis and have been carefully placed on the report in the context of your analysis. Let's use the choices on the menu for the oneway analysis to find out if there are differences in profits among company types.

You can already see some of these differences in the graph. Let's further quantify the differences in profits by company type.

From the red triangle, select **Means and Std Dev**. (See Figure 5.23.)

Figure 5.23 Select Means and Std Dev

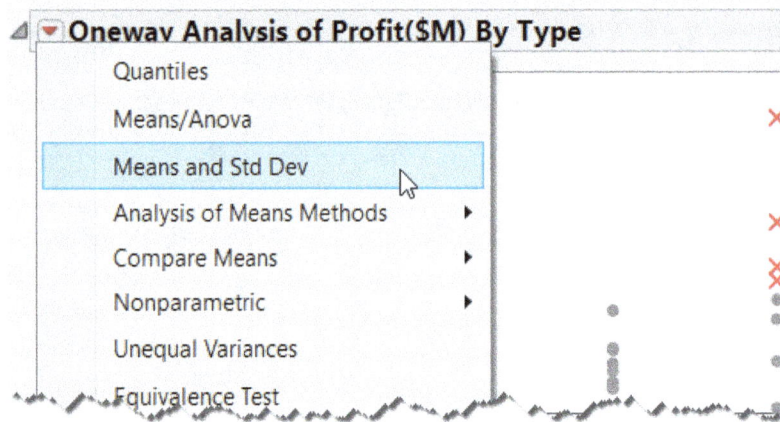

An additional table appears below your graph, and blue mean error bars and standard deviation lines appear on your graph. The blue lines on the graph correspond to mean lines, mean error bars, and standard deviation lines. These values also appear in the table as numbers. (See Figure 5.24.)

Figure 5.24 Means for Drugs and Oil

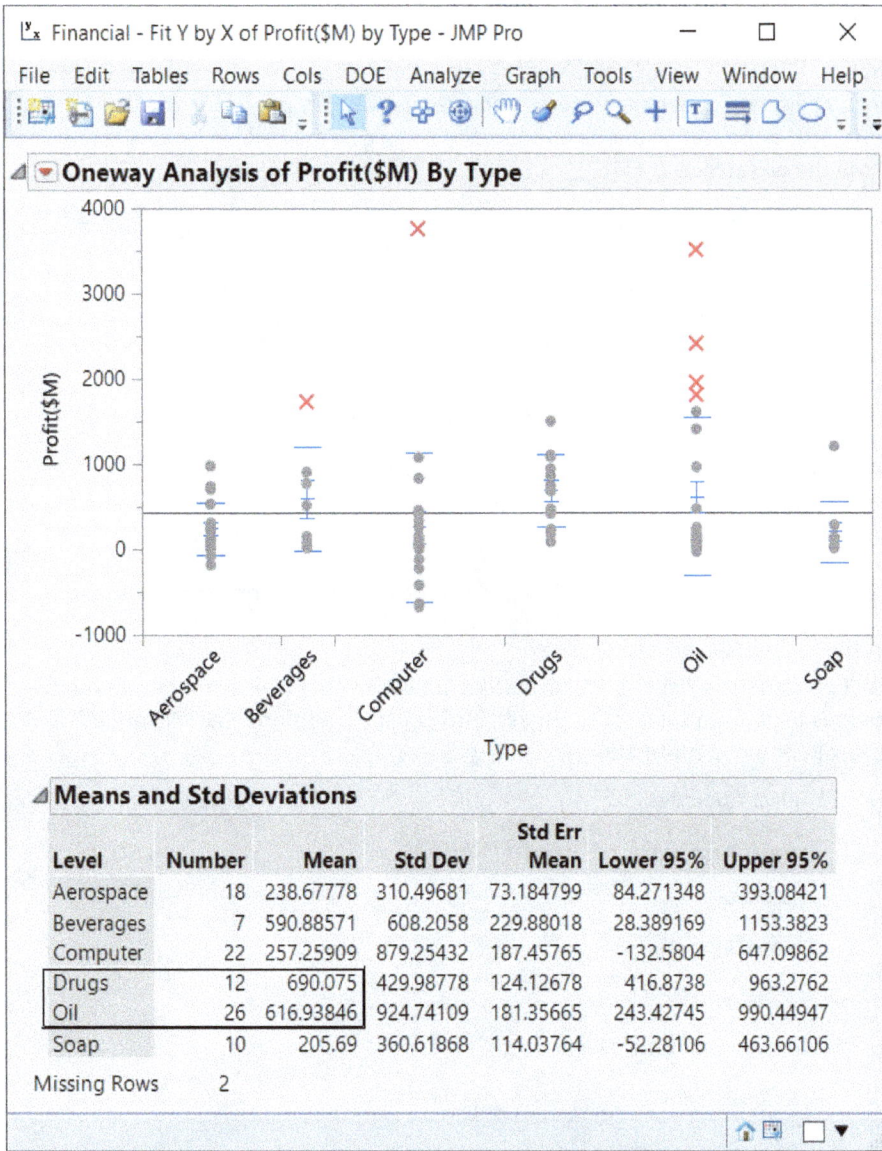

Level	Number	Mean	Std Dev	Std Err Mean	Lower 95%	Upper 95%
Aerospace	18	238.67778	310.49681	73.184799	84.271348	393.08421
Beverages	7	590.88571	608.2058	229.88018	28.389169	1153.3823
Computer	22	257.25909	879.25432	187.45765	-132.5804	647.09862
Drugs	12	690.075	429.98778	124.12678	416.8738	963.2762
Oil	26	616.93846	924.74109	181.35665	243.42745	990.44947
Soap	10	205.69	360.61868	114.03764	-52.28106	463.66106
Missing Rows	2					

Using the table, we can now see that the highest average profits (mean) were obtained from the 12 companies represented in the drugs category. The second highest average profits were obtained from the oil category, and there were 26 companies represented.

Let's review what we have learned so far. In Distribution, we learned that companies with the highest profits are likely to be in the oil category. From Fit Y by X, we learned that the highest average profits are found in the drug category and the next highest average profits are found in the oil category. Remember, you are trying to accumulate just enough knowledge to make your decisions. If you are keeping score, Oil is showing up as a good performer by at least two measures. The most extreme profits are coming from some of these oil companies and overall averages are also high for Oil.

Did you also notice the number of oil companies represented? This is another vote for Oil to be represented in our selection of company stocks because many highly profitable stocks are from this category. The risk associated with one oil company going down is hedged by having many profitable ones in the oil category.

Note
This is true when the companies in the category are not highly correlated on the fundamentals. Seek professional advice when making investment decisions.

If you are keeping score, Oil looks better by three measures now.

We don't really have numerical measurements of range yet to help answer the question: "How big and in what direction (negative and positive) are the differences in profit between the types of business sectors?"

Where might we go to answer this question? Yes, it's in the red triangle for the oneway result.

1. Select the red triangle next to **Oneway**.
2. Select **Quantiles**. (See Figure 5.25.) *Quantiles* are values that divide an ordered set of continuous data (from smallest to largest) into equal proportions. See "Quantiles" in Appendix B for more information.

Figure 5.25 Select Quantiles

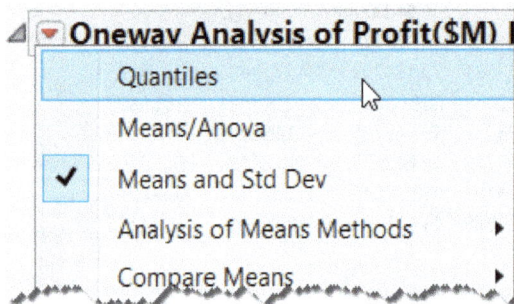

A Quantiles table appears. This table appears in the context of your initial analysis, and box plots are added to your graph illustrating the quantiles. (See Figures 5.26a and 5.26b.)

Figure 5.26a Oneway Means and Quantiles

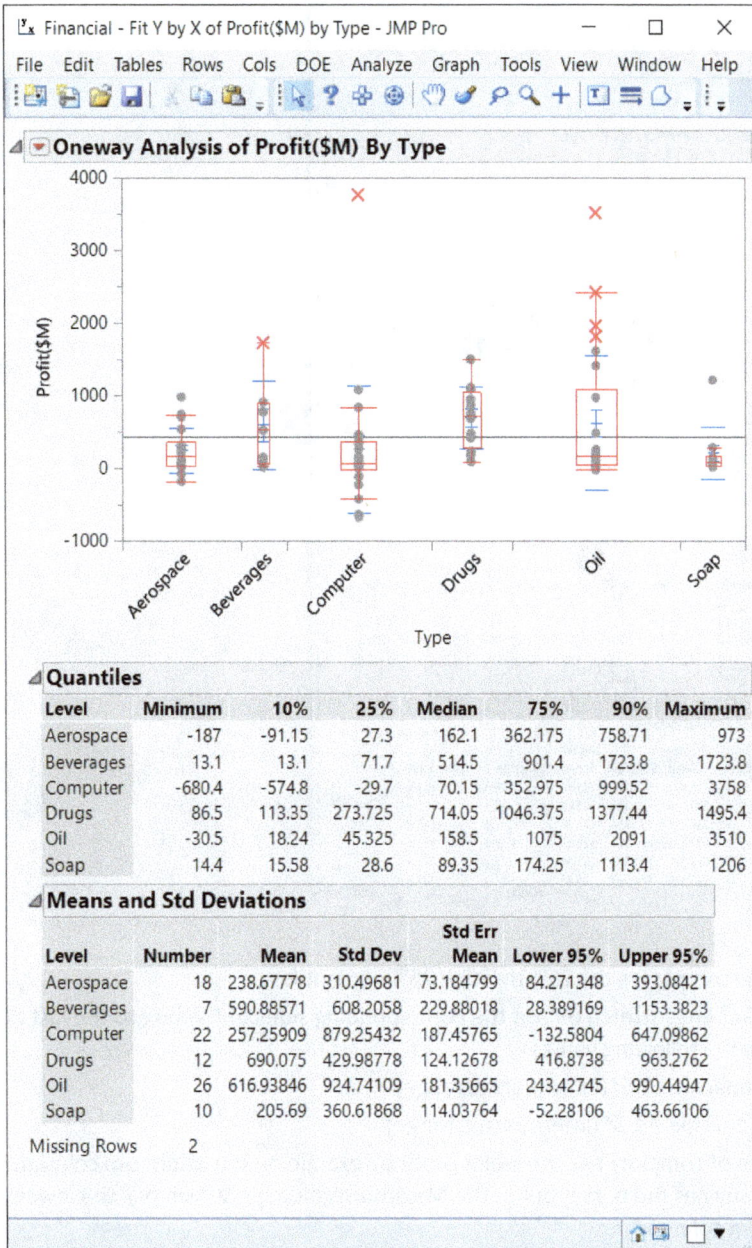

Inside the window:

Financial - Fit Y by X of Profit($M) by Type - JMP Pro — □ ×

File   Edit   Tables   Rows   Cols   DOE   Analyze   Graph   Tools   View   Window   Help

Oneway Analysis of Profit($M) By Type

Quantiles

Level	Minimum	10%	25%	Median	75%	90%	Maximum
Aerospace	-187	-91.15	27.3	162.1	362.175	758.71	973
Beverages	13.1	13.1	71.7	514.5	901.4	1723.8	1723.8
Computer	-680.4	-574.8	-29.7	70.15	352.975	999.52	3758
Drugs	86.5	113.35	273.725	714.05	1046.375	1377.44	1495.4
Oil	-30.5	18.24	45.325	158.5	1075	2091	3510
Soap	14.4	15.58	28.6	89.35	174.25	1113.4	1206

Means and Std Deviations

Level	Number	Mean	Std Dev	Std Err Mean	Lower 95%	Upper 95%
Aerospace	18	238.67778	310.49681	73.184799	84.271348	393.08421
Beverages	7	590.88571	608.2058	229.88018	28.389169	1153.3823
Computer	22	257.25909	879.25432	187.45765	-132.5804	647.09862
Drugs	12	690.075	429.98778	124.12678	416.8738	963.2762
Oil	26	616.93846	924.74109	181.35665	243.42745	990.44947
Soap	10	205.69	360.61868	114.03764	-52.28106	463.66106

Missing Rows	2

Figure 5.26b Oneway Means and Quantiles with Call-out Numbers

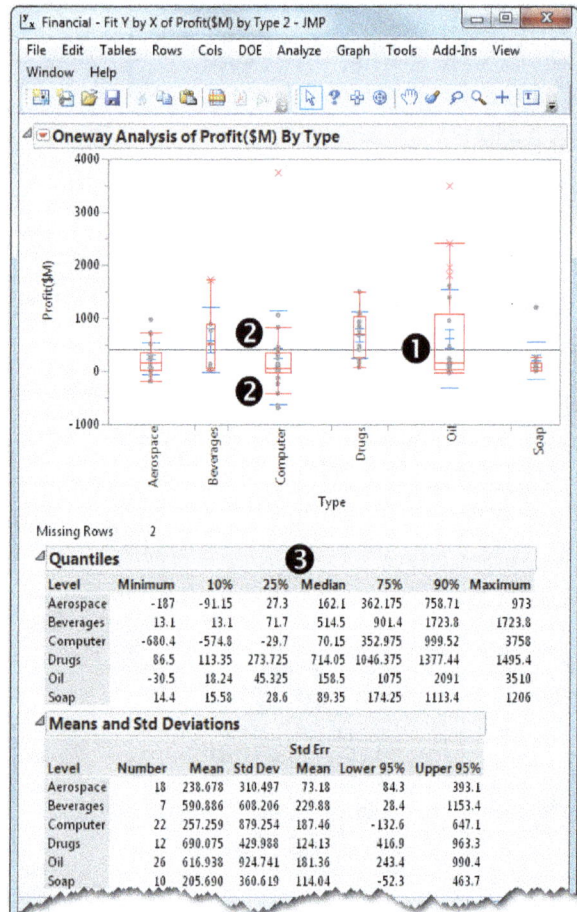

❶ The box encompasses the 25th through 75th percentiles.

❷ The whisker lines stretch out on the corresponding sides of the box to the last data point between the following values:

> 1st quartile - 1.5*(interquartile range)
>
> 3rd quartile + 1.5*(interquartile range)

❸ What type of company has the worst profit by examining the **Minimum** column? What type of company has the best profit in the **Maximum** column? Which has the lowest value for the **Maximum** column?

You can make the differences easier to see. Try this:

1. Right-click the **Quantiles** table.
2. From the submenu, select **Sort by Column**. (See Figure 5.27.)

Figure 5.27 Sort by Column

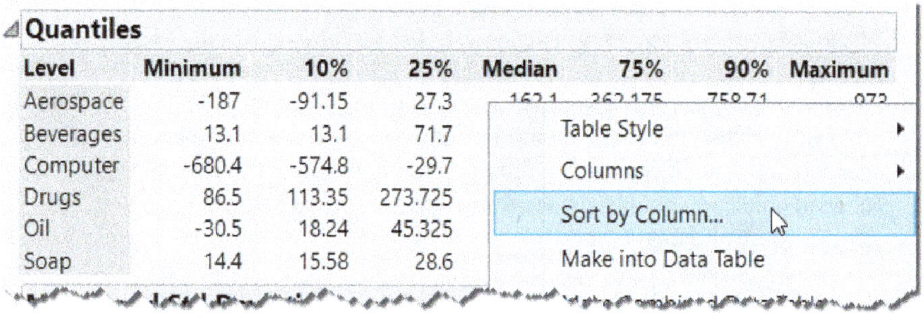

Level	Minimum	10%	25%	Median	75%	90%	Maximum
Aerospace	-187	-91.15	27.3	162.1	262.175	758.71	973
Beverages	13.1	13.1	71.7		Table Style		▶
Computer	-680.4	-574.8	-29.7		Columns		▶
Drugs	86.5	113.35	273.725				
Oil	-30.5	18.24	45.325		Sort by Column...		
Soap	14.4	15.58	28.6		Make into Data Table		

◢ **Quantiles**

3. Select **Minimum** and click **OK**. (See Figure 5.28.)

Figure 5.28 Sort Column by Minimum

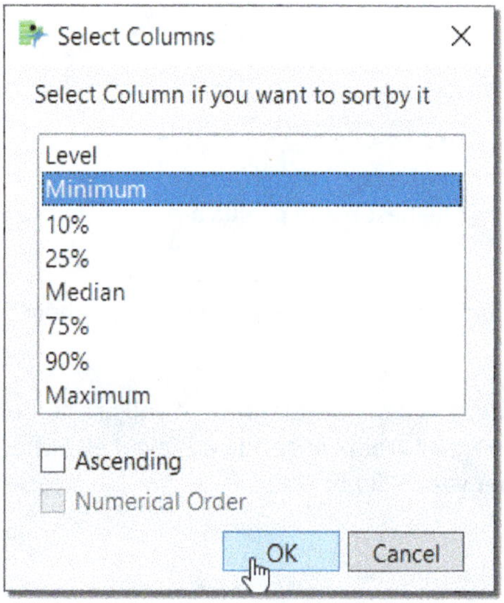

Select Columns ✕

Select Column if you want to sort by it

Level
Minimum
10%
25%
Median
75%
90%
Maximum

☐ Ascending
☐ Numerical Order

OK   Cancel

4. The Profit table now appears sorted by the **Minimum** column. (See Figure 5.29.)

Figure 5.29 Quantiles Sorted by Minimum

◢ **Quantiles**

Level	Minimum	10%	25%	Median	75%	90%	Maximum
Drugs	86.5	113.35	273.725	714.05	1046.375	1377.44	1495.4
Soap	14.4	15.58	28.6	89.35	174.25	1113.4	1206
Beverages	13.1	13.1	71.7	514.5	901.4	1723.8	1723.8
Oil	-30.5	18.24	45.325	158.5	1075	2091	3510
Aerospace	-187	-91.15	27.3	162.1	362.175	758.71	973
Computer	-680.4	-574.8	-29.7	70.15	352.975	999.52	3758

You can now see that the worst loss (or least profitable) is in the Computer category at -680.4. Can you find the biggest profits in the **Maximum** column? Yes, it's also in the Computer category at 3758. They are circled for you. (See Figure 5.29.)

Now you can start asking yourself about your risk tolerance. Review the oneway graph that you produced, the mean and standard dev, and the Quantiles table. If you are conservative, what types of companies would you select to assure profits? Wouldn't you choose only companies that show no negative profits? Which ones are those? They are Drugs, Soap, and Beverages.

Higher profits appear in other categories, but more of those companies show higher variability and losses and, therefore, present more risk.

5. Finally, save the data table **Finance.jmp**. You will need this data table for the next chapter.

Summary of 5.3: What You Learned by Comparing a Continuous Column to a Categorical Column

You might have noticed that the highest mean (average) profits are found overall in the 12 companies in the drug category and the next highest mean profits are found in the 26 companies in the oil category.

You found that the biggest loser was in the computer category as well as the most profitable one.

You found that choosing company types depends on risk tolerance. If you are conservative, you might select Drugs, Soap, and Beverages. Higher profits might be achieved elsewhere, but they are in company types where there is higher variability or risk.

5.4 Comparing Two Continuous Columns

As shown in Figure 5.20, the Fit Y by X platform does different analyses depending on the modeling role of the columns entered into the Y and X roles in the dialog box. Section 5.3 examined the case where the column in the Y roles is continuous and the column in the X role is categorical. There, Fit Y by X invokes the Oneway platform for the analysis. In the case where both columns are continuous, Fit Y by X conducts a Bivariate analysis.

Continuing with use of the Financial.jmp data table, we can examine the relationship between Profit and Sales.

1.   Select **Analyze ▶ Fit Y by X**.
2.   Assign **Profit** to the **Y, Response** role and **Sales** to the **X, Factor** role.
3.   Select **OK**.

The result shown in Figure 5.30 is a scatterplot of the data.

Figure 5.30 Bivariate Plot of Profit by Sales

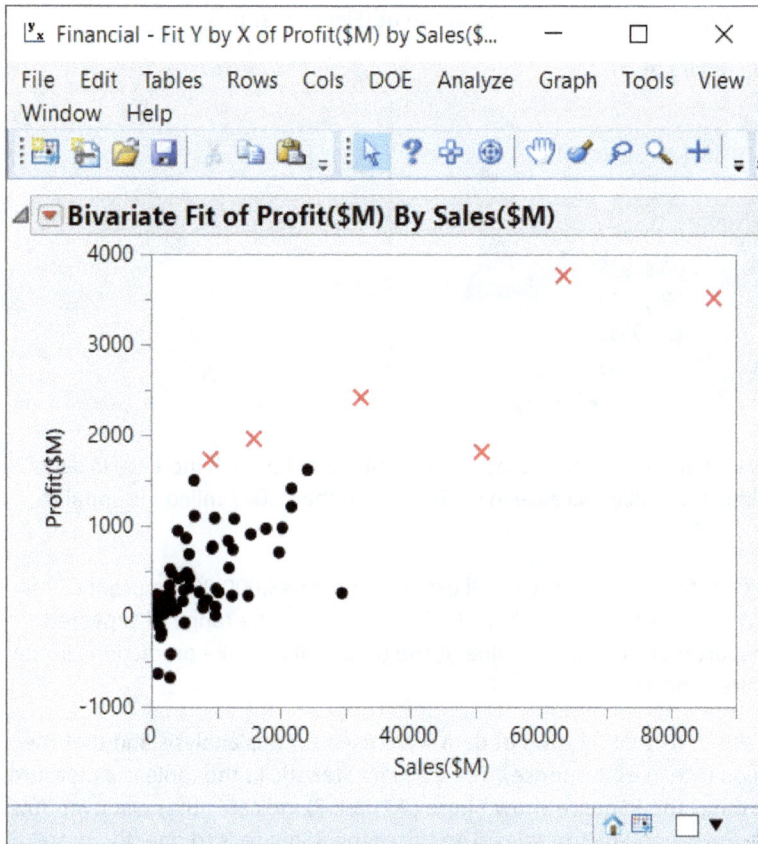

The scatterplot shows what we might have expected. Higher sales seem to indicate higher profit. To define the relationship between profit and sales, you can fit a line to the data.

4. Click on the red triangle and select **Fit Line**. (See Figure 5.31.)

Figure 5.31 Select Fit Line from the Red Triangle Menu

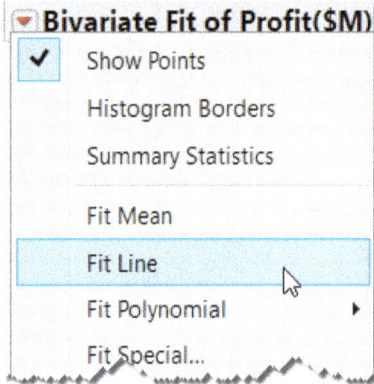

▾ **Bivariate Fit of Profit($M)**

✔	Show Points
	Histogram Borders
	Summary Statistics
	Fit Mean
	Fit Line
	Fit Polynomial ▸
	Fit Special...

This adds several tables of statistical information to the results. Concentrating on the first two, we see the equation of the line of fit and a summary of the fit (Figure 5.32).

Figure 5.32 Equation and Summary of Fit

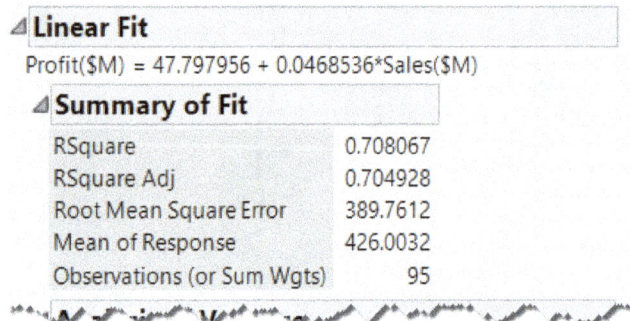

◢ **Linear Fit**

Profit($M) = 47.797956 + 0.0468536*Sales($M)

◢ **Summary of Fit**

RSquare	0.708067
RSquare Adj	0.704928
Root Mean Square Error	389.7612
Mean of Response	426.0032
Observations (or Sum Wgts)	95

The equation tells us that, within the range of sales in this data table, for each increase in sales of $1 million, we see an almost $47,000 increase in profits. (Note that 0.047 million is equal to 47,000.)

The equation also tells us that, for $0 in sales, we still expect almost $48,000,000 in profits. This intercept value is often not very interpretable when "0" falls outside of the range of expected values for sales, but it is important to helping the line fit the data well to make predictions in the range of sales values that are expected.

The Summary of Fit table shows us that 95 rows of data were used for this analysis and that the average profit is $426 million (Mean of Response). The RSquare statistic in the table is a measure of the fit of the line to the data. The RSquare shown here (0.708067) indicates that just over 70% of the variability in profits is explained by the sales. The closer the RSquare is to one, the more perfect the fit of the data to the line. In the next chapter we will introduce the use of more than one X variable in the regression model which can result in a better fit of the data and a higher RSquare.

Examining the scatter plot shown in Figure 5.30 more closely, there appear to be a few companies with very high sales compared to the others. You might consider using the local data filter to exclude these points from the analysis and examine the effect of the exclusions on the equation of the line and the RSquare statistic.

5. From the red triangle, select **Local Data Filter**.
6. Select **Sales** as a filter column, then click on the + button to add it as a filter
7. Use your cursor to move the right blue line to exclude the companies with very high sales (over $40 billion) from the analysis. (See Figure 5.33.)

Figure 5.33 Bivariate Fit with Local Data Filter

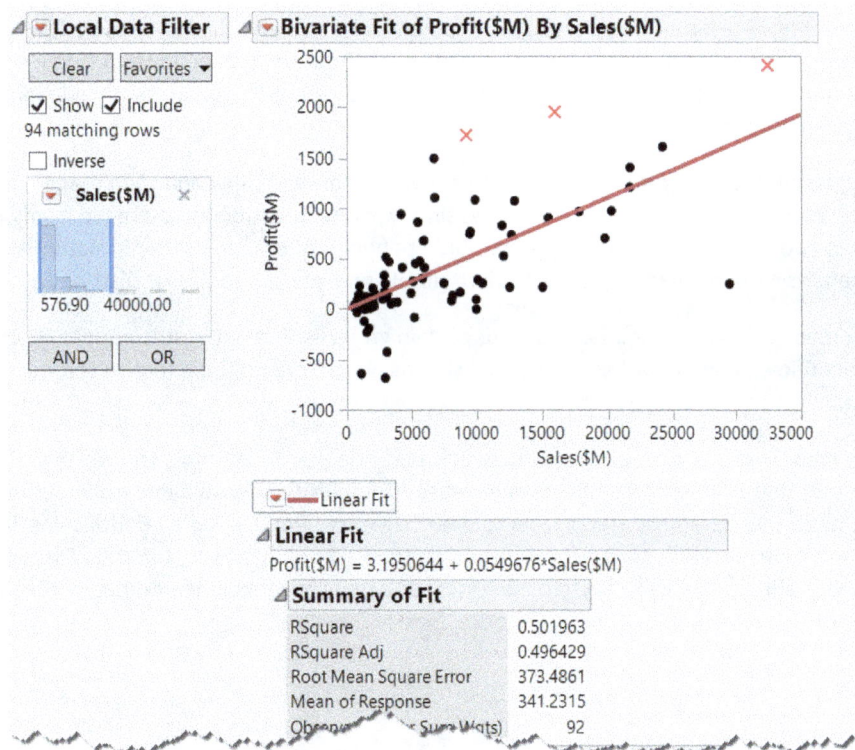

Examine the differences between the linear equation and the RSquare for various setting in the Local Data Filter. For example, in Figure 5.33 you see that three companies have been excluded resulting in a steeper slope for the line and a much lower RSquare statistic. You might also use the Local Data Filter to examine the differences in the line for each of the Types of companies represented in the data.

Summary of 5.4: What You Learned by Comparing Two Continuous Columns

When using the Fit Y by X platform to compare two continuous columns, the simple linear regression analysis provides an equation for a line fit to the data. This equation provides a slope that defines the magnitude of the increase (or decrease) in the "Y, Response" column as the "X, Factor" column increases.

In this data, we found (as expected) an increase in profits as sales increased. We also discovered that the relationship might be defined differently if unusual observations (rows of data) are excluded from the analysis. We used the Local Data Filter to explore how the relationship changes as we include or exclude certain values. Finally, we considered, using our own common sense about what we need from our model, whether to include or exclude the extreme values and which equation to report.

5.5 Summary

This chapter has presented an approach to problem-solving that is unique to JMP. The approach underscores the progressive nature of problem solving that tends to build from simple descriptions of one column of data to relationships between two columns of data, leading to a better understanding of the data.

Learning about data using JMP tends to start slowly, increase rapidly, and then reach understanding. Also, the process does not go simply in one direction as you move from one column to two columns. As discoveries are made between two columns, confirmation and further analysis can be made by going back to simple one-column tools like Distribution.

Marking rows with colors and markers helps certain groups stand out in subsequent analyses. Visual identification with markers and colors speeds discovery and the effective communication of results.

Because many real-world problems involve multiple variables or columns, you will learn in the next chapter that JMP easily handles this increased complexity. We will build upon our example and the basic analyses introduced in this chapter to explore multivariable relationships. In the next chapter, we introduce several tools including the Partition platform, the Data Filter, and the Prediction Profiler to help you explore and discover deeper insights in your data.

Chapter 6: Problem-Solving with Multiple Columns

The previous chapter focused on problem-solving when you have one or two columns of interest. This chapter builds on the previous chapter's framework for problem-solving and introduces a method to understand multiple column and group relationships.

We introduce advanced analytics including Partition, Fit Model, and Prediction Profiler to help us understand the multi-column/variable relationships in the data. Most real-world problems involve more than one- or two-column relationships, so these methods will usually be needed for any real data analysis problem.

We also introduce the Data Filter for on-the-spot slicing of the data. The Data Filter enables easy inference testing and hypothesis development by excluding, hiding, and marking selected observations across any number of column ranges or groups.

The Predictive Modeling platforms include, among other methods, the tree-based Partition model, which allows us to explore making partitions in the data to explain a response. The Fit Model platform includes several modeling personalities, including Multiple Regression. Unlike the two-column analysis (one Y and one X) we introduced in the last chapter, Partition and Multiple Regression enable you to use additional X columns. We briefly introduce these topics at the end of the chapter, along with a very powerful model visualization tool, the Prediction Profiler.

6.1 Introduction

As we explained in the previous chapter, the transformation of data into information can be advanced using a few basic JMP platforms including Distribution and Fit Y by X, which are found in the Analyze menu. Table 6.1 provides a review of the tool name, how many columns it supports, and its common statistical identity. The process of discovery starts with univariate/one-column analysis using the Distribution platform to explore the single-column/variable properties and proceeds to bivariate/two-column relationships using the Fit Y by X platform. As in life itself, these analyses can often lead to more complex questions involving more than two columns, which is the subject of this chapter.

In Table 6.1, we add Fit Model and Predictive Modeling to the items on the Analyze menu column. Partition and Multiple Regression are associated with these menus and are added to the table framework. These methods support multivariable or multi-column relationships.

Table 6.1 Tool Name, Columns Supported, and Statistical Terms

Analyze Menu	How Many Columns	Statistical Terminology
Distribution	Single column	Univariate methods, histograms, box plots, quantiles, descriptive statistics, and more
Fit Y by X	Two columns	Bivariate, contingency, logistic, oneway ANOVA, nonparametrics, linear regression, and more
Fit Model	Multi-Column	Data mining, multiple regression, logistic regression, stepwise variable reduction, and more
Predictive Modeling	Multi-Column	Data mining, partitioning, neural networks, and more

Remember, JMP's goal for this menu framework is to introduce you to methods in a logical order, enabling you to learn about your data progressively without forcing you to first understand statistical jargon or the conditions required to implement them. The process of digging deeper into your data is built into the menu structure.

6.2 Comparing Multiple Columns

We now want to know whether there are important multiple-column relationships (looking at more than two columns) that might be present in the data. Complex products or systems often have these relationships. In fact, sometimes to see the forest for the trees, you need these methods so that you can focus on only those columns that drive your learning from the data.

Example 6.1: Financial

We will be using the **Financial.jmp** data file to illustrate the steps in this chapter. The data is from companies in the Fortune 500, selected from the April 23, 1990 *Fortune* magazine issue. This data includes columns for:

- **Type:** type of company
- **Sales($M):** yearly sales in millions of dollars
- **Profit($M):** yearly profits in millions of dollars
- **#emp:** number of employees at time of measurement

- **Profits/emp:** profits per employee in thousands of dollars
- **Assets($Mil.):** assets in millions of dollars
- **Sales/emp:** sales per employee in thousands of dollars
- **Stockholder's Eq($Mil.):** stockholder's equity in millions of dollars

You can access this data at **Help ▶ Sample Data Library ▶ Financial**.

> Note
>
> Statistical modeling in practice is both art and science, and simply adding more complexity to your model is not always better. Real-world models often include more than two columns, and sometimes adding too much complexity will undermine the usefulness of the model when it is used on new data. There are tools built into JMP to help you decide how much complexity to include in a model. JMP Pro includes many state-of-the-art platforms and tools to validate your models and generalize them for new data and prediction.

You should have the Financial.jmp data table open from the previous chapter. If you do not, you can open it at: **Help ▶ Sample Data Library ▶ Financial.jmp**. (See Figure 6.1.)

Figure 6.1 Financial Data Table

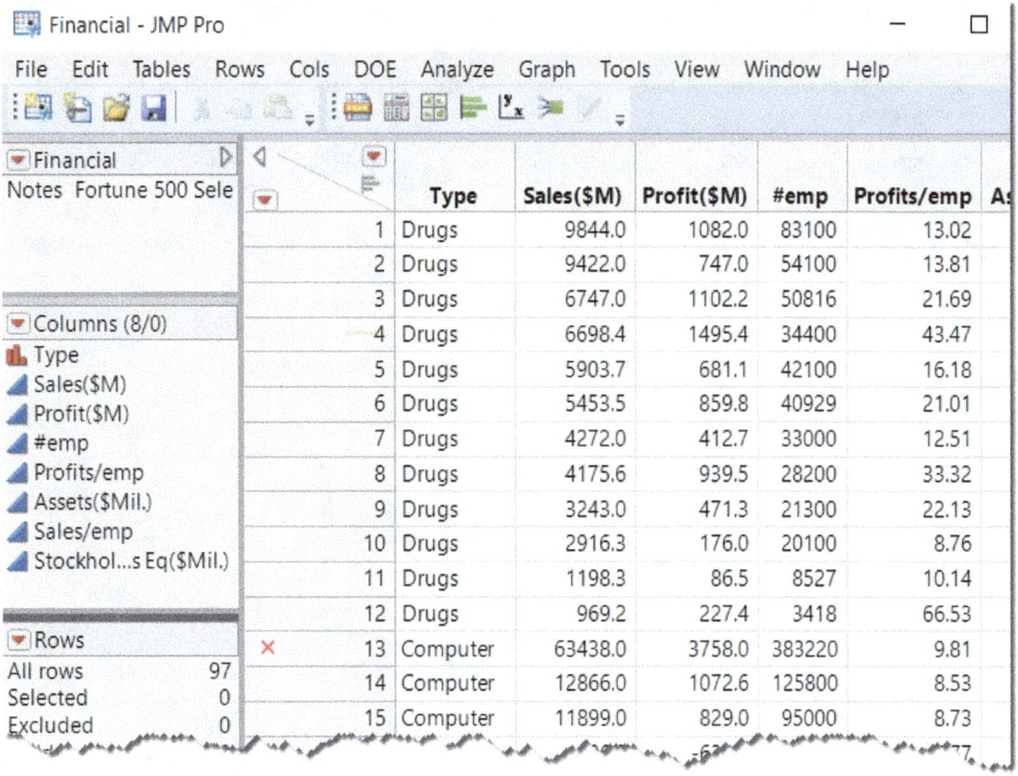

	Type	Sales($M)	Profit($M)	#emp	Profits/emp
1	Drugs	9844.0	1082.0	83100	13.02
2	Drugs	9422.0	747.0	54100	13.81
3	Drugs	6747.0	1102.2	50816	21.69
4	Drugs	6698.4	1495.4	34400	43.47
5	Drugs	5903.7	681.1	42100	16.18
6	Drugs	5453.5	859.8	40929	21.01
7	Drugs	4272.0	412.7	33000	12.51
8	Drugs	4175.6	939.5	28200	33.32
9	Drugs	3243.0	471.3	21300	22.13
10	Drugs	2916.3	176.0	20100	8.76
11	Drugs	1198.3	86.5	8527	10.14
12	Drugs	969.2	227.4	3418	66.53
13	Computer	63438.0	3758.0	383220	9.81
14	Computer	12866.0	1072.6	125800	8.53
15	Computer	11899.0	829.0	95000	8.73

Preparing Data by Grouping

Before we perform an analysis, let's create a new column that simply identifies each company as being profitable or not. (We will, in effect, be making a new profit column that indicates whether it meets a certain profit threshold and will be treated as nominal.) Our intention is to identify profitable companies based on all the other columns. We want to select companies whose profits are sufficient to provide us with solid returns, so we want to identify companies whose profits are above $10 million and store them in a new column. The $10 million mark is an arbitrary breakpoint (it could be any value), but it is based on the idea that companies with higher profits in general will probably be larger and less volatile. For purposes of illustration, we will use $10 million as a reasonable and conservative strategy.

> **Note**
>
> We do not need to change the profit column to a nominal type to use Partition. Output columns like profit can be either continuous, nominal, or ordinal in the Partition platform as well as any input columns. We made this change only to simplify the interpretation of our analysis for illustration purposes.

Let's make our new Profit column by deriving what we need from the existing **Profits($M)** column:

1. Click the **Rows** menu.
2. Select Row Selection.
3. Click **Select Where...** (Figure 6.2). This generates the **Select Rows** window. (See Figure 6.3.)

Figure 6.2 Select Where to Generate the Select Rows Window

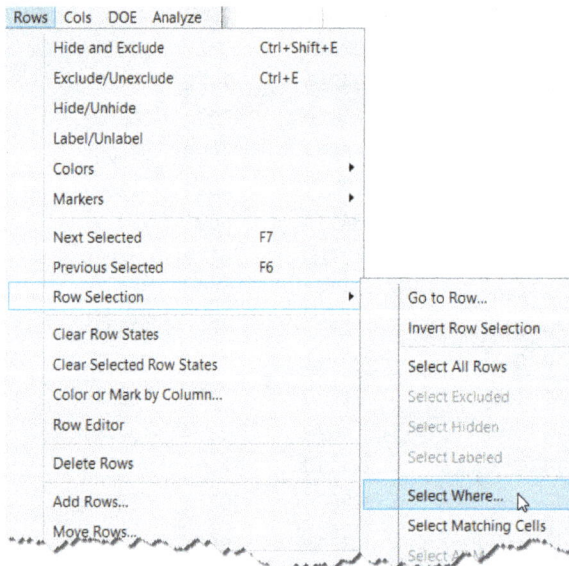

Figure 6.3 Select Where Profit is Greater Than $10M

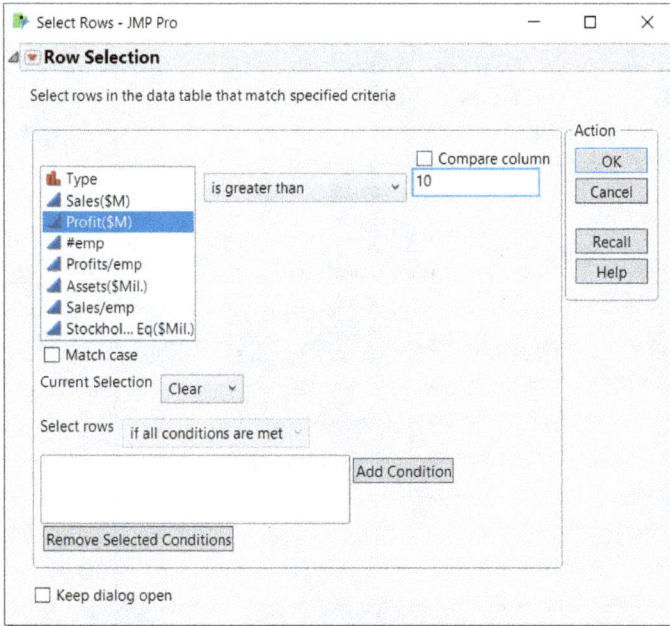

4. Select **Profits($M)**.
5. Select **is greater than**.
6. Enter **10**. (The data is expressed in millions.)
7. Click **OK**.

   Rows that meet the condition of Profits being greater than $10 million are now highlighted in the data. (See Figure 6.4.)

Figure 6.4 Selected Rows are Highlighted

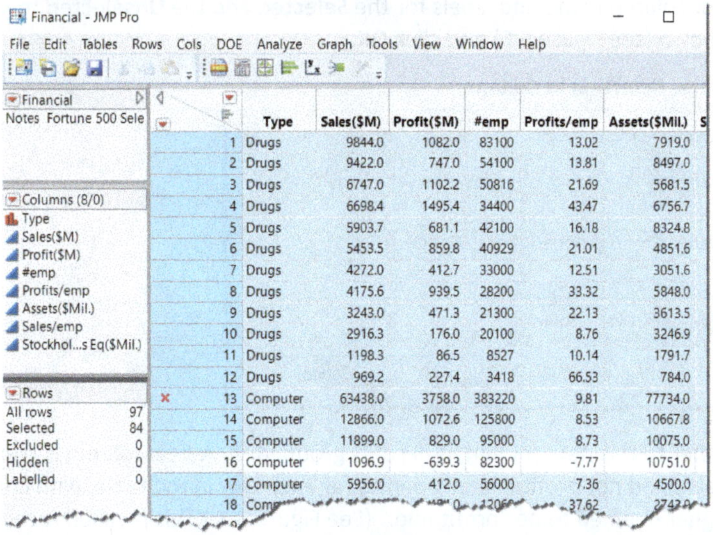

8. Now select **Rows ▶ Row Selection ▶ Name Selection in Column...** (Figure 6.5).

    Figure 6.5 Select Name Selection in Column

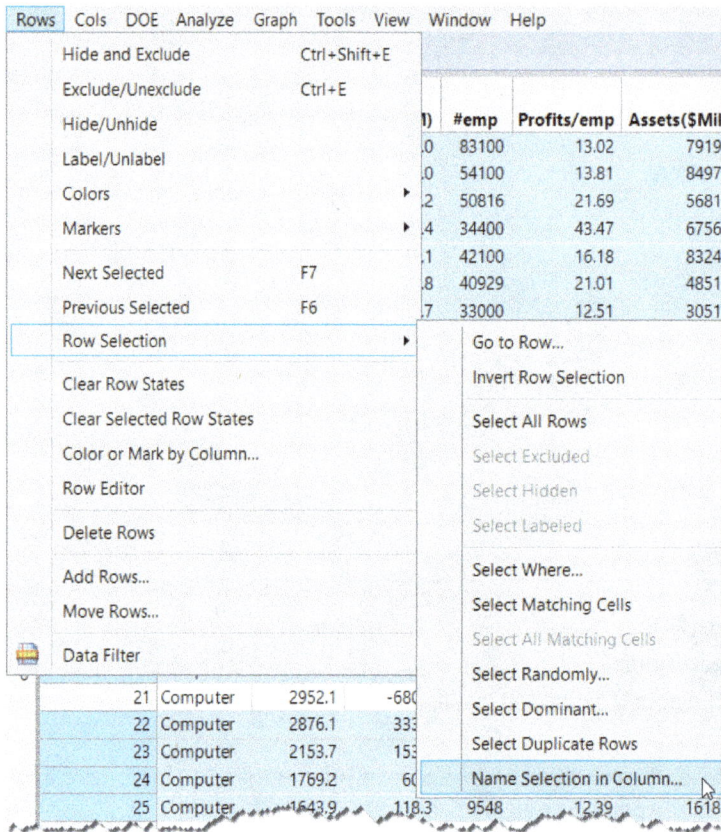

9. Type the new column name and labels for the **Selected** and the **Unselected** rows exactly as shown (see Figure 6.6) and click **OK**.

    Figure 6.6 Name Selection in Column

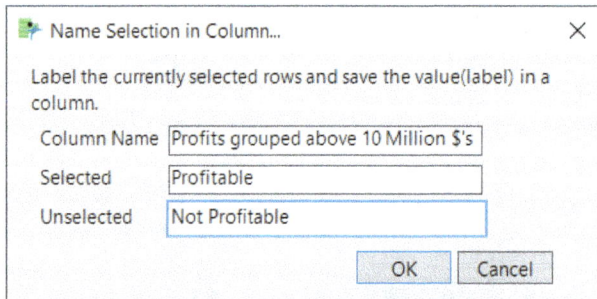

We will henceforth refer to this new column as "Profits grouped...". A new column is created that identifies profitable and not profitable companies for each row in the table, with companies at $10 million and higher deemed to be "profitable." (See Figure 6.7.) This is called *recoding the*

data. In this case, we are making two groups where one group is labeled "Profitable" above a certain threshold and one group "Not Profitable" below that threshold. Not having to write any code to do this is very handy. To learn about another method to accomplish data recoding using the Recode command on the Cols menu, see Section 2.4.

Figure 6.7 New Profits Grouped Column

Profits grouped above 10 Million $'s
Profitable
Profitable
Profitable
Profitable
Profitable
Profitable
Profitable
Profitable
Profitable
Profitable
Profitable
Profitable
Profitable
Profitable
Profitable
Not Profitable
Profitable
Profitable
Not Profitable
Profitable
Not Profitable
Profitable

Let's move on with the analysis.

Consider the data table in Figure 6.8. More than one column might be related to the **Profits grouped** column: for example, **Sales($M)**, **#emp**, and **Assets($M)** may explain or influence a company's profitability. With more than two columns to investigate, we need to move farther down the Analyze menu. There are many methods to investigate this type of relationship. Let's look at one of these with a tool called **Partition**.

Figure 6.8 Grouped Profits

	Type	Sales($M)	Profit($M)	#emp	Profits/emp	Assets($MIL.)	Sales/emp	Stockholder's Eq($Mil.)	Profits grouped above 10 Million $'s
12	Drugs	969.2	227.4	3418	66.53	784.0	283.56	569.4	Profitable
13	Computer	63438.0	3758.0	383220	9.81	77734.0	165.54	38509.0	Profitable
14	Computer	12866.0	1072.6	125800	8.53	10667.8	102.27	8035.7	Profitable
15	Computer	11899.0	829.0	95000	8.73	10075.0	125.25	5446.0	Profitable
16	Computer	1096.9	-639.3	82300	-7.77	10751.0	13.33	3881.8	Not Profitable
17	Computer	5956.0	412.0	56000	7.36	4500.0	106.36	1985.0	Profitable
18	Computer	5284.0	454.0	12068	37.62	2743.9	437.85	1485.7	Profitable
19	Computer	3078.4	-424.3	28334	-14.97	2725.7	108.65	1130.8	Not Profitable
20	Computer	2959.3	252.8	31404	8.05	5611.1	94.23	1428.3	Profitable
21	Computer	2952.1	-680.4	18000	-37.80	1860.7	164.01	411.5	Not Profitable
22	Computer	2876.1	333.3	9500	35.08	2090.4	302.75	1171.6	Profitable
23	Computer	2153.7	153.0	8200	18.66	2233.7	262.65	1176.0	Profitable
24	Computer	1769.2	60.8	10200	5.96	1269.1	173.45	661.8	Profitable
25	Computer	1643.9	118.3	9548	12.39	1618.8	172.17	989.1	Profitable
26	Computer	1517.9	-230.2	•	•	•	•	•	Not Profitable
27	Computer	1382.3	0.3	2900	0.10	1076.8	476.66	442.5	Not Profitable
28	Computer	1324.3	-119.7	13740	-8.71	1040.2	96.38	522.1	Not Profitable
29	Computer	1014.0	47.7	9100	5.24	977.0	111.43	337.1	Profitable
30	Computer	990.5	20.9	8578	2.44	624.3	115.47	169.7	Profitable
31	Computer	873.6	79.5	8200	9.70	808.0	106.54	629.8	Profitable
32	Computer	855.1	31.0	7523	4.12	615.2	113.66	243.2	Profitable
33	Computer	784.7	89.0	4708	18.90	955.8	166.67	594.3	Profitable
34	Computer	709.3	41.4	5000	8.28	468.1	141.86	200.8	Profitable
35	Oil	86656.0	3510.0	104000	33.75	83219.0	833.23	30244.0	Profitable
36	Oil	50976.0	1809.0	67900	26.64	39080.0	750.75	16274.0	Profitable
37	Oil	32416.0	2413.0	37067	65.10	25636.0	874.52	9180.0	Profitable

Note

You might find these multi-way relationships by simply (but laboriously) producing distributions or examining output from Fit Y by X. We use Partition because it is a powerful and quick method for finding these relationships when you have more than two potential relationships in your data. Further, relationships among the predictors can be found using Partition; these are called interactions. In general, Partition results are usually easier to interpret than looking at distributions and examining output from Fit Y by X.

Mining Data Using Partition

The Partition platform enables you to determine which columns in a data table most influence or predict the outcome of another column (profits, in our example). It achieves this by searching all of your selected X columns and finding a set of splits or subgroups within them whose values best predict your Y value. These splits (or partitions) of the data are done recursively (starting with the best predictor), forming a tree of decision rules.

In our example, **Profits grouped** is our column of interest (Y Column) and we want to determine which of the remaining columns (X Column) in our data table most influence or predict Profit. (See Figure 6.9.)

The Partition method introduced here is one of a set of methods that is sometimes termed *data mining* because you are mining relationships in your data to find which columns most relate to or predict profits.

Figure 6.9 Determine Which Columns Affect Profit Grouped (Outputs Versus Inputs)

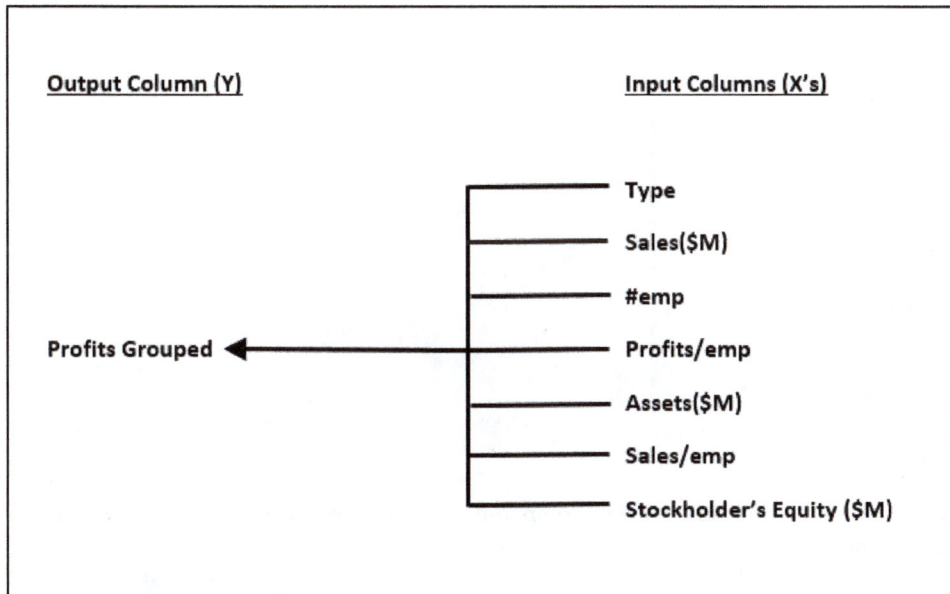

> **Note**
>
> The Partition technique is introduced here to enable quick exploration of your data. This topic is an advanced analytical technique. We recommended that you familiarize yourself with the underlying concepts by using the question mark tool to call up the relevant chapters in the documentation. Because these tools often require greater sophistication in their interpretation, we recommend that you seek out experienced data analysts for assistance when necessary.

To use Partition, we will move farther down the Analyze menu to the **Predictive Modeling** menu. (See Figure 6.10.)

Figure 6.10 Analyze Modeling Menu

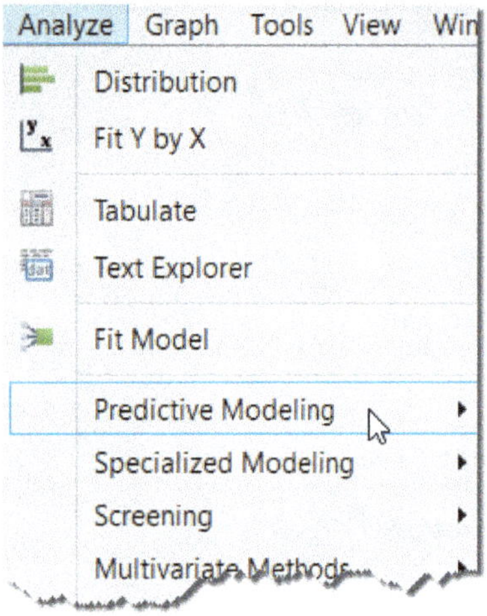

Because we have already explored profits compared with the type of company, we might now ask other questions.

Framing Our Analysis

Questions That Involve Multiple Columns

Are there relationships between profits and other columns?

What are the best predictors of profit?

What column conditions contribute to the biggest differences found between company profits?

How can we use the information in our many columns to choose the most profitable companies?

These are just a few of the questions that can be answered using the Modeling platforms.

1. Select **Analyze ▶ Predictive Modeling ▶ Partition** (Figure 6.11).

   Figure 6.11 Analyze Modeling Partition

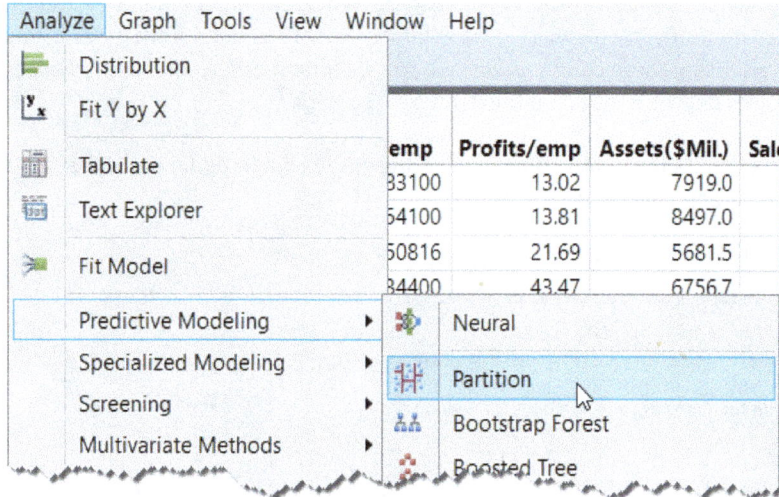

2. Select **Profits grouped** for **Y, Response.** (See Figure 6.12.)

   Figure 6.12 Partition Window Selections

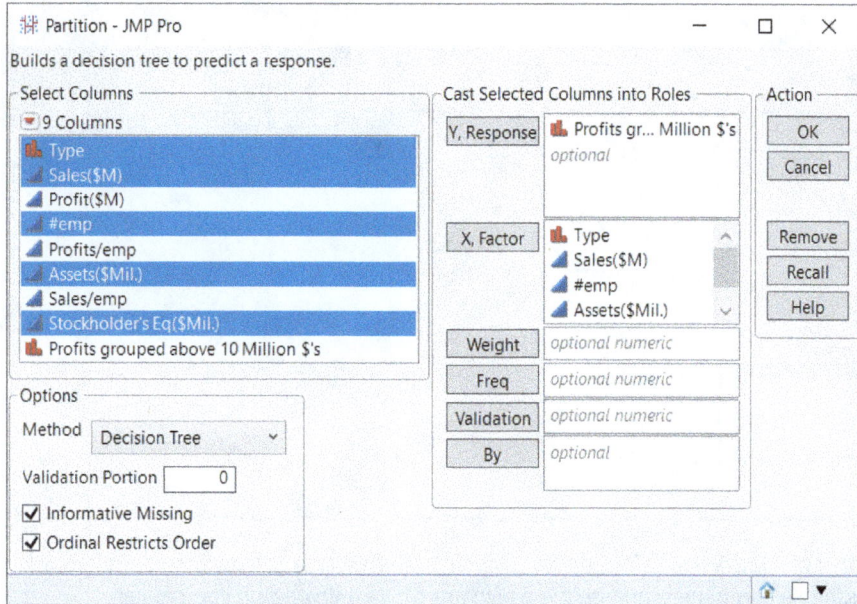

3. And for **X, Factor**, select **Type, Sales($M), #emp, Assets($Mil.),** and **Stockholder's Eq** (see Figure 6.12).

   Click **OK**.

> **Note**
>
> We left out the column Sales/emp above because the column contains a calculated ratio of two other columns already used in the analysis. When exploring data using Partition and other tree-based methods, try to avoid using columns that are derived from other X columns used in the analysis. Using these columns will lead to a condition called collinearity, which is when two variables describe the same thing.

You now see profits for all the companies represented by a (mostly) random scattering of dots in the partition graph (see Figure 6.13).

Figure 6.13 Profits Before Split

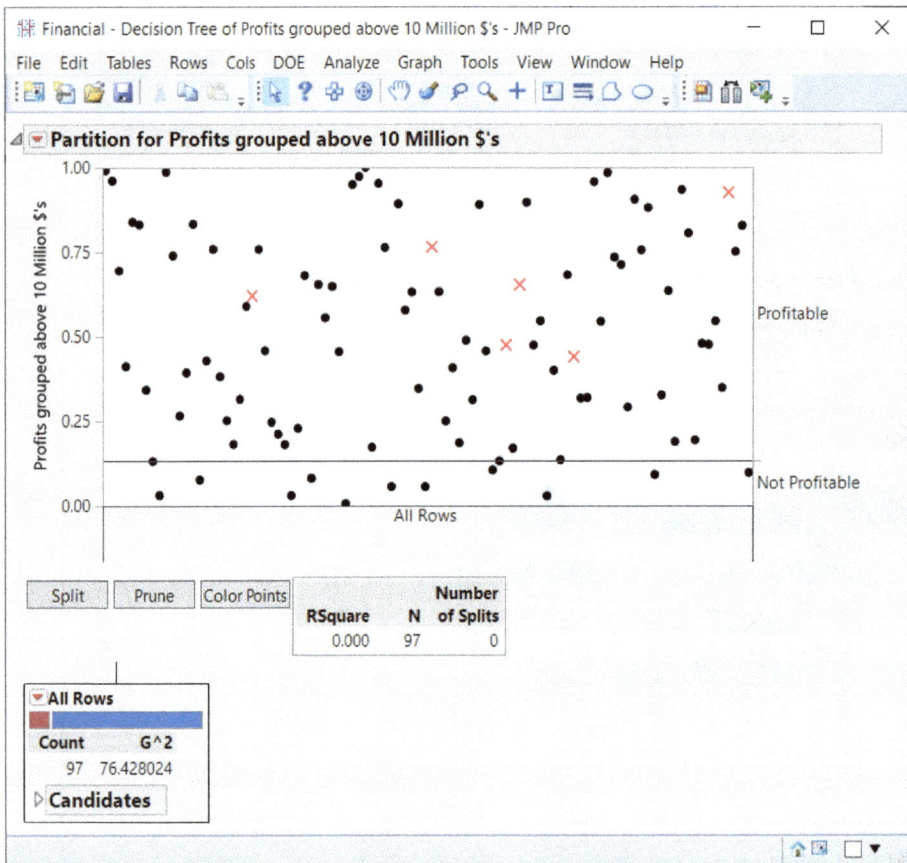

The horizontal black line that is perpendicular to the Y axis is the dividing point, above which are the Profitable companies, and below which are the Not Profitable companies.

Specifically, the companies that are above the line are those that we have labeled as "Profitable" according to the $10 million cutoff, and those below are "Not Profitable, having profit less than $10 million. So, looking at where the line crosses the Y axis on the scatterplot, about 13% of the companies are not profitable and about 87% are profitable. Your graph might look slightly different because the dots representing the data points are randomly arranged in the plot (or, in statistical terms, jittered).

You can see those very high values (outliers) to which we assigned a red X marker from the previous chapter. We saved these in the table as row properties previously. If you don't see the red Xs, it is okay. We can go forward without them.

Now let's try something.

4. Click **Color Points**. It is the button next to **Split** and **Prune**. (See Figure 6.13.)

   Companies determined to be profitable (by our definition) are now displayed as blue. Red points represent those companies that are not profitable. Remember that the point of this analysis is to understand the multiple-column relationships between profits and the other columns in the data table, so seeing these contrasted by color is helpful.

   If you still have the row states applied from the previous chapter, you can see that the X markers (for our high outliers on profit) have all turned blue, because they are all in the Profitable group.

5. Select **Split** and new output appears (Figure 6.14).

Figure 6.14 First Split of Profits

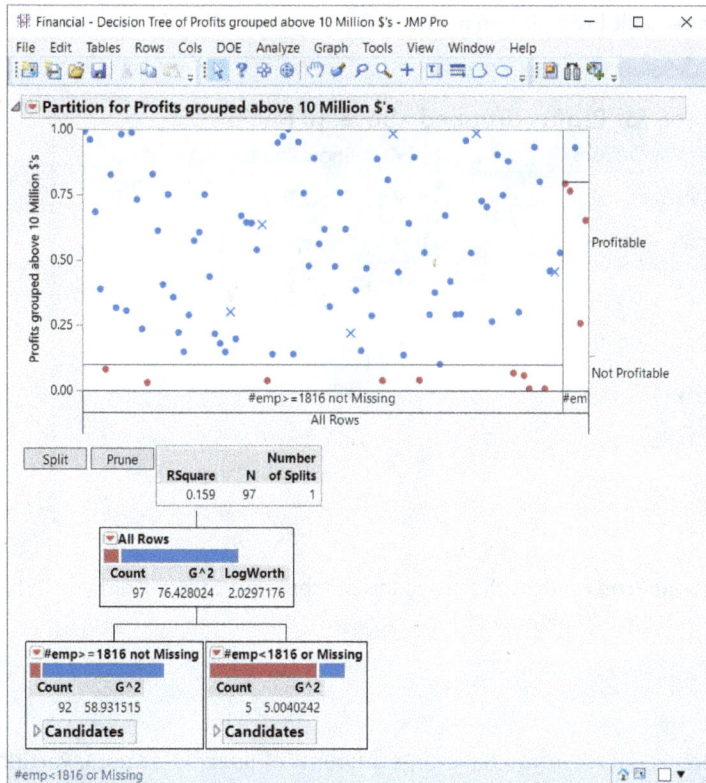

What happened? Partition went through all the columns you selected as Xs and found that the best predictor of **Profits grouped** was number of employees (**#emp**). Remember, you also included a half-dozen other columns in your model, but the first to be selected was the #emp column. The Partition algorithm selected this #emp column and made the split at 1,816 employees. This tells us that the companies with greater than or equal to 1,816 employees have a much different rate of being Profitable than do the companies with less than 1,816 employees. In fact, splitting the data into the two groups ("#emp >= 1816" and "#emp < 1816") is the BEST split to make in order to get the biggest difference in probabilities of Profitable between the two categories. That is how the Partition algorithm found this column and made the split at 1,816 – it searched all the columns and all of the places that it could make splits on all of those columns, and this split makes the biggest difference in predicting "Profitable" versus "Not Profitable."

Partition works similarly when we have a continuous Response, like the actual Profit amount. In that case, we find a split that pushes the average profit in the two groups as far apart as possible, instead of pushing apart the probability of group membership. Want to know more about the analytical method used by Partition? Go to **Help ▶ JMP Documentation Library ▶ Predictive and Specialized Models ▶ Partition Models**.

Let's express the splits as probabilities. Go to the red triangle:

6. Select **Display Options**.
7. Select **Show Split Prob** (Figure 6.15).

Figure 6.15 Select Show Split Prob

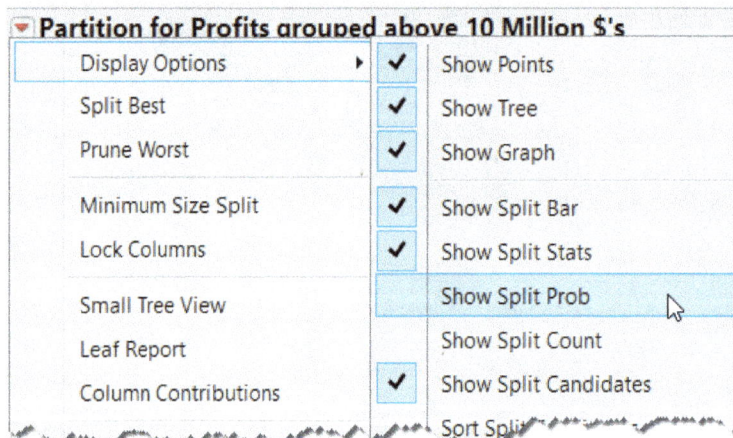

▼ Partition for Profits grouped above 10 Million $'s		
Display Options ▶	✔	Show Points
Split Best	✔	Show Tree
Prune Worst	✔	Show Graph
Minimum Size Split	✔	Show Split Bar
Lock Columns	✔	Show Split Stats
Small Tree View		Show Split Prob
Leaf Report		Show Split Count
Column Contributions	✔	Show Split Candidates
		Sort Split

Turning on **Show Split Prob** renders the groups as percentages that will aid you in finding the most and least profitable combinations.

Partition trees are useful when you have large amounts of unexplored data. Partition trees are also flexible to your column modeling types. The output column (**Profits grouped** in our example) can be either continuous, nominal, or ordinal, AND the input columns (in our example, **#emp** and **Type** among others) can be any combination of continuous, nominal, and ordinal. However, you should be cautious on drawing conclusions from partition trees when the data sets are small, sparse, or messy, especially as you continue to split. Starting in JMP 11, enhancements were made to many modeling methods, including Partition, to use the information value of missing data in improving model accuracy. Methods to measure the usefulness of a partition model are also supported in JMP. To learn more about the Partition platform, go to **Help ▶ JMP Documentation Library ▶ Predictive and Specialized Models ▶ Partition Models**.

Figure 6.16 Probabilities for Each Split

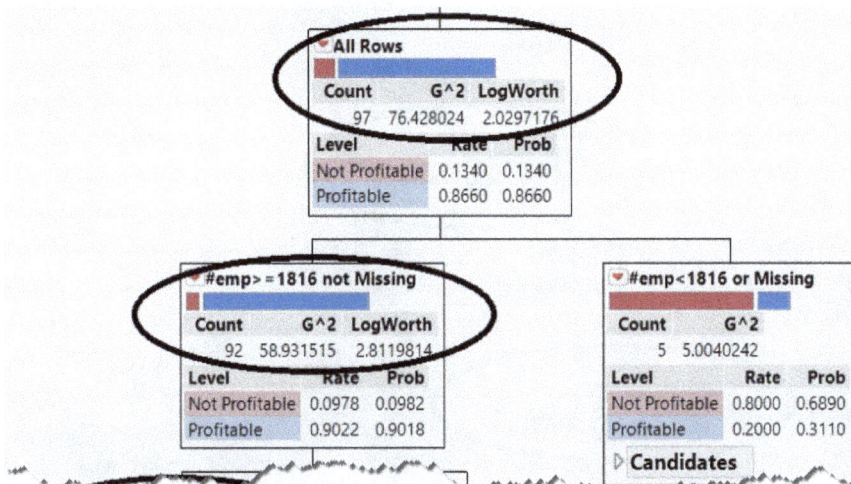

The partition tree output now also shows the probabilities for each split. (See Figure 6.16.) Out of the total group of companies (before splitting) we found about 13% of them are not profitable and about 86% of them are profitable.

Now, after making some splits, where is most of the blue? (Reminder: Blue is profitable!) It is mostly on the left branch of the tree. These blue bars have been circled for you. It shows that 90% of these companies are profitable and therefore about 10% are not profitable. This branch defines companies with employees greater than or equal to 1,816. On the right side, the branches define companies where the number of employees was less than 1,816 or we were missing the information about the number of employees. You see more red on the right side of the tree. (Reminder: Red is unprofitable by our definition.) So, we conclude that the best single predictor of profit is the **#emp** column, where about 90% of the companies that have more than 1,816 employees meet our criteria for profitable companies. In comparison, for smaller companies (and companies who failed to report the number of employees), only about 31% of the companies are profitable.

Note that in the output, JMP has taken Missing or not Missing into account in the model. The information value of missing data can result in improved models. Go to **Help ▶ JMP**

Documentation Library ▶ Predictive and Specialized Models ▶ Partition Models and the "Informative Missing" section to learn more.

Based on what we just learned from the first split, we probably want to look for companies with more than 1,816 employees if we want the best chance of identifying Profitable companies. Now what additional information can we learn by splitting our tree again?

Let's continue growing the tree.

8. Select **Split** again. Another branch appears on the report (Figure 6.17).

Figure 6.17 Probabilities for Each Split After Second Split

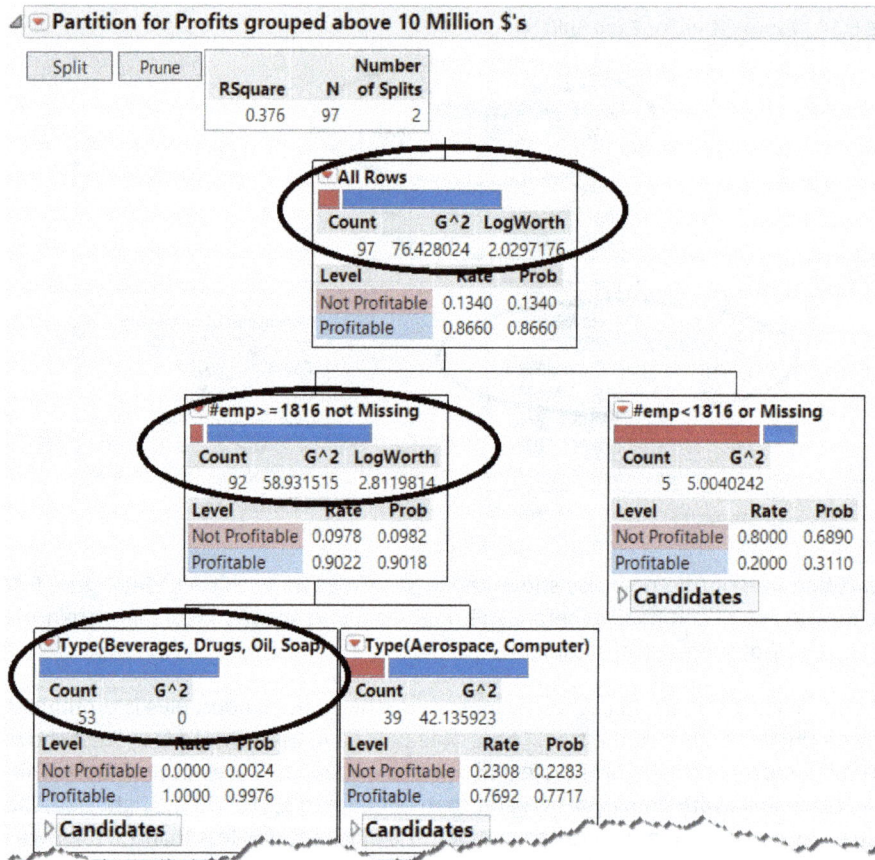

Where are the most profitable groupings after your last split? (See Figure 6.17.) Hint: Follow the blue. On what side of the tree did the split happen? Can you find the combination where >99% of the companies are profitable? We find these where the number of employees is greater than 1,816 AND the Type is Beverages, Drugs, Oil, and Soap. The best branches in the tree are circled for you.

You have now identified the most likely combination of predictor values for profitable companies so far. Let's make that easier to see in the Leaf Report.

9. From the red triangle for partition, select **Leaf Report** (Figure 6.18).

Figure 6.18 Select Partition Leaf Report

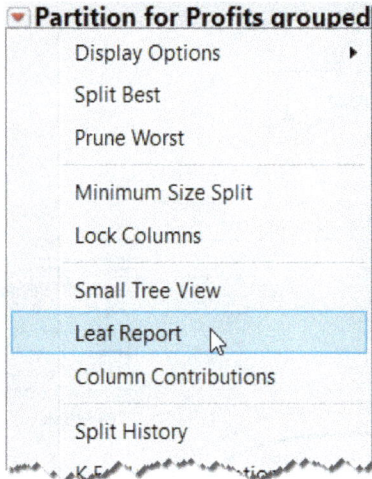

The Leaf report assembles your discoveries in a table format that mirrors the tree format we just covered. (See Figure 6.19.) Looking at the Leaf report and reading the **Leaf Label** from left to right, we can see which split combinations are most or least desirable. What combination looks the most risky so far? It appears to be where the number of employees is less than 1,816 (or we are missing the information about how many employees the company has). Almost 69% of companies in that group do not achieve our definition of profitable.

Figure 6.19 Leaf Report

Leaf Label	Not Profitable	.2 .4 .6 .8	Profitable	.2 .4 .6 .8
#emp>=1816 not Missing&Type(Beverages, Drugs, Oil, Soap)	0.0024		0.9976	
#emp>=1816 not Missing&Type(Aerospace, Computer)	0.2283		0.7717	
#emp<1816 or Missing	0.6890		0.3110	

What group looks most profitable?

The most profitable group appears to be where the number of employees is greater than or equal to 1,816 and the type of companies are Beverages, Drugs, Oil, and Soap. In our example, 99% of companies in this group are profitable. Remember this group because we will be using it in the Using Data Filter section a bit later.

10. Back to our tree, click **Split** again.

Where did the next split happen? (See Figure 6.20). We have found another leaf where companies have a very high probability of being Profitable.

Figure 6.20 Results Showing Third Split

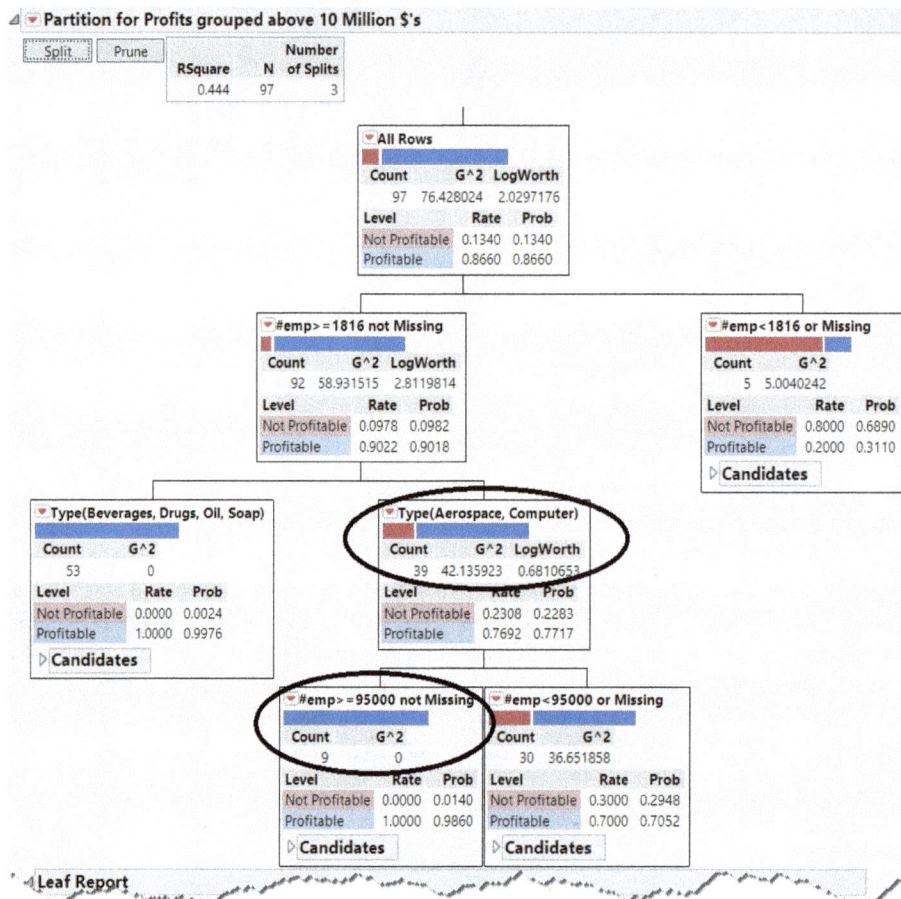

The combination of Aerospace and Computer companies and Number of Employees greater or equal to 95,000 contains nine companies that are profitable. All of the companies in that combination are Profitable. This profitable set has been circled for you. (See Figure 6.20.)

You can now see this last split in the updated Leaf report that appears in Figure 6.21.

Figure 6.21 Third Split Leaf Report

⊿ **Leaf Report**

Response Prob

Leaf Label	Not Profitable	.2 .4 .6 .8	Profitable	.2 .4 .6 .8
#emp>=1816 not Missing&Type(Beverages, Drugs, Oil, Soap)	0.0024		0.9976	
#emp>=1816 not Missing&Type(Aerospace, Computer)&#emp>=95000 not Missing	0.0140		0.9860	
#emp>=1816 not Missing&Type(Aerospace, Computer)&#emp<95000 or Missing	0.2948		0.7052	
#emp<1816 or Missing	0.6890		0.3110	

Response Counts

Leaf Label	Not Profitable		Profitable	
#emp>=1816 not Missing&Type(Beverages, Drugs, Oil, Soap)	0		53	
#emp>=1816 not Missing&Type(Aerospace, Computer)&#emp>=95000 not Missing	0		9	
#emp>=1816 not Missing&Type(Aerospace, Computer)&#emp<95000 or Missing	9		21	
#emp<1816 or Missing	4		1	

In Figure 6.21 at the top (Response Prob), we see that, for the third leaf in the list, defined by #emp is greater than or equal to 1816 or the company did not report employee counts and the type of company is Aerospace, Computer, and the number of employees is less than 95,000 and employee count was not missing, the probability of expecting a company to be Profitable is 0.7052. From the bottom (Response Counts), we see that there are 21 Profitable companies in this group, and nine Not Profitable companies. You might notice that the listed Probabilities are not exactly the Number Profitable/Total Number in that group. JMP is using an adjustment for calculating these probabilities that works better, in practice, for estimating the true probabilities rather than just the in-sample probabilities. (In other words, just because for the second leaf in the list, nine of nine companies were Profitable, it does not guarantee that every company in the world that falls into that leaf's characteristics will be Profitable. So, we adjust the probability slightly away from 100% to account for the chance of just randomly sampling nine Profitable companies when we chose nine companies.)

How do we decide when to stop splitting the tree? This is a very important question! The best way to answer it is to go back to the beginning of this analysis and split our data into two groups: one group of data to build the Partition, and one group to test the resulting model. To learn more about this important idea go to **Help** ▶**JMP Documentation Library** ▶ **Predictive and Specialized Models** ▶ **Partition Models** ▶ **Validation**.

6.3 Filtering Data for Insight

Drilling down or filtering your data is an important objective in data analysis and can be accomplished in several ways with JMP. In this section, we will show you three distinct ways this can be done with our example including commands from the Tables menu, the Data Filter platform, and Lasso tool. You will find that each filtering approach provides unique benefits and is suited to the context of the analysis that you are performing.

Using a Table Command to Extract a Subset

Now let's identify those nine companies:

1. Select the red triangle within the tree node where **#emp>=95000 not Missing**
2. Select **Select Rows.** (See Figure 6.22.)

Figure 6.22 Select Rows on the Partition Node

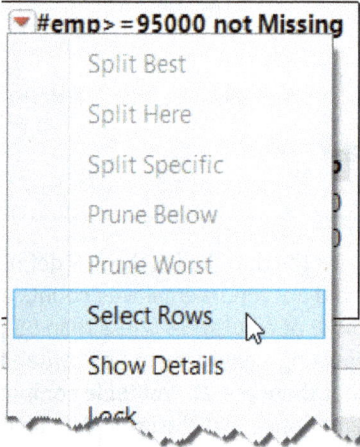

3. Select the **Tables** menu.
4. Select **Subset** (Figure 6.23).

Figure 6.23 Tables Subset

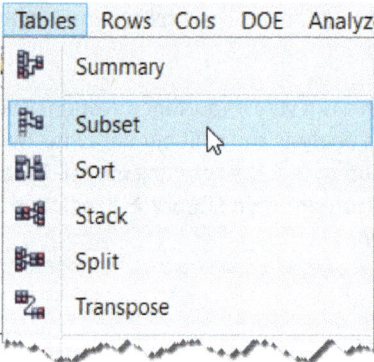

5. In the window, select **Selected Rows** and **All Columns** as shown, and click **OK.** (See Figure 6.24.)

Figure 6.24 Tables Subset Selected

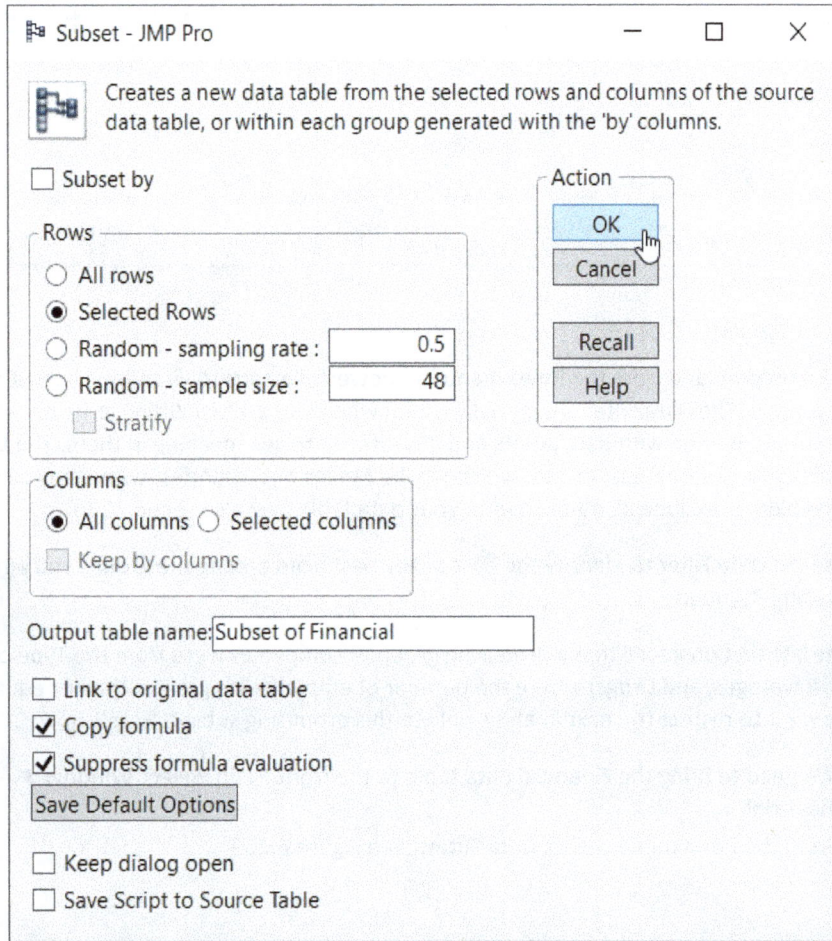

A new data table appears. (See Figure 6.25.) You have identified a subset of desirable and profitable companies that are from the computer and aerospace categories and that have sales above $10 million. These companies can now be submitted for further study and consideration for the investment portfolio. This subset step saves you time from finding them in the data table.

Figure 6.25 New Data Table Showing Profitable Subset

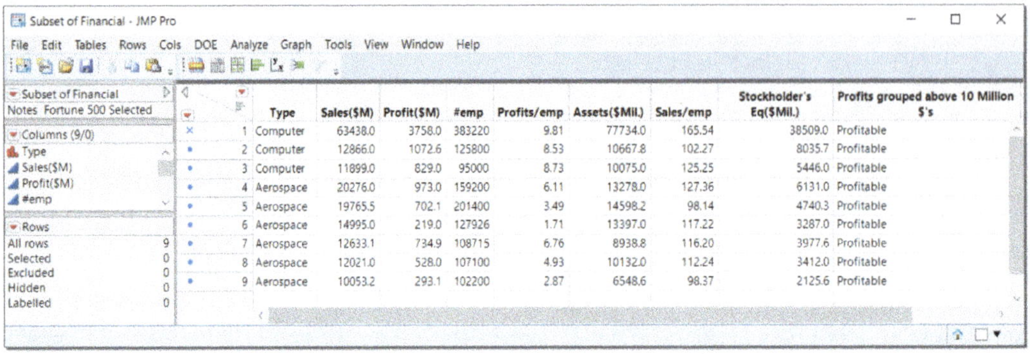

	Type	Sales($M)	Profit($M)	#emp	Profits/emp	Assets($Mil.)	Sales/emp	Stockholder's Eq($Mil.)	Profits grouped above 10 Million $'s
1	Computer	63438.0	3758.0	383220	9.81	77734.0	165.54	38509.0	Profitable
2	Computer	12866.0	1072.6	125800	8.53	10667.8	102.27	8035.7	Profitable
3	Computer	11899.0	829.0	95000	8.73	10075.0	125.25	5446.0	Profitable
4	Aerospace	20276.0	973.0	159200	6.11	13278.0	127.36	6131.0	Profitable
5	Aerospace	19765.5	702.1	201400	3.49	14598.2	98.14	4740.3	Profitable
6	Aerospace	14995.0	219.0	127926	1.71	13397.0	117.22	3287.0	Profitable
7	Aerospace	12633.1	734.9	108715	6.76	8938.8	116.20	3977.6	Profitable
8	Aerospace	12021.0	528.0	107100	4.93	10132.0	112.24	3412.0	Profitable
9	Aerospace	10053.2	293.1	102200	2.87	6548.6	98.37	2125.6	Profitable

Using Data Filter

The Data Filter command from the Rows menu is a powerful means to visualize a subset of your data interactively. The Data Filter is especially useful when you are visualizing large data tables where graphs are packed with data points and it is difficult to see meaning in them. The Data Filter enables you to easily select rows (which can be ranges or categories) within any column of interest and hide or exclude all other rows in your data table (see also section 2.5).

We will use the Data Filter to identify the best of the best from among the groups you already identified using Partition.

Early in the last section, recall that a promising group of companies were from the Type column (Soap, Oil, Beverages, and Drugs) where the number of employees is greater than or equal to 1,816. We want to restrict the next analysis only to this promising subset. Here's how:

1. We need to bring the **Financial** data table to the front again. Select **Window ▶ Financial**.
2. From the Rows menu, select **Data Filter**. (See Figure 6.26.)

Figure 6.26 Select Rows Data Filter

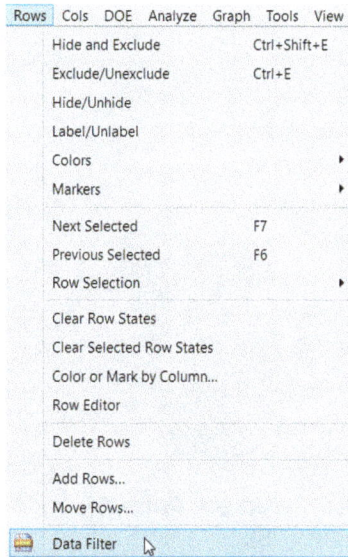

The Data Filter window appears (Figure 6.27).

Figure 6.27 Data Filter Window

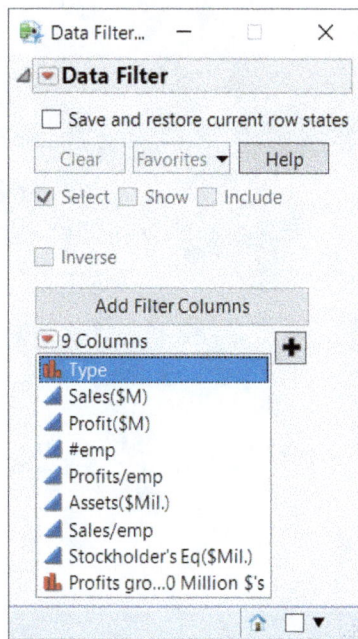

3. Select **Type** and click , which adds it to the filter. A change to the Data Filter appears, showing the different company types. (See Figure 6.28.)

Figure 6.28 Data Filter Type

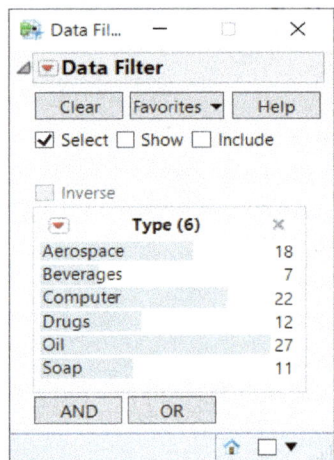

4. Now, press and hold the **CTRL** key and select **Beverages, Drugs, Oil,** and **Soap**. Click the **AND** button. (The results are shown in Figure 6.29).

Figure 6.29 Data Filter Beverages, Drugs, Oil, and Soap

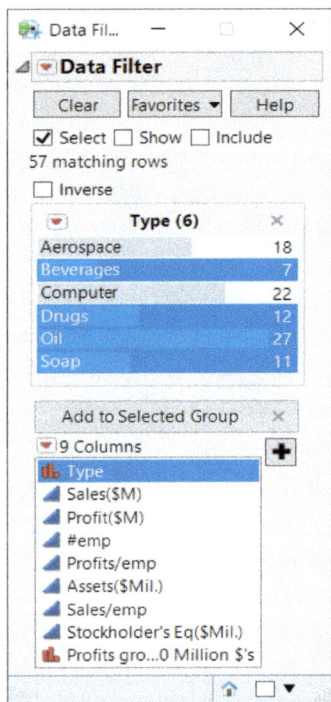

5. Select the **#emp** column and click ➕. (The results with **#emp** slider control are shown circled in Figure 6.30).

Figure 6.30 Select Number of Employees

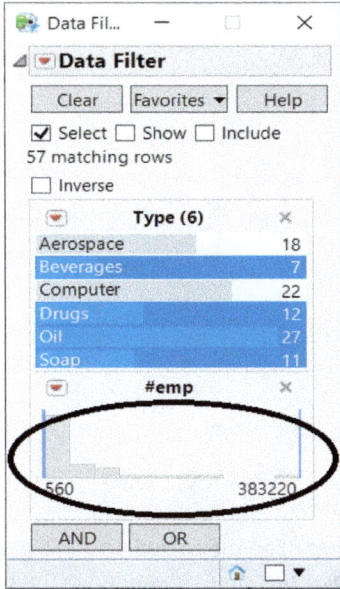

A histogram appears under the number of employees showing the values of the column to be between 560 and 383,220. Recall that the dividing point discovered in the Partition platform earlier favored companies with a number of employees greater than or equal to 1,816.

6. Click and enter **1816** where you see 560 on the left side of the histogram. Click away from the edited value. (See Figure 6.31.)

Figure 6.31 Data Filter

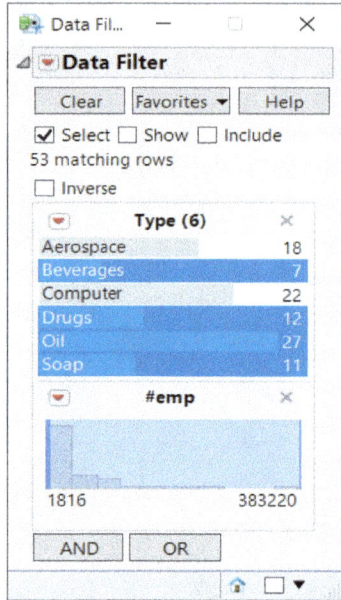

7. Now select the **Show** and **Include** check boxes. (See the circled area in Figure 6.32.) This restricts the analysis to those same profitable companies identified in the partition and leaf report earlier in this chapter.

Figure 6.32 Select Show and Include

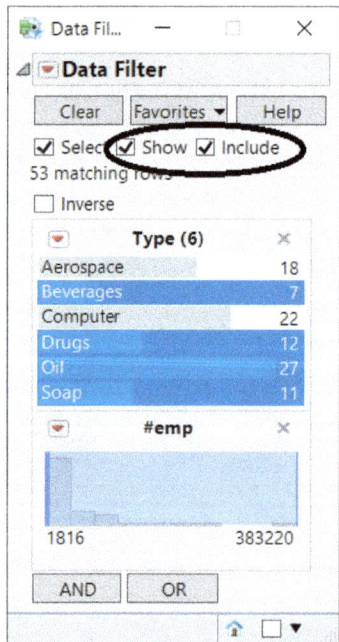

8. Let's look at the result of this exercise by returning to the data table. Select **Window ▶ Financial** (Figure 6.33).

Figure 6.33 Bring Financial Window to Front

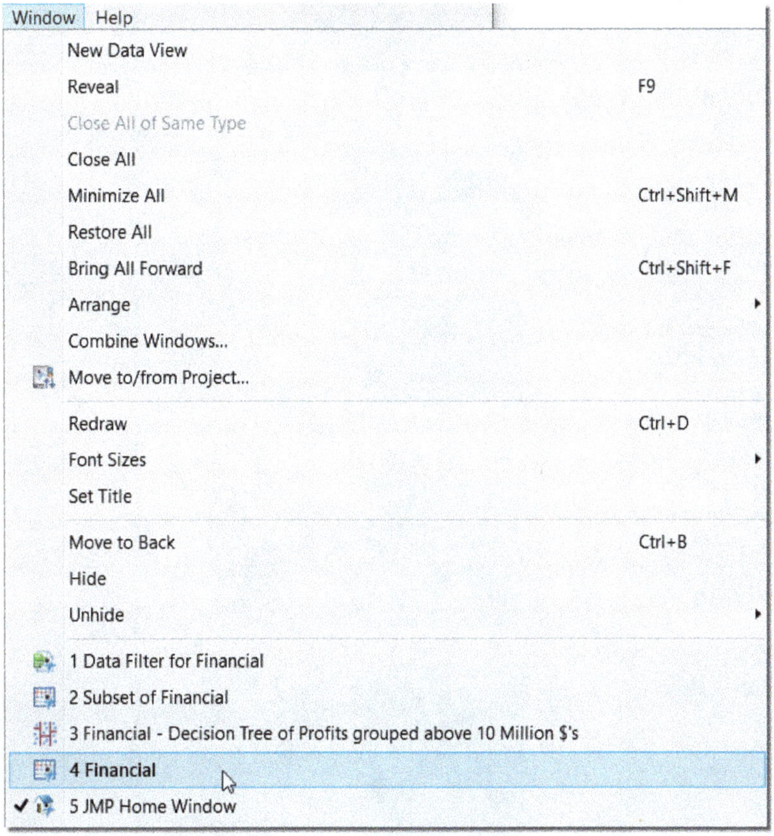

Window	Help	
	New Data View	
	Reveal	F9
	Close All of Same Type	
	Close All	
	Minimize All	Ctrl+Shift+M
	Restore All	
	Bring All Forward	Ctrl+Shift+F
	Arrange	▶
	Combine Windows...	
	Move to/from Project...	
	Redraw	Ctrl+D
	Font Sizes	▶
	Set Title	
	Move to Back	Ctrl+B
	Hide	
	Unhide	▶
	1 Data Filter for Financial	
	2 Subset of Financial	
	3 Financial - Decision Tree of Profits grouped above 10 Million $'s	
	4 Financial	
✓	5 JMP Home Window	

Take a look at the data table and notice what has happened. It now shows groups of rows that are selected and included for profitable rows. (See Figure 6.34.)

Figure 6.34 Data Table Showing Rows Hidden and Excluded

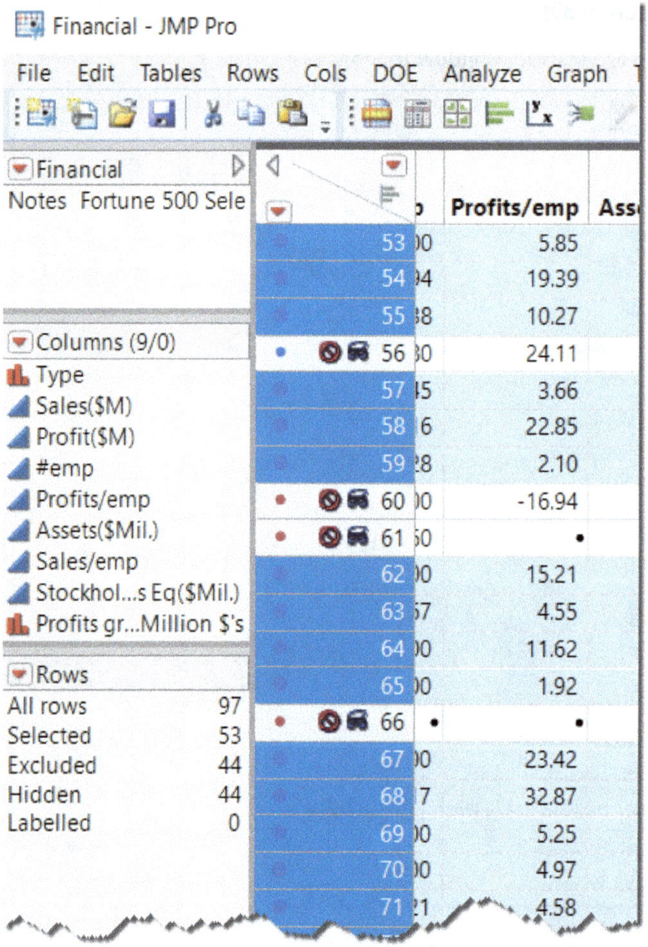

As we covered in Section 2.5, the rows that are hidden (and that will not appear in graphs) feature a mask icon in the row, while rows that are excluded (not used in any analysis) feature a circle icon with a strike-through symbol.

Rows that are still included and selected (highlighted) are those that meet the criteria for selection that we established from our Partition example. They are now selected with the Data Filter.

The Data Filter is a terrific tool for exploring your data visually by enabling you to see subsets of your data in any JMP graph. These subsets can be derived from a prior analysis as demonstrated here or can simply be used to restrict your analysis to any subset you want. As you will see in the next few pages, the Data Filter will be used to further subset the most desirable companies in our example.

Now let's save the Data Filter for the future:

9. From the red triangle on the Data Filter, select **Save Script** ▶ **To Data Table...** (Figure 6.35). Select **OK** to complete the action.

Figure 6.35 Save Script to Data Table

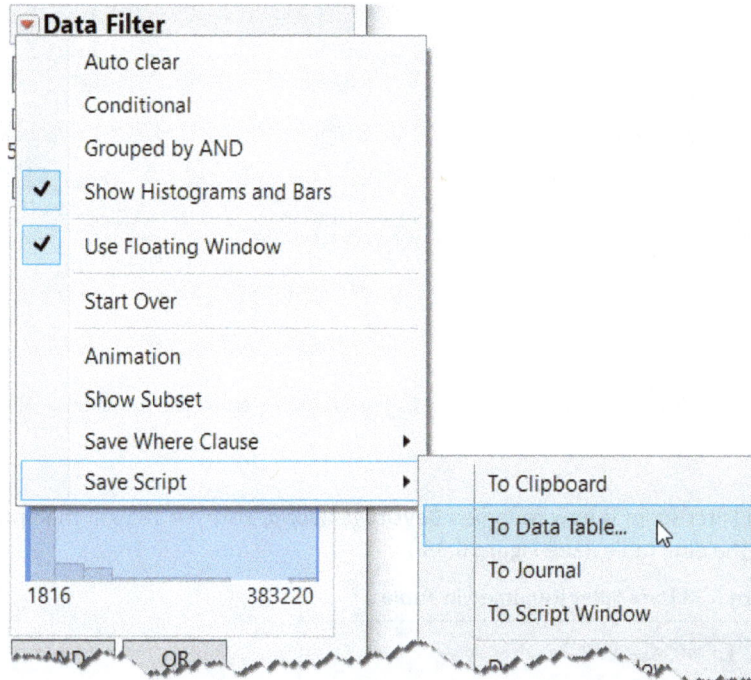

In the upper left panel of the data table, a new item appears named **Data Filter**. It is circled for you. (See Figure 6.36.) The **Data Filter** item has created a JMP script that stores the steps needed to reproduce the subset that you selected.

Figure 6.36 Data Filter Script in Data Table

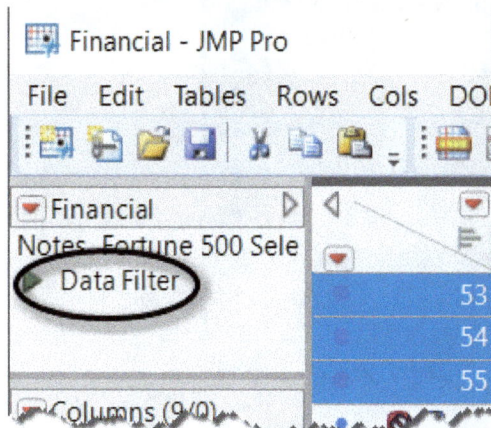

Let's rename it something more descriptive so that we can remember what it does.

10. Right-click on **Data Filter** and select **Edit.** Rename it **BevDrugOilSoap, EMP >= 1816** in the window that appears. (See Figure 6.37.)

Figure 6.37 Data Filter Script Renamed

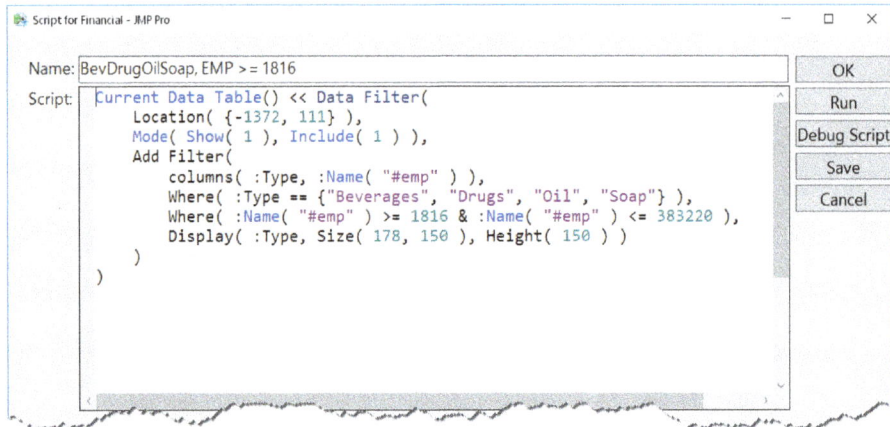

11. Click **OK**.

The Data Filter script is now renamed **BevDrugOilSoap, EMP > = 1816** in the upper left panel of the data table. (See Figure 6.38.)

Figure 6.38 Data Filter Renamed in Table

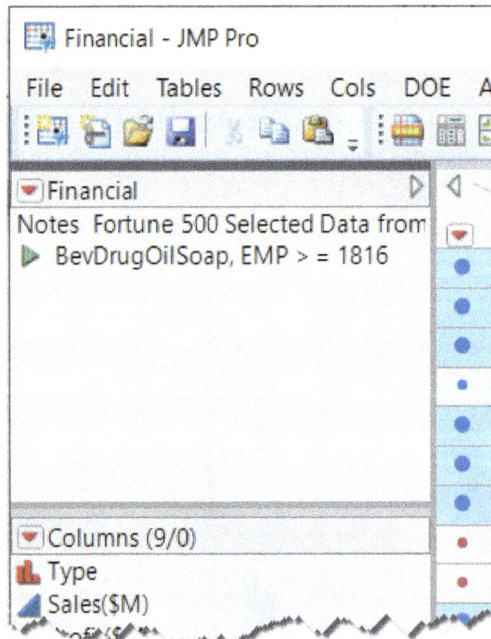

Now we can reuse the filter if we want to apply this criterion to the data. It's easy with the new script that applies the filter. You simply click the ▶ in the upper left panel of the data table.

You have now filtered your data to only the best performers, so our task is to select only the very best from this group for our portfolio. There are many methods to finding high performers from among the groups that we selected. We will go back and use the Fit Y by X platform (that we introduced in the previous chapter) along with the Lasso tool to visually select individual points. We find that using Lasso in this context easily enables visual selection of points that meet our performance requirements. The goal is to select the most profitable companies by industry type. In this example, we will use the continuous representation of profits, **Profits($M)**, as it allows us to distinguish with more precision among this group of high performers.

Recall that we also identified profitable companies from the Aerospace and Computer types in the partition Analysis and we used the subset command to extract those. Because these companies also had the requirement of having sales over $10 million, we could have used a conditioning statement (such as an "and" or "or" statement) in the Data Filter or merged that group with the one illustrated above to create one data table. To simplify this illustration, we have omitted these steps. To learn about how to merge data tables, see **Help ▶ JMP Documentation Library ▶ Using JMP ▶ Reshape Data** and the topic **Concatenate**.

12. Select **Analyze ▶ Fit Y by X**. (See Figure 6.39.)

    Figure 6.39 Analyze Fit Y by X

    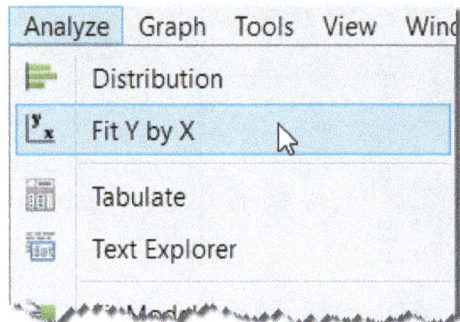

13. In the launch window, select **Profits($M)** as the **Y, Response** and select **Type** for the **X, Factor**. (See Figure 6.40.)

Figure 6.40 Fit Y by X Profit by Type

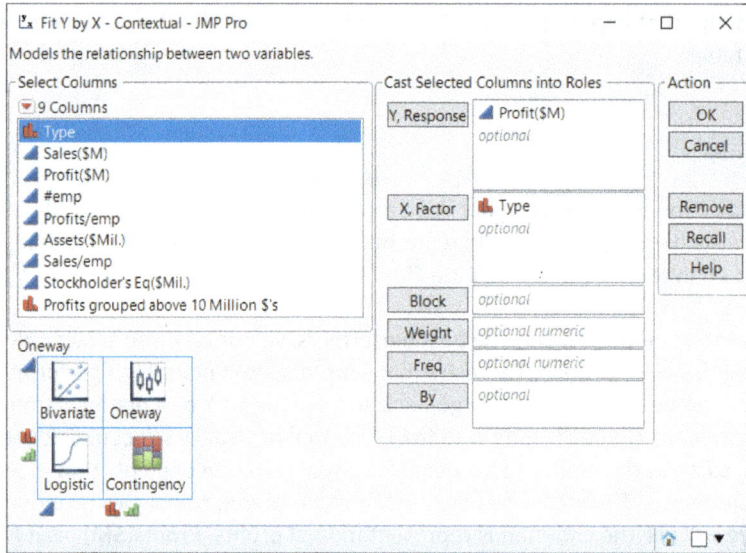

14. Click **OK**.

The Oneway graph (Figure 6.41) shows profits from one of the best performing groups that you identified in your partition analysis and selected using the Data Filter. All of the other companies have been excluded and hidden from this analysis. It is from these included companies that we want to identify the very best of these performers.

Figure 6.41 Oneway ANOVA Report

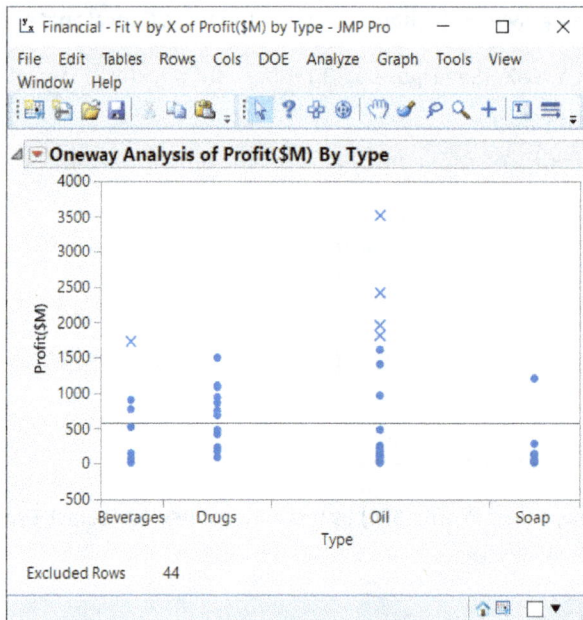

15. From the red triangle, select **Means and Std Dev**. (See Figure 6.42.)

Figure 6.42 Select Means and Std Dev from Red Triangle

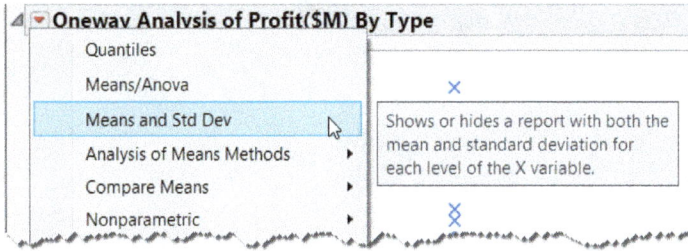

The result appears in Figure 6.43a.

Figure 6.43a Oneway Means Std Dev Lines

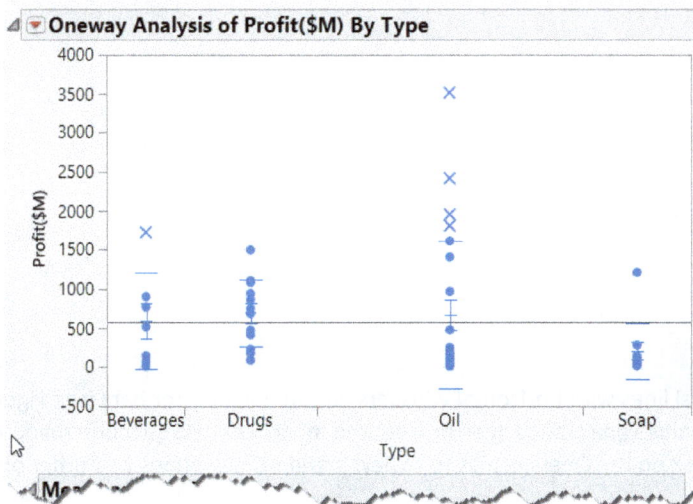

We want to select companies that are exceptional performers. Let's define that by looking for companies that are at least one standard deviation above the mean within each Type. Notice that the wide blue horizontal lines that appear in each type mark the values at the mean plus and minus one standard deviation. (See Figure 6.43b.)

Figure 6.43b Mean Error Bars

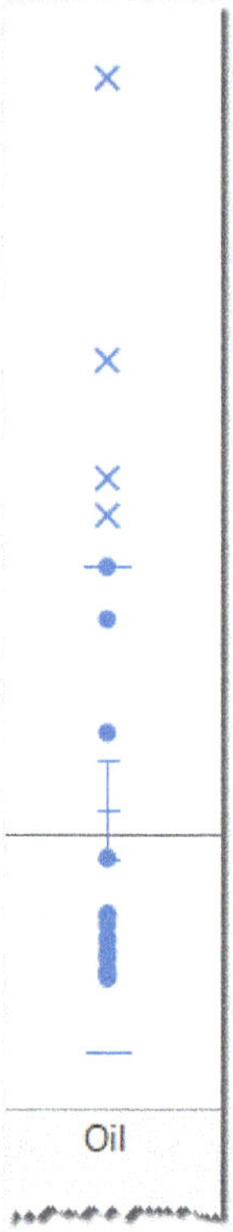

The smaller blue vertical lines with horizontal whiskers denote mean error bars (see Figure 6.43b), indicating the values one standard error from the mean. Use the question mark tool from the Tools menu and click on the Oneway plot to review standard deviations and other options.

We are interested in the companies performing better than one standard deviation above the mean Profit within each Type (in other words, above the blue horizontal lines for each Type) because they are exceptional performers within their type in terms of profit. Let's try a new method to select them.

Using Lasso to Select Individual Points

1. From the toolbar, select the **Lasso** tool. (See Figure 6.44.)

Figure 6.44 Select Lasso Tool

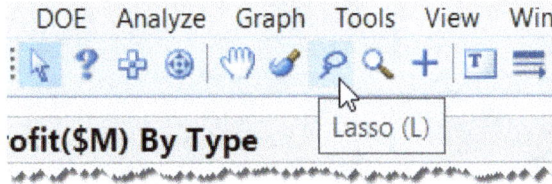

2. Now move the cursor (which should look like a little lasso) to the Oneway result. Left-click and draw a circle around the points above the top whiskers of the error bars (see Figure 6.45), making sure that you close the circle around the points before releasing the button.

Figure 6.45 Lasso Around Points

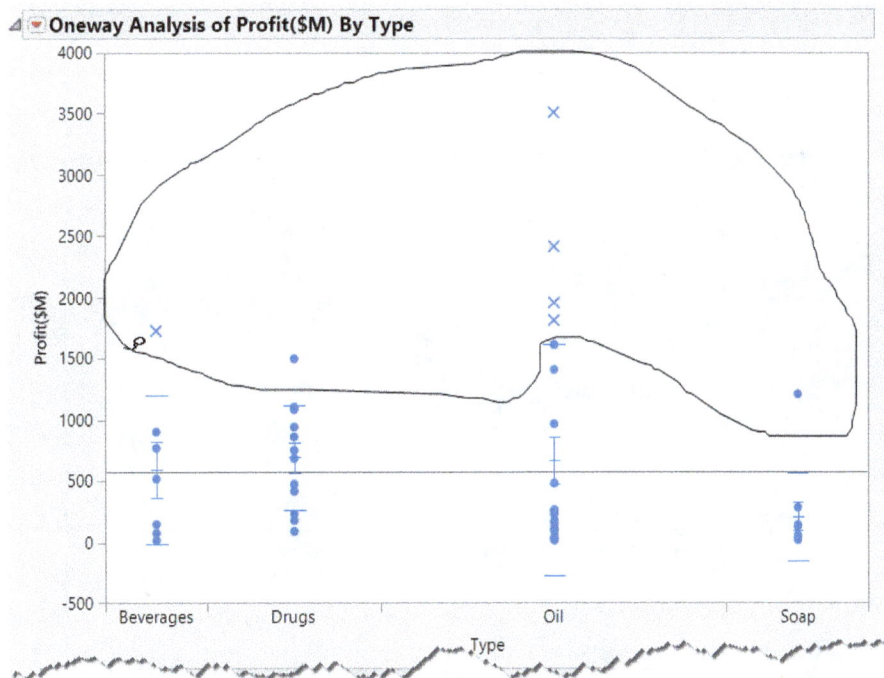

There should be about seven selected rows in your table. It's okay if the count is not exact. Now we just need to extract these very best historical performers (with respect to their Type) from the rest.

Note
To help make these easier to see, the size of the graph has been increased by clicking and dragging the lower right corner to enlarge the graph.

3. From the **Tables** menu, select **Subset**. (See Figure 6.46.)

Figure 6.46 Tables Subset

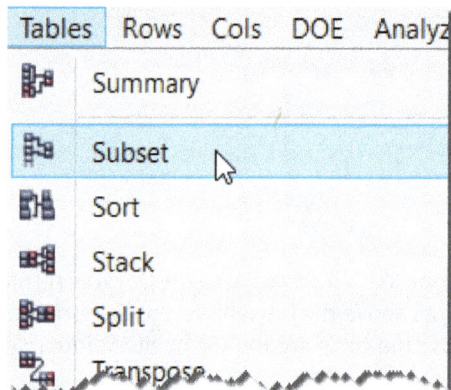

4. In the launch window, select **Selected Rows** and **All Columns**. (See Figure 6.47.)

Figure 6.47 Select Selected Rows and All Columns

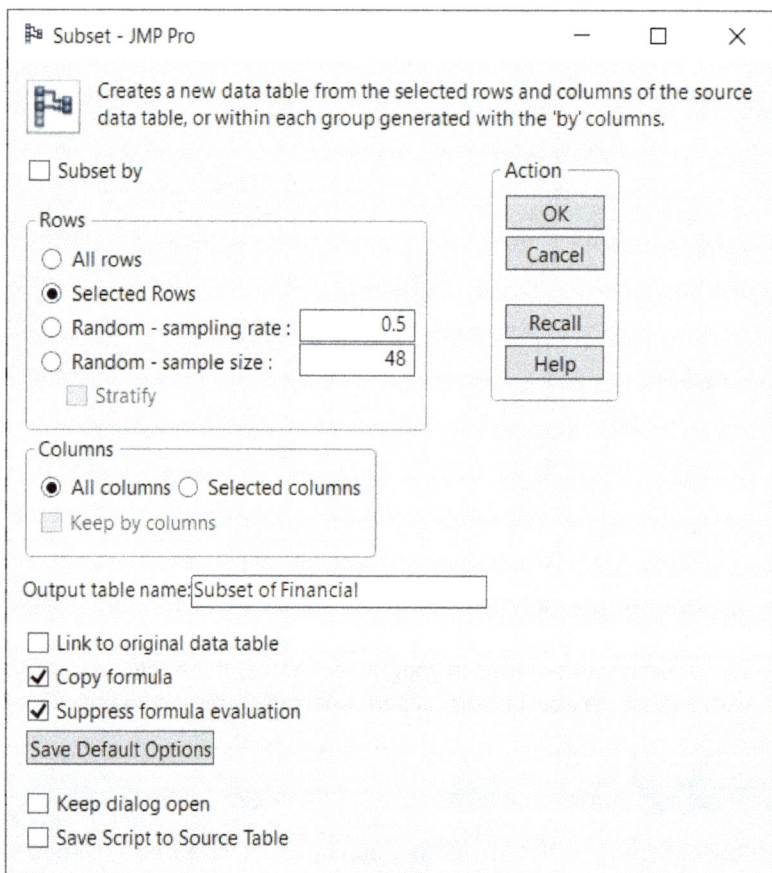

5. Click **OK**.

The resulting data table (Figure 6.48) contains the seven most promising companies out of the 97 original companies. The most profitable companies are mostly oil companies, though soap, drugs, and beverage companies each made a showing among the most profitable. Note that we have not necessarily chosen the seven most profitable companies – rather, we have chosen the seven companies that, within their Types, are more profitable than the one standard deviation cutoff. This might mean that we have chosen a Soap company that technically performs with slightly less Profit than an Oil company that we rejected; however, this can be a benefit in diversifying our choices. You could also just choose the seven most profitable companies, regardless of Type, if you are not concerned about diversifying among several company Types. We now have some potential portfolio selections for further study or investment.

> **Note**
>
> You might not get exactly seven. This is okay because some of the best performing companies might be challenging to grab using the lasso tool.

Figure 6.48 Data Table Showing Most Profitable Companies

	Type	Sales($M)	Profit($M)	#emp	Profits/emp	Assets($Mil.)	Sales/emp	Stockholder's Eq($Mil.)	Profits group
1	Drugs	6698.4	1495.4	34400	43.47	6756.7	194.72	3520.6	Profitable
2	Oil	86656.0	3510.0	104000	33.75	83219.0	833.23	30244.0	Profitable
3	Oil	50976.0	1809.0	67900	26.64	39080.0	750.75	16274.0	Profitable
4	Oil	32416.0	2413.0	37067	65.10	25636.0	874.52	9180.0	Profitable
5	Oil	15905.0	1953.0	26600	73.42	22261.0	597.93	6562.0	Profitable
6	Soap	21689.0	1206.0	79300	15.21	16351.0	273.51	6215.0	Profitable
7	Beverages	9171.8	1723.8	20960	82.24	8282.5	437.59	3485.5	Profitable

6.4 Model Fitting, Visualization, and What-If Analysis

In this section, we will extend many of the JMP skills and concepts that you have learned thus far into the insight you will need to articulate your findings effectively. We will now combine our models with the Data Filter in order to better understand the data. We will also introduce the Prediction Profiler for interactive visual what-if analysis.

Consider for a moment a thought experiment. Some old radios have an analog tuner dial on them. Imagine you turn on the radio and hear noisy static. You slowly turn the tuner dial through the frequencies until you hear a transmission. Sometimes you might tune past the station, but you slowly tune back to obtain the optimal signal. Your ear hears the difference between the signal and the background noise. The Profiler and Data Filter (as described in the previous section) can be used as just such a tuner, but instead of using your ears, you use your eyes. The next section shows you how.

Let's test the idea that a positive relationship exists between profits and the number of employees in all of our companies. To do this, we need to use two things you already learned: Fit Y by X and the Data Filter.

1. Select **Rows ▶ Clear Row States**. (See Figure 6.49.)

Figure 6.49 Clear Row States

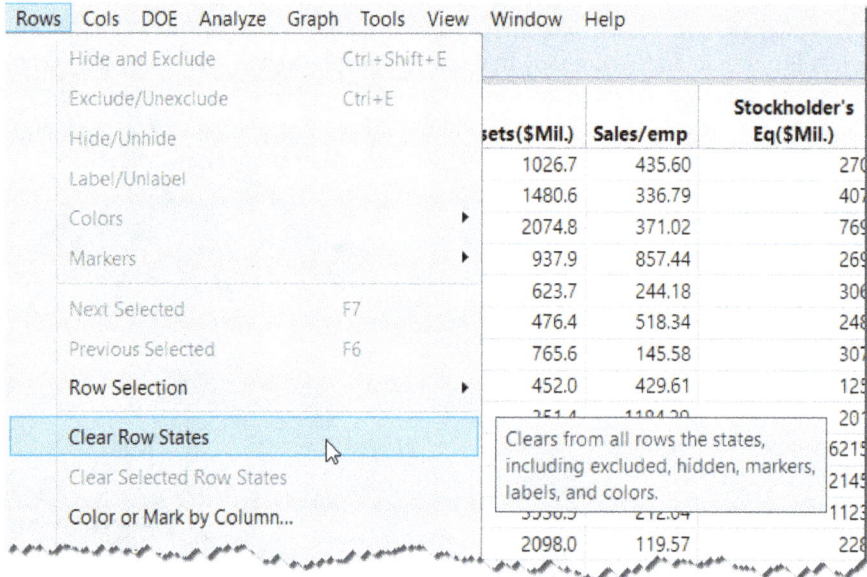

Rows	Cols	DOE	Analyze	Graph	Tools	View	Window	Help

		Stockholder's
Hide and Exclude	Ctrl+Shift+E	
Exclude/Unexclude	Ctrl+E	
Hide/Unhide		sets($Mil.) Sales/emp Eq($Mil.)
Label/Unlabel		1026.7 435.60 270
		1480.6 336.79 407
Colors	▶	2074.8 371.02 769
Markers	▶	937.9 857.44 269
		623.7 244.18 306
Next Selected	F7	476.4 518.34 248
Previous Selected	F6	765.6 145.58 307
Row Selection	▶	452.0 429.61 129
		251.4 1194.20 20
Clear Row States		Clears from all rows the states, 6215
Clear Selected Row States		including excluded, hidden, markers, 2145 labels, and colors.
Color or Mark by Column...		1123
		2098.0 119.57 228

This clears the row selections, the markers, and the hidden and excluded rows from the last section.

Now let's build a simple model of the relationship between profits and number of employees.

2. Select **Analyze ▶ Fit Y by X**. Select **Profit($M)** as the **Y, Response** role and **#emp** as the **X, Factor** role. (See Figure 6.50.)

Figure 6.50 Fix Y by X Profit Employees

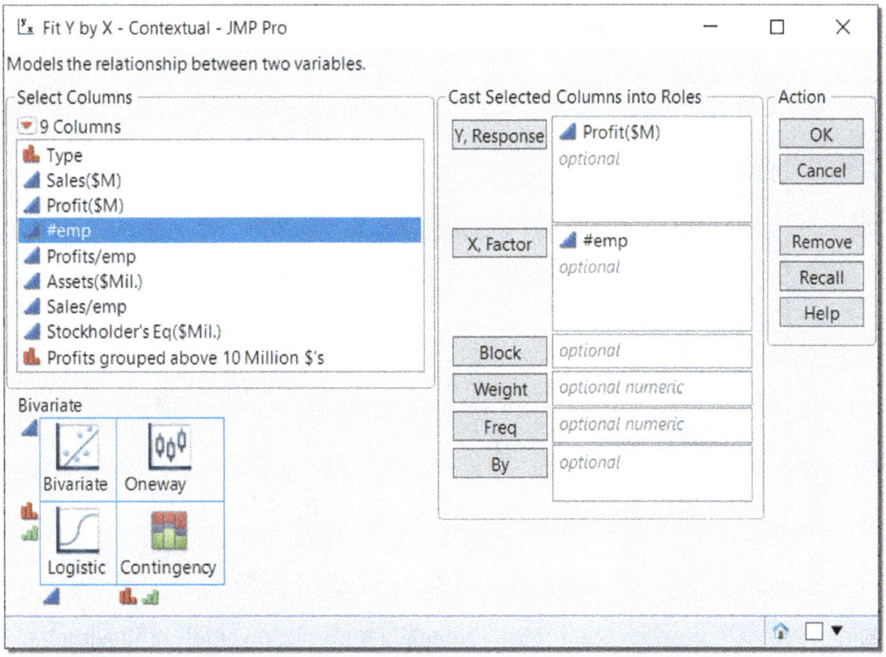

3. Click **OK**. A bivariate fit appears (Figure 6.51).

Figure 6.51 Bivariate Plot

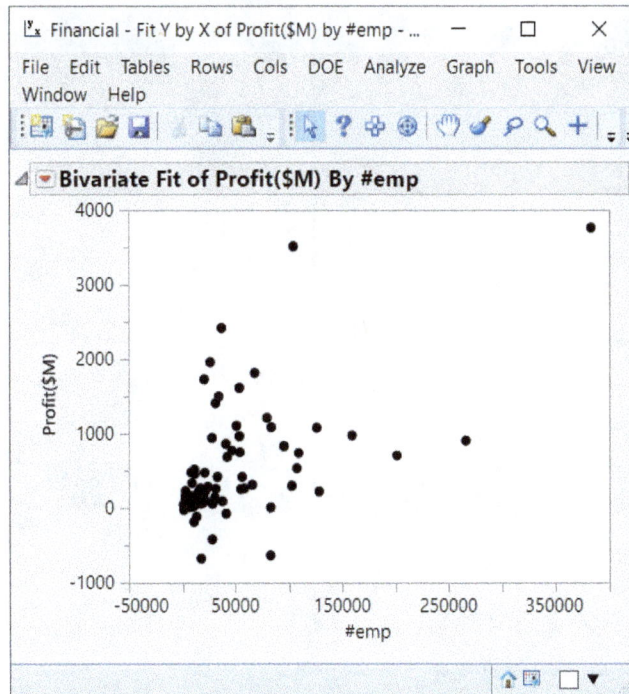

From the red triangle for the bivariate fit, select **Fit Line** and **Fit Mean**. (See Figure 6.52.)

Figure 6.52 Select Fit Line

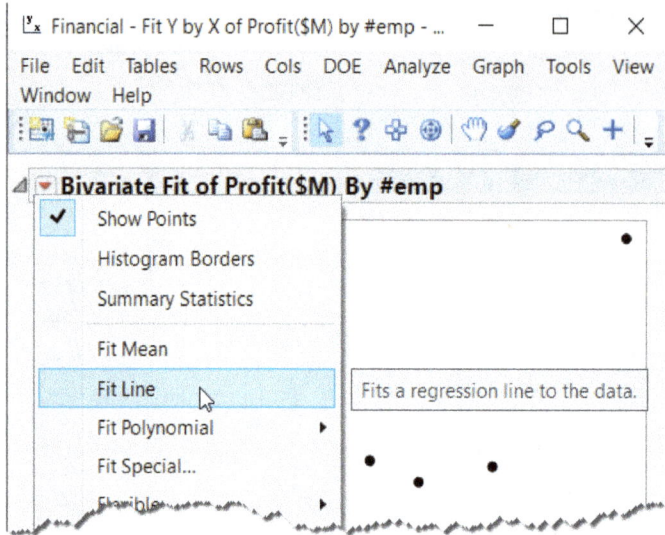

A red line appears in the data that extends generally from the lower left of the graph to the upper right of the graph. A green line appears that represents the mean of the response or average profit. (See Figure 6.53.)

Figure 6.53 Bivariate Profits by Employee with Fit Line

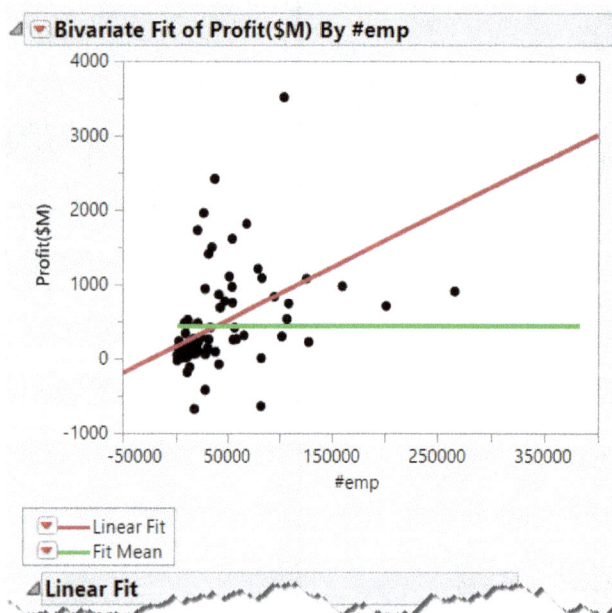

This red line shows that, in general, as the number of employees increases, so do the profits. Notice how sparse the data is for companies above 150,000 employees. Because there are few companies or rows of data above 150,000 employees, it might suggest that the relationship is not as well defined in that region. An unusually large and profitable company in the upper right is influencing the slope of the line as well.

> **Note**
>
> This is a simple or linear least squares regression fit, which fits a straight line through the points in a manner that balances the differences between those points above and below the line.

Let's see whether graphics can help us interpret this relationship:

4. Select the red triangle item next to the line labeled Linear Fit and select **Confid Shaded Fit** (Figure 6.54). A red shaded boundary around the line appears. (See Figure 6.55.)

Figure 6.54 Conf Shaded Fit

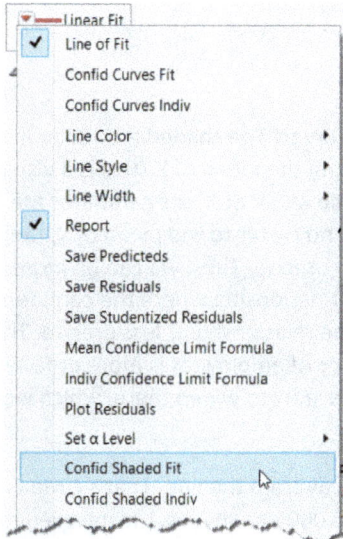

Figure 6.55 Conf Shaded Fit Displayed

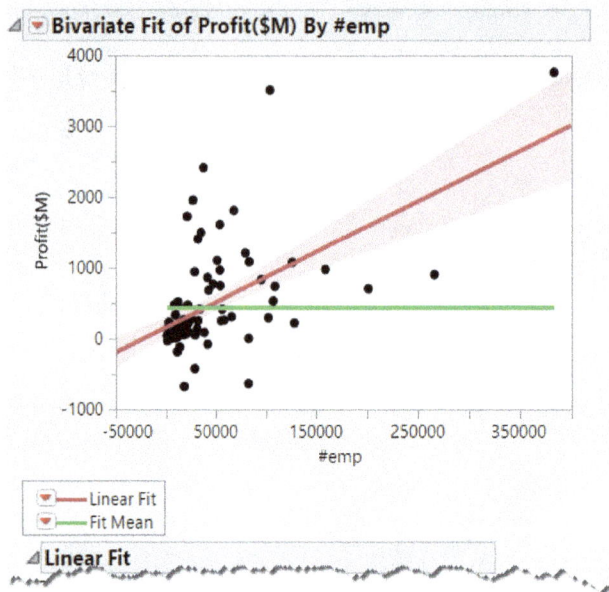

The curves appearing on the graph are 95% confidence curves. The shaded area around the line gets wider where data is sparser and the farther away from the mean of X. This is a visual indication that where data is sparse and far away from the mean of X, the estimates are less precise. The converse is also true: where data is denser and closer to the mean of X, the confidence curves are narrower, and estimates are more precise. Thus, we can be more confident that the line better represents a more precise relationship where the confidence curves are narrower than where they are wider where the relationship is less precise. This might indicate that the relationship between profits and number of employees is more precise up to about 150,000 employees because the confidence curves start to widen there, which would indicate the relationship is less precise.

Notice, too, that a lot of the data points are pretty far away from the line. That's curious. It appears especially true for companies with more than 50,000 employees. Maybe even more than one pattern is present. How might we explore this?

Let's start by exploring if the positive relationship between profits and number of employees holds across all company types. The Data Filter acts like a tuner as it toggles through the company types. Here's how.

5. Select **Rows ▶ Data Filter**. (See Figure 6.56.)

Figure 6.56 Select Rows Data Filter

Rows	Cols	DOE	Analyze	Graph	Tools	View	Window	Help

Hide and Exclude	Ctrl+Shift+E		
Exclude/Unexclude	Ctrl+E		St
Hide/Unhide		sets($Mil.)	Sales/emp
Label/Unlabel		1026.7	435.60
		1480.6	336.79
Colors	▶	2074.8	371.02
Markers	▶	937.9	857.44
		623.7	244.18
Next Selected	F7	476.4	518.34
Previous Selected	F6	765.6	145.58
Row Selection	▶	452.0	429.61
		351.4	1184.29
Clear Row States		16351.0	273.51
Clear Selected Row States		9379.0	265.44
Color or Mark by Column...		3536.5	212.04
Row Editor		2098.0	119.57
		•	•
Delete Rows		1213.1	277.58
		969.6	206.19
Add Rows...		362.6	128.11
Move Rows...		325.2	216.97
		335.3	103.90
Data Filter		335.2	144.23

3.39

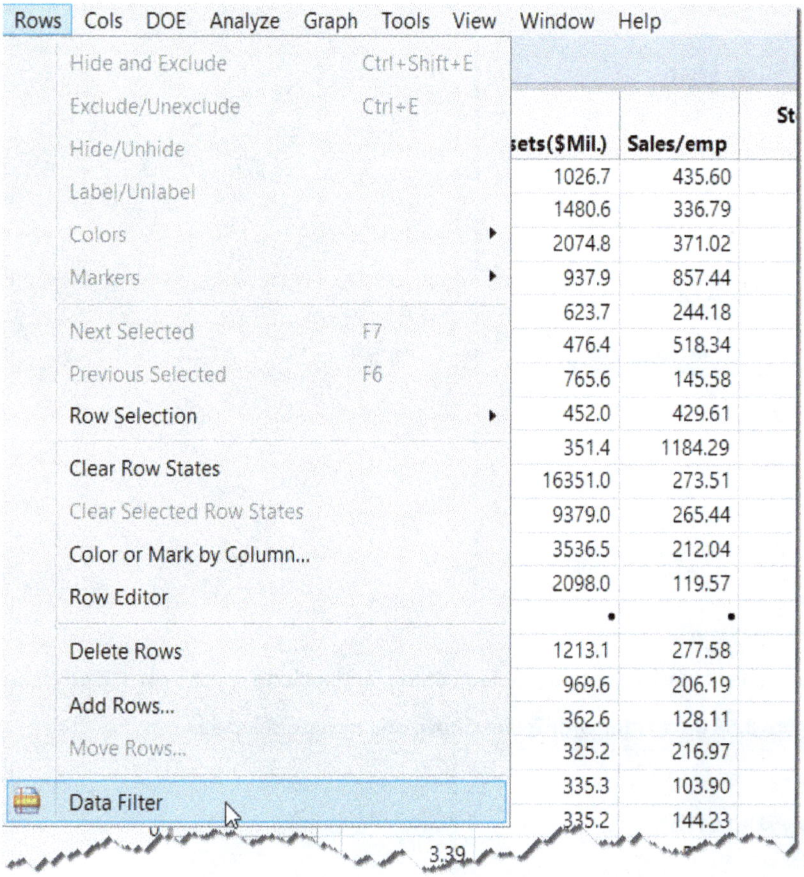

6. In the Data Filter window, select the **Type** column (see Figure 6.56) and click ➕.

Figure 6.57 Select Data Filter Type

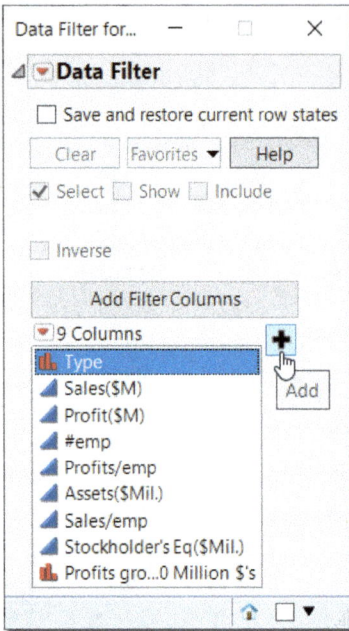

7. Put a check mark next to the **Select**, **Show**, and **Include** check boxes. (See Figure 6.58.)

Figure 6.58 Data Filter with Select, Show, and Include Selected

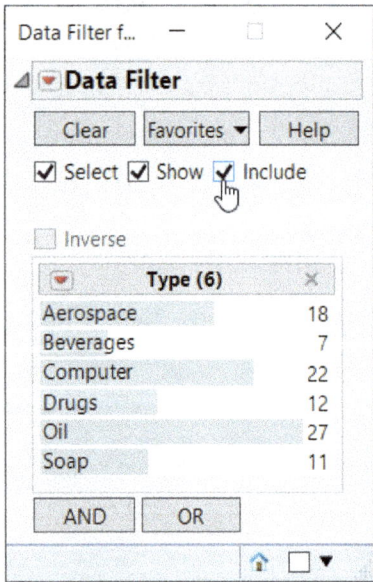

8. Now return to your Fit Y by X results. From the red triangle, select **Redo ▶ Automatic Recalc**. (See Figure 6.59.)

Figure 6.59 Automatic Recalc

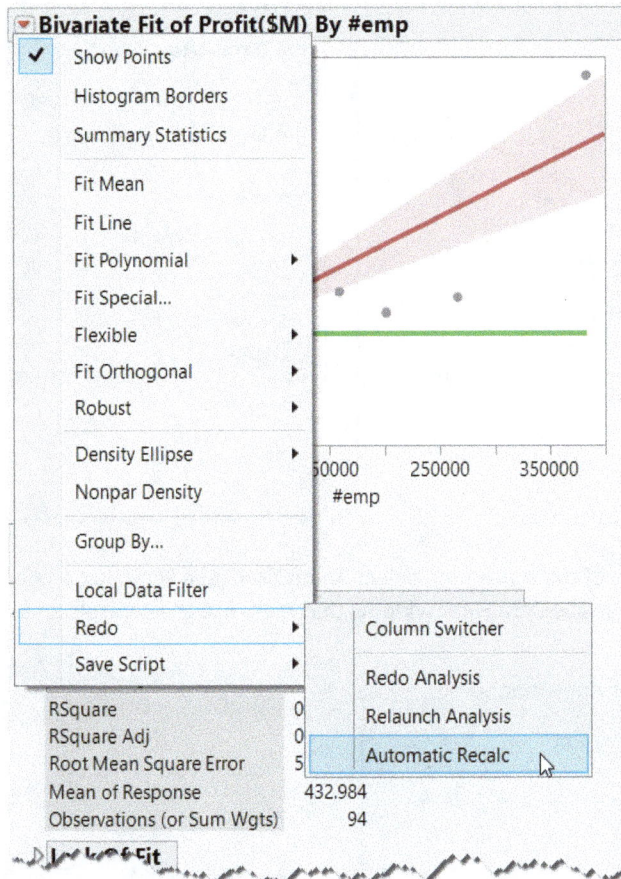

Automatic Recalc updates the scatterplot and fit based on the new row states controlled by the Data Filter. Alternatively, you could leave Automatic Recalc off and re-fit the trend line iteratively for multiple views, but this takes more time to do.

Now you should see two floating windows. (See Figure 6.60.)

Figure 6.60 Bivariate Fit Y by X and Data Filter

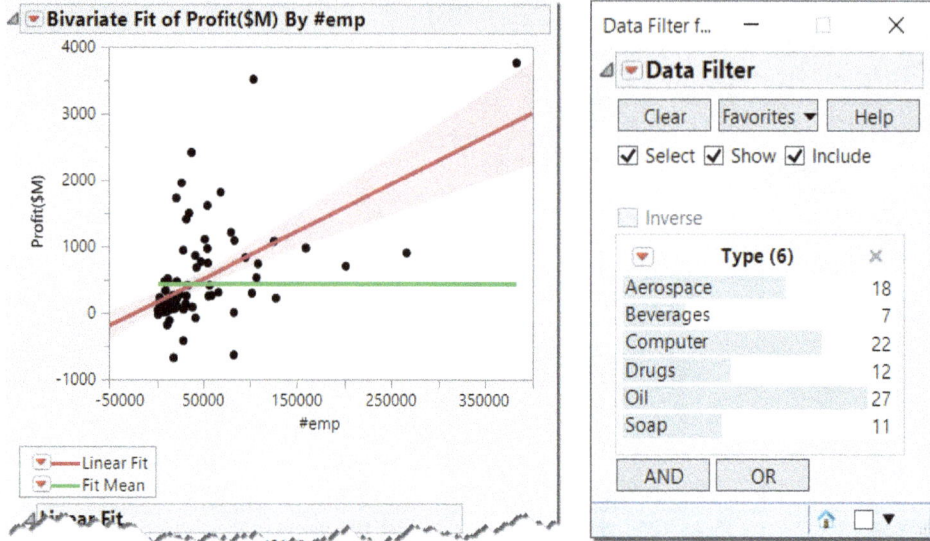

9. From the red triangle of the Data Filter, select **Animation**. (See Figure 6.61.) Animation controls now appear in the Data Filter window that looks like the controls on a DVD player.

Figure 6.61 Data Filter Animation

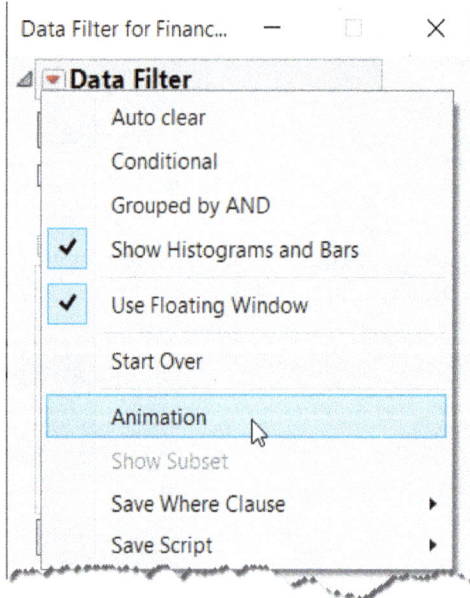

10. Click on (the step forward control). (See Figure 6.62.)

Figure 6.62 The Step Forward Control

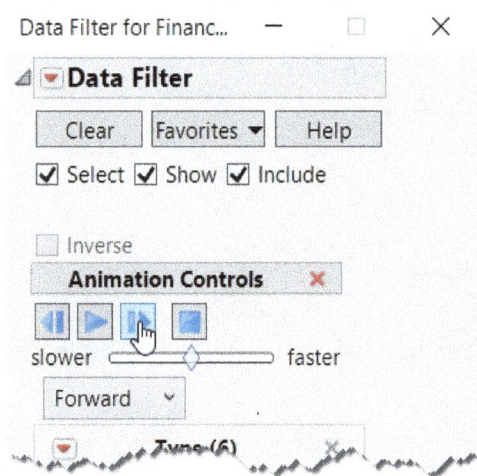

Each time you click the step forward control, you filter on just one of company groups within the **Type** column, and you will see these highlighted in the Data Filter window.

Now click sequentially through the company types and notice what happens to the graph each time you click. Toggle through several times. Can you see the one type that is different from the rest? Figure 6.63 includes the series of graphs that you should be seeing as you toggle through each company type.

Figure 6.63 Results of Clicking Through Company Type

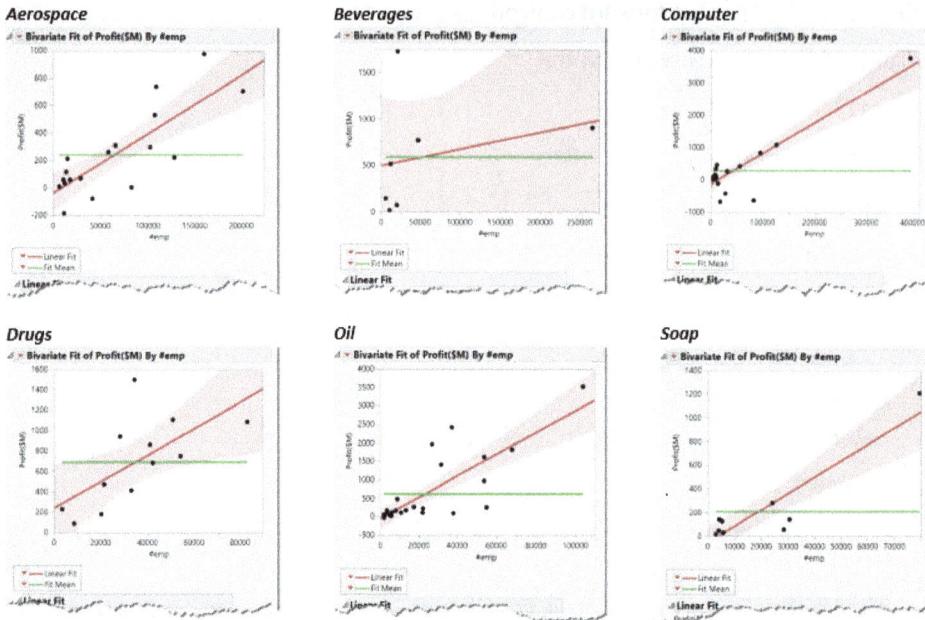

Which graph is most different from the rest? Which graph has a fit line that is almost flat? You might have noticed that Beverages shows a line that is much flatter than the others, which indicates that the number of employees or size of the company has less influence on the profitability. Notice that, in general, the other company types show a reasonably strong correlation between profits and number of employees.

Notice also that Beverages is the sparsest fit. It has only seven data points widely dispersed, so the confidence bands also flare out widely. Also notice that the green mean line for Beverages is completely inside the confidence curves. This is another indication that the correlation for Beverages is not strong. At least for the beverage company type, we can conclude that the greater number of employees does not predict greater profits in this sparse sample.

11. Select **Rows ▶ Clear Row States** for next section.

Conducting What If Analysis

The Prediction Profiler is a different type of graphical tuner for your data and provides a clear picture of your model. The Prediction Profiler is an interactive graph that produces estimates of your Y column of interest (profits, in our example) subject to your predictors, or X, columns (such as number of employees from the last example). The interactive feature enables you to drag and change the settings of any column to see the estimated effect on the other columns. The following example shows you how the Prediction Profiler is used in the context of a Linear Model.

The advantage of the Prediction Profiler is that it lets you try what-if scenarios dynamically and get immediate estimates on any column of interest. Very cool!

The simplest version of the prediction profiler can be seen in the Fit Model platform when fitting a line as we did with the Fit Y by X platform.

12. To create a profiler, you first need to create a model to describe the relationships in your data. Using the same Financial data table with the row states cleared, select **Analyze ▶ Fit Model**. (See Figure 6.64.)

Figure 6.64 Analyze Fit Model

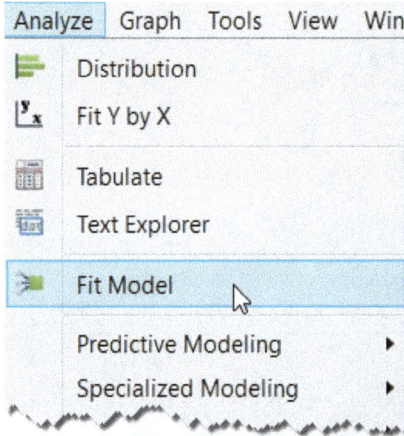

13. In the Fit Model window, select **Profit($M)** as the **Y** column. Then select **#emp** and click on the **Add** button to place it in the Construct Model Effects window. (See Figure 6.65.)

Figure 6.65 Fit Model Window

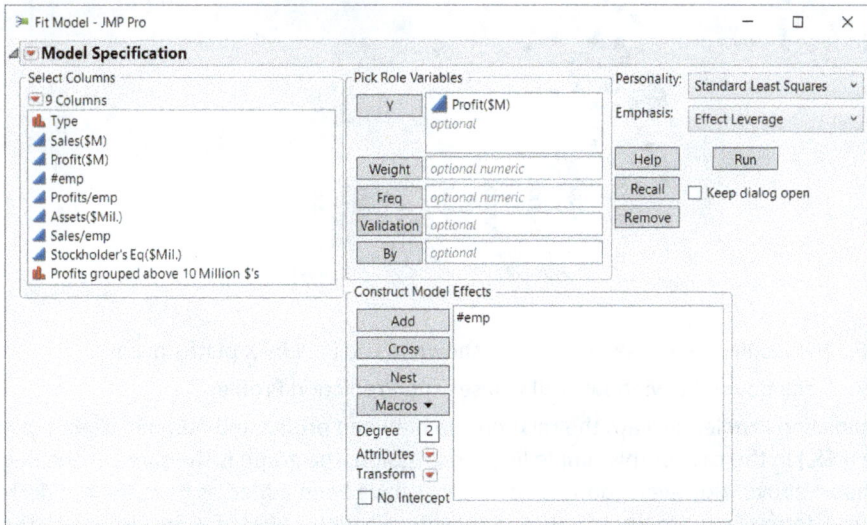

14. Select the pop-down menu for **Emphasis** and change it to **Effect Screening**. (See Figure 6.66.) This just changes the format of the results shown in the output window (Figure 6.67).

Figure 6.66 Fit Model with Effect Screening Emphasis

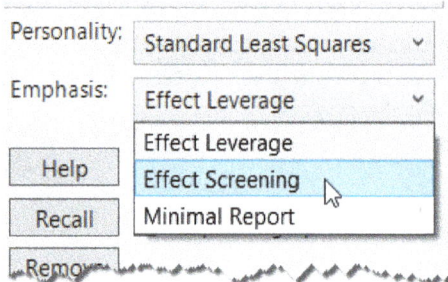

15. Click **Run**.

Figure 6.67 Fit Model Screening Results

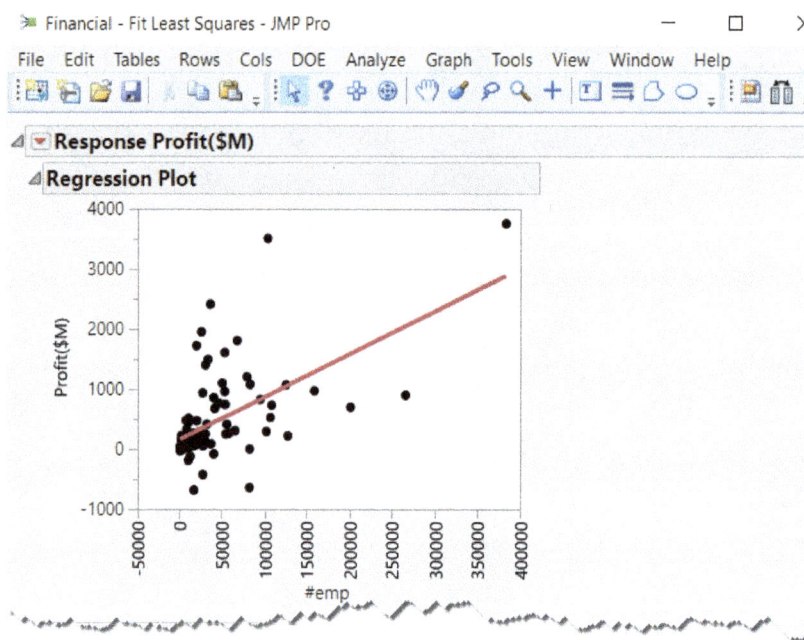

16. The results are the same as those shown in the Fit Y by X platform earlier.
17. Scroll down the window until you see the Prediction Profiler.

The Prediction Profiler displays the relationship between profits and number of employees. (See Figure 6.68.) In the case of this simple linear regression, the graph is the same as the Regression Plot shown above. However, confidence intervals have been added as have the red dotted lines. These red dotted lines are interactive and provide use with a kind of X-ray vision into the relationships between the columns.

Figure 6.68 Prediction Profiler

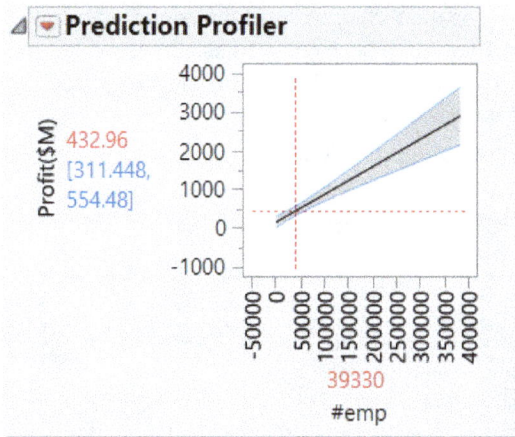

As you drag the red dotted line for number of employees to 100,000, notice the estimate for profits increases to around 865. (See Figure 6.69.) You might find it difficult to get the exact value when dragging the line. This is okay. You can also click on the red value and enter the new desired value (100,000 in this case).

Figure 6.69 Adjusted Prediction Profiler

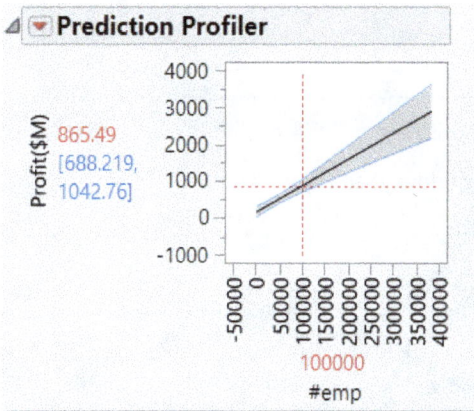

Look at the angle of the fitted line in the plot. The angle of the fitted line gives us clues as to the relationship between the predictor (number of employees in this case) and the profit. A steeper line suggests that small changes in X (number of employees) will have significant changes in Y (profit). If the line were flatter, this would indicate that changes in X will not change Y very much (if at all).

While the Prediction Profiler is interesting for this simple linear regression case, its real power is for cases where there are multiple predictor variables because it enables you to see the effect of all of the X variables at the same time. Recall the earlier example that used the data filter to look the relationship between the number of employees and profit for each of the types of companies. This can also be done with the prediction profiler in the Fit Model platform.

18. Select **Analyze ▶ Fit Model**.

19. In the Fit Model window, select **Profit($M)** as the **Y** column. Hold the **control** key down on your keyboard, then select **#emp** and **Type**. With both columns selected, click on the **Macros** button and select **Full Factorial**. This adds both columns and the interaction between them to the Construct Model Effects window.

20. Change the emphasis to **Effect Screening**. (See Figure 6.70.)

Figure 6.70 Fit Model Window

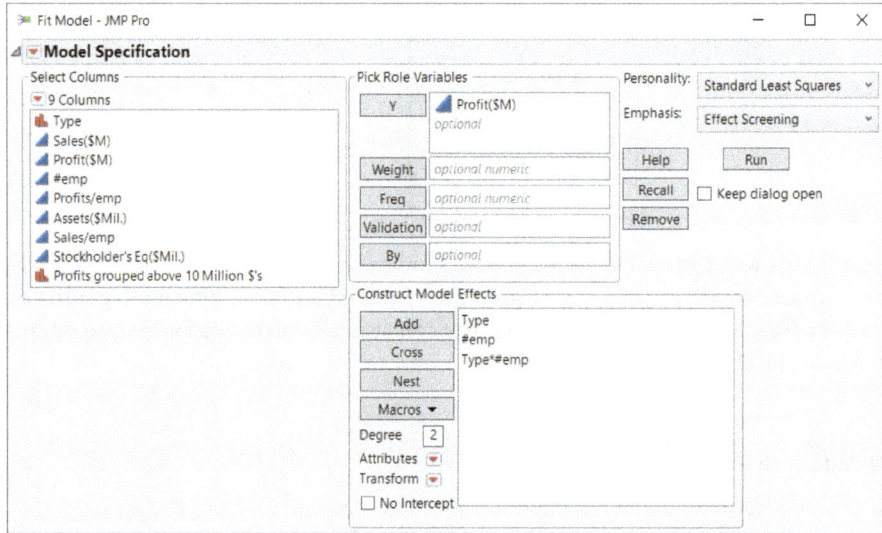

21. Click the **Run** button.

Examine the Regression Plot in the results window. (See Figure 6.71.)

Figure 6.71 Regression Plot in Fit Model Results

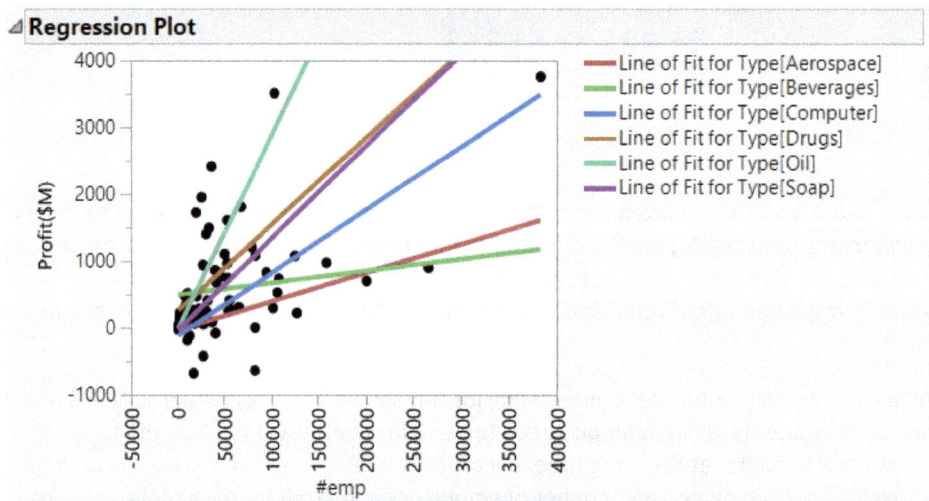

As you can see in the Regression Plot, a separate line is fit for each type of company. This becomes even more clear in the Prediction Profiler. (See Figure 6.72.)

Figure 6.72 Prediction Profiler with Type = Aerospace Selected

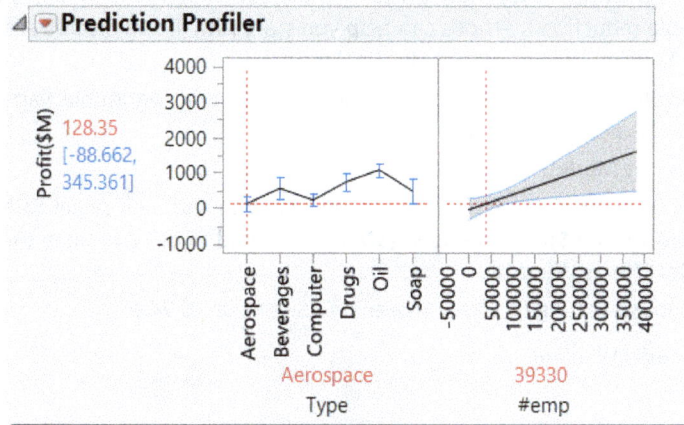

As you click on each of the company types in the left panel, you can see the right panel change to reflect the relationship between profit and number of employees for the company type selected. (See Figures 6.72 and 6.73.)

Figure 6.73 Prediction Profiler with Type = Computer Selected

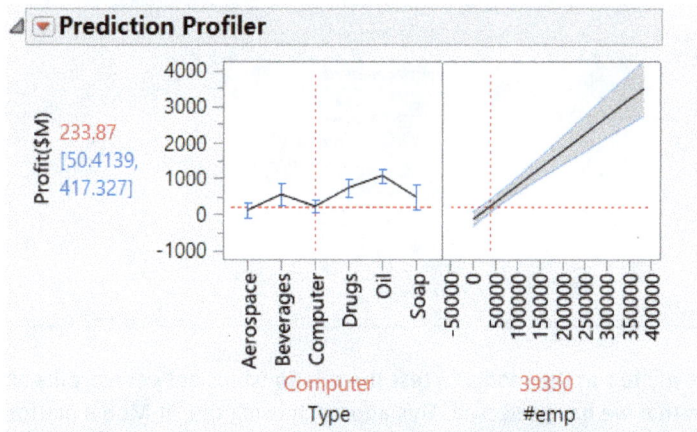

Note

There is some statistical advantage to fitting a single model (as is done here) as opposed to the earlier approach using the data filter. While discussion of this is beyond the scope of this book, the reader is encouraged to consult a statistics text.

Previously we used a full factorial model, which allowed the slope of #emp to change for the different company Types. We could also include interactions, or even more complicated effects from these columns here, but we will instead stick to a simpler model for the sake of a simpler interpretation. In practice, you will probably want to try models with more interactions because these interactions often truly exist, and you cannot discover the evidence for them unless you include them in the model. Stepwise and other variable reduction methods can help you start with a complicated set of predictors and narrow it down to a simpler final model.

Here, we begin with a simple model and we will look at several continuous variables that might impact Profit.

1. Select **Analyze ▶ Fit Model**.

2. In the Fit Model window, select **Profit($M)** as the **Y** column and then select **Sales($M)**, **#emp**, **Assets($Mil.)**, and **Stockholders Eq($Mil.)**. Click **Add** to place them in the **Construct Model Effects Window**.

3. Change the **Emphasis** setting to **Effect Screening.** (See Figure 6.74.)

Figure 6.74 Fit Model Window

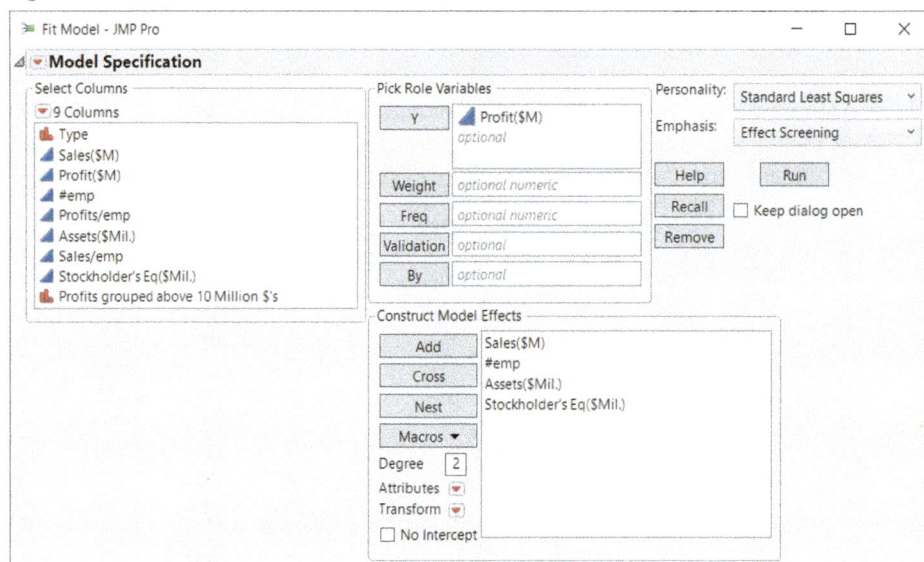

Now we are ready to run the model to test the relationships between profits and the other columns that we have selected. This approach using the Fit Model platform uses a method called *multiple regression*. We could choose to add the type of company to this model, but we will exclude it for now because we will use it later with the data filter to investigate differences.

4. Click **Run** and examine the results window. (See Figure 6.75.)

Figure 6.75 Fit Model Results

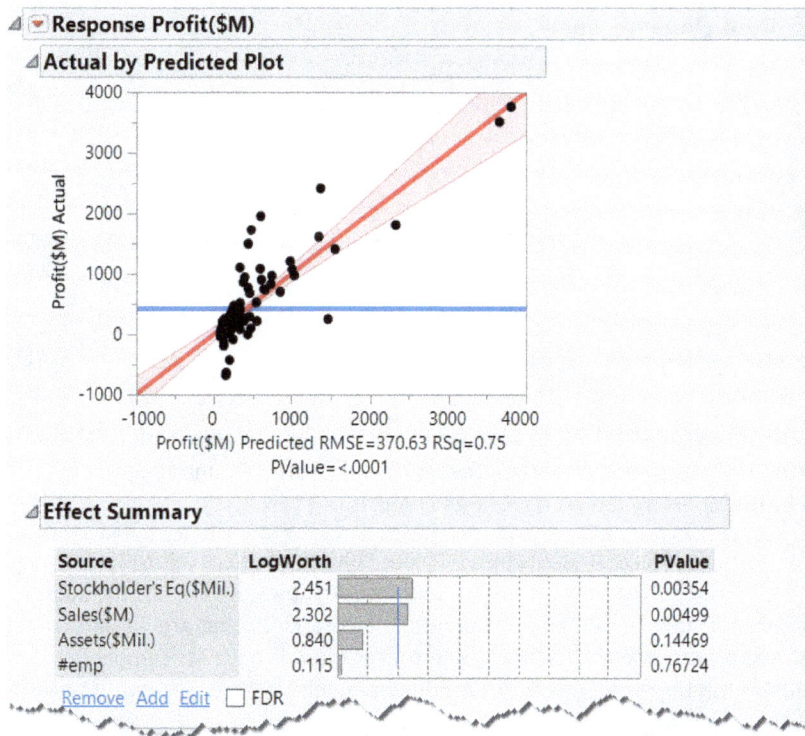

The Actual by Predicted Plot indicates there is a strong, positive relationship between profits and the columns you chose because of the angle of the red fit line relative to the blue mean line. Just as in the earlier example, the blue mean line is not within the confidence bands, indicating a strong relationship. There are additional technical indicators of the quality of the relationships in other report items, including those available from the red triangle menu under Regression Reports. You can turn some of those reports on from the menu and use the question mark tool (**?**) to learn more about the results.

5. Scroll down the window until you see the Prediction Profiler. (See Figure 6.76.)

    As mentioned before, the Prediction Profiler enables you to see visual relationships between profits and sales, number of employees, assets, and stockholder equity (X columns) as depicted by the fitted solid black lines.

Figure 6.76 Prediction Profiler for multiple regression model

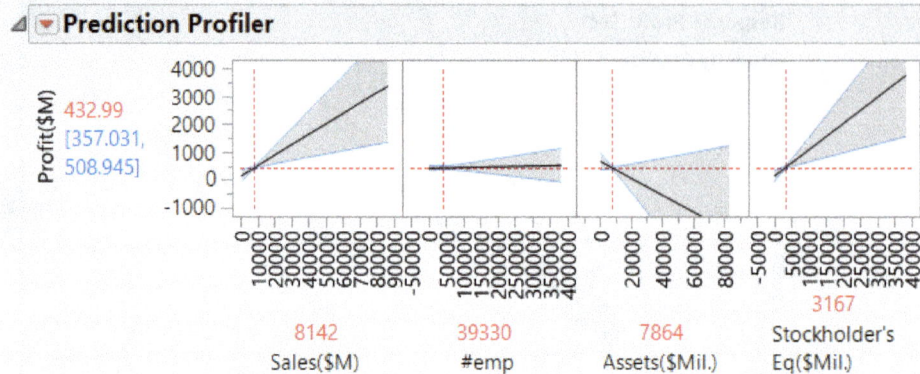

The angle of the fitted line in the plots with the Prediction Profiler give us clues about their influence on Profit in the presence of the other variables. A steeper line suggests that small changes in X (sales for example) will have large changes in Y (profit).

With sales, for example, this is an indication that our column of interest (profits) changes a lot when we move the red dotted vertical line associated with **Sales($M).** Thus, the steeper the line, the greater effect a change has on our Y column of interest. Lines that are flat or nearly flat (such as number of employees), mean that changes in these columns have little impact on our Y column (profits). With the Profiler, you can conduct dynamic, interactive what-if analyses.

We previously learned that profits varied among some company types. However, the model we just created did not include the **Type** column. Let's explore how each type of company might impact profits in this model. Here we will use the Data Filter to accomplish this; however, we could instead add the **Type** column to the model.

1. Select **Rows ▶ Data Filter**. (See Figure 6.77.)

Figure 6.77 Rows Data Filter

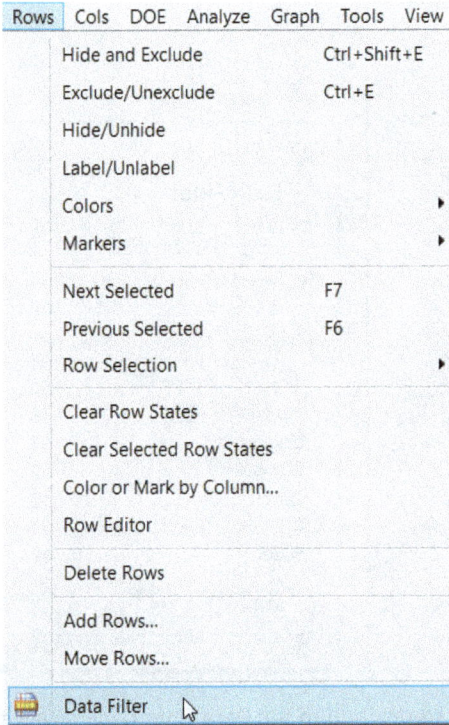

Rows	Cols	DOE	Analyze	Graph	Tools	View

Hide and Exclude	Ctrl+Shift+E
Exclude/Unexclude	Ctrl+E
Hide/Unhide	
Label/Unlabel	
Colors	▶
Markers	▶
Next Selected	F7
Previous Selected	F6
Row Selection	▶
Clear Row States	
Clear Selected Row States	
Color or Mark by Column...	
Row Editor	
Delete Rows	
Add Rows...	
Move Rows...	
Data Filter	

2. In the Data Filter window, select **Type** and click ➕. Select **Show** and **Include**. (See Figure 6.78.)

Figure 6.78 Data Filter Type and Show and Include

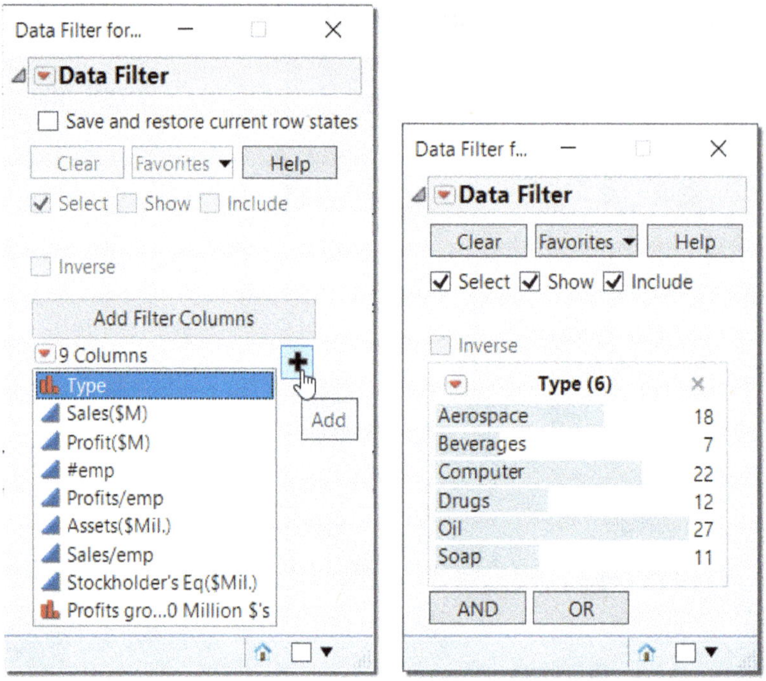

We need the report window to respond to the Data Filter dynamically using automatic recalculation. To do this, follow these steps:

3. From the red triangle (within the report window that contains the Prediction Profiler), select **Redo ▶ Automatic Recalc**. The Prediction Profiler now responds to changes you make in the Data Filter.

4. In the Data Filter window, click on the red triangle and select **Animation**. (See Figure 6.79.)

Figure 6.79 Data Filter Animation

5.   Click on ⏯ (the step forward control). (See Figure 6.80.)

Figure 6.80 Data Filter Animation Step Control

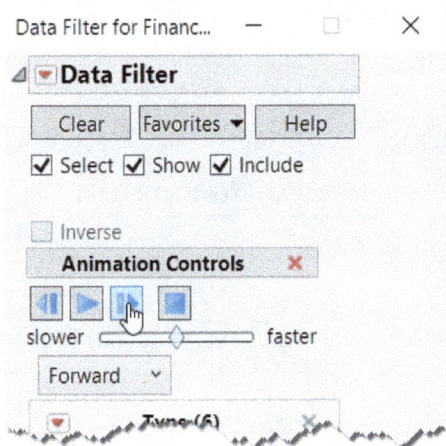

Each time you click the step forward control, you are filtering the Fit Model analysis to just one of the company groups within the **Type** column, and you will see these highlighted in the Data Filter window.

6.   Now click the step forward control again (which toggles through the company types). Watch what happens to the Prediction Profiler with each click as it steps through the company types.

Have you noticed that some of the fitted lines in the profiler flip directions as the company type changes? How would you interpret these changes? We will summarize the first few of the profilers that you see as you click through the company types. We will also provide a brief interpretation of each profile. Within each type, we can still drag the red dotted lines to investigate the sensitivity analysis or relationships within the model for that company type.

Aerospace

As the number of employees and stockholder equity increases, profits increase. (See Figure 6.81.) As Sales and Assets increase, profits decrease. Only Stockholder's Equity is likely to have a statistically significant relationship, however, because this is the only column for which the confidence bands do not include the straight flat horizontal line. We can confirm the statistical significance by looking at the *p*-value in the model report. The other relationships, having confidence bands that include the horizontal line, indicate that, although they do slope slightly in the positive or negative relationship, that slope is not big enough, compared to the general variability in the data, to be definitive. We are not confident that the true relationship, in new but similar data, would have that same direction for the slope.

Figure 6.81 Prediction Profiler for Aerospace

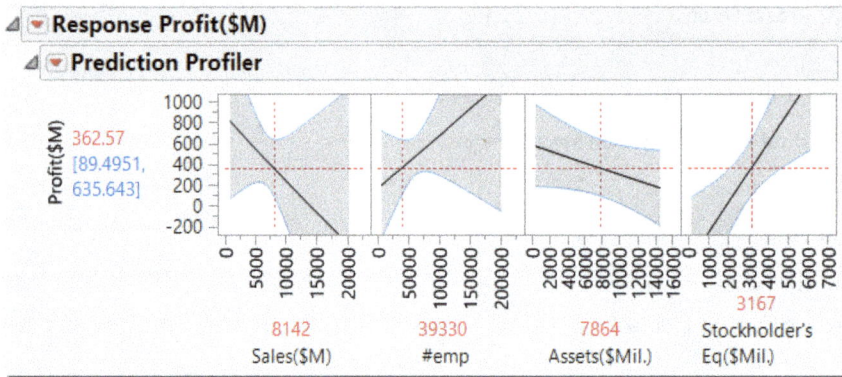

Beverages

The confidence bands are so far away from the fit that we must be very cautious about trusting the profiler for Beverages. (See Figure 6.82.)

Figure 6.82 Prediction Profiler for Beverages

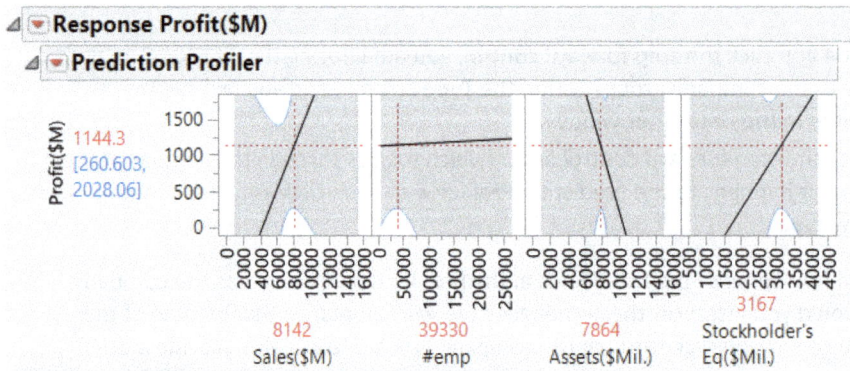

Computer

As sales increase, profits increase (but only just barely statistically significantly). (See Figure 6.83.) As the number of employees increases, profits decrease slightly in these data (but not statistically significantly, so we shouldn't assume this relationship for new data). As assets increase, profits decrease steeply. As stockholder's equity increases, profits increase (but not statistically significantly).

Figure 6.83 Prediction Profiler for Computer

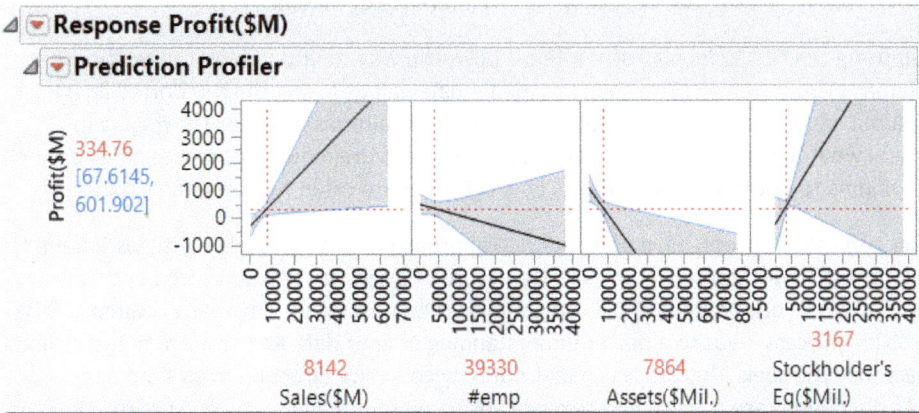

Soap

As sales increase, profits increase. (See Figure 6.84.) Because the line (or trace) for employees is flat, a change in the number of employees does not change profits.

As assets increase, profits decrease. As stockholder's equity increases, profits increase.

Figure 6.84 Prediction Profiler for Soap

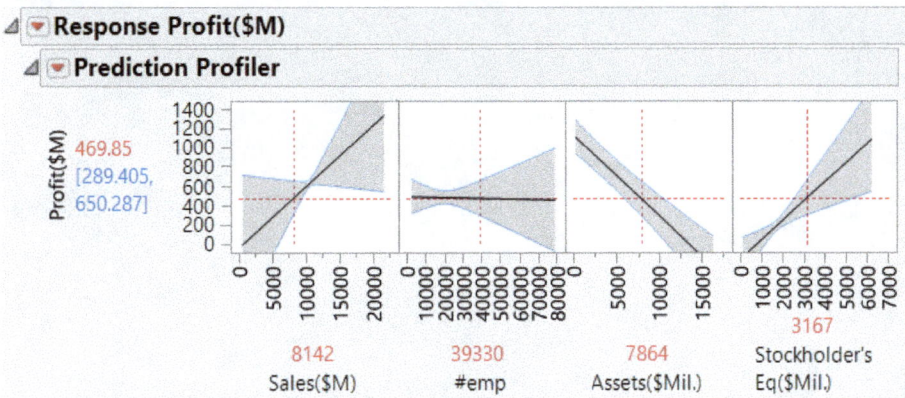

Now it's your turn. On your computer screen, try reading the Prediction Profiler for Oil using what you have learned.

6.5 Summary

Most real-world problems are complex and involve multiple columns. The Partition platform is a flexible tool for solving many types of problems that involve multiple columns and rapidly identifies the key relationships in the data.

Tools like Data Filter and Lasso enable quick identification and extraction of interesting subsets of data. The Data Filter also provides an exploration method using animation that lets you tune in to just the slice of the data that best supports an inference or hunch.

The Profiler using the Fit Model platform offers a powerful way to understand the relationships among columns in your models. The ability to manipulate values (using the red dotted lines) within a column and immediately see their effect on other columns provides the means to conduct visual what-if analyses. With the Data Filter, you can drill down to subcategories or ranges of columns to discover those nuggets of insight that are often hidden at first glance.

This chapter and the Chapter 5 have presented an approach to problem-solving that is unique to JMP. The approach underscores the progressive nature of discovery that tends to build from simple descriptions of one column of data to complex relationships among many columns. This problem-solving process leads to a better understanding of your data and, in turn, to the insights and answers that you seek. This process might not only go in one direction from simple to complex. As discoveries are made with the more advanced multi-column tools, confirmation and further analysis can be made with the simpler ones, Distribution and Fit Y by X, from where our journey began.

Chapter 7: Sharing Graphs

This chapter focuses on common ways you can customize and share JMP graphs with others whether you are presenting those findings yourself, placing them in a document, sending a file, or sharing them on the web. We also provide some advice for effectively doing so.

One of the most important things a JMP user can do is communicate results (data, statistics, and graphs) clearly and accurately. Most people recognize that it can be easy to manipulate results or create a graph that does not tell the whole story. The first section of this chapter is devoted to some principles of effective communication with graphs.

If you want to annotate a graph with comments or change colors or other settings before moving it into another document, JMP provides several convenient tools introduced in Section 7.2.

We will work with three types of graphical output in this chapter:

1. Fixed or static graphs (Section 7.3).
2. Animated graphs, which perform like movies. Simply press go and watch (Section 7.4).
3. Dynamic or interactive graphs, which enable you to manually click on points to highlight (Section 7.5).

Moving *static* graphs into other applications, such as Microsoft Word and PowerPoint, is a straightforward operation in JMP. We illustrate some options in Section 7.3 that enable you to edit the results in other applications.

Moving *dynamic* and *animated* graphs into HTML and onto the web is now a simple operation which we will walk through in Sections 7.4 through 7.6.

Further, you might want to combine graphs into an interactive dashboard and share it with others. We cover the essentials of these features in Section 7.7.

7.1 Presenting Graphs Effectively

Presenting statistical graphics in a report or presentation is a common task for many JMP users. As such, it is critical that you do this well. By *well*, we mean that you rely upon the data to tell the story, be succinct, and, most importantly, present an accurate and valid interpretation (visual, numeric, verbal). It helps to keep the following in mind when presenting statistical graphics:

- Understand your data and its limitations. Time spent with your data before you present it prepares you for the questions that may emerge.
- Focus on the important factors. (See Note.) What columns are core to the problem or question at hand?
- Provide the simplest expression of the data to convey the most complete description of it.
- Present graphs accurately. Do not reduce ranges in graphs to magnify changes or increase ranges to hide trends.
- Let graphs speak for themselves and avoid using unnecessary background colors or dimensions that distract from the core information that you are conveying. Make sure the graph can be seen by your audience.

> Note
>
> How do you determine which variables or columns are most important? The methods described in Chapter 6 can help you identify important columns of interest. For example, the Partition platform (Section 6.1) can help you find out which columns are the key factors affecting another variable.

An important feature of JMP is that it helps you follow these principles and commonly accepted standards. JMP can provide many different types of appropriate graphs of your data, so how do you know which is the most effective way to communicate your information? You will find that using common sense goes hand-in-hand with effective presentation of data and graphical integrity.

In recent years, there have been a number of books on the subject of presenting quantitative graphs and visualizing data effectively. Perhaps the best-known of these are books by Edward Tufte, whose beautifully crafted works have won wide acclaim and illustrate the core principles of presenting quantitative information effectively. These books and those by Stephen Few and others (see the Bibliography) are highly recommended if you use graphs extensively.

7.2 Customizing Graphs for Presentation

There are many features in JMP that provide you with the ability to customize graphs for presentation. Sometimes you would like to annotate a graph by pointing out a key attribute of the graph. Or, you might want to use color or markers to enhance other columns and attributes, as described in Section 2.6.

This section introduces the common tools needed to prepare your graphs and results before moving them into another application or document.

Most of JMP's formatting options appear by simply right-clicking on the area that you would like to customize (for example, on a graph or its axis). When customizing your graphs, it is often easier to make changes within JMP before pasting or exporting them into another application. The first step, however, is to look at a powerful set of tools called the JMP Toolbar that we will begin with in this section.

Example 7.1 SAT by Year

We will be using the **SATByYear.jmp** data table to illustrate the concepts in this chapter. SAT by Year is data representing SAT and ACT test performance by college-bound students in the United States covering the years 1992 to 2004. The tests serve as an important admission metric for colleges and universities.

The data consists of 17 columns and 408 rows of data and includes the following columns:

- **State:** Students' state of residence
- **Expenditures (1997):** State expenditures per pupil
- **Student/Faculty ratio (1997):** Ratio of students to faculty
- **Salary (1997):** Mean salary of teachers (in thousands)
- **%Taking (2004):** Percent taking the SAT exam in 2004
- **Population:** Population of the state
- **% Taking (1997):** Percent of students taking the SAT in 1997
- **ACT Score (2004):** Mean test scores by state in 2004
- **ACT % Taking (2004):** Percent taking ACT in 2004
- **ACT Score (1997):** Mean test scores by state in 1997
- **ACT %Taking (1997):** Percent taking ACT in 1997
- **Year:** Year of the tests
- **SAT Verbal:** Mean SAT Verbal test score by state
- **SAT Math:** Mean SAT Math test score by state
- **Region:** Region in which state resides

You can access this data at **Help ▶ Sample Data ▶ Open the Sample Data Directory ▶ SATByYear.jmp**.

Using the JMP Toolbar

When JMP graphs are generated, the Toolbar contains several tools that help you copy, annotate, and share them (Figure 7.1).

Figure 7.1 JMP Toolbar

Once a tool is selected, your mouse is transformed into its corresponding function. If these do not appear on your home window, select **View ▶ Toolbars ▶ Tools**. Many of these tools have familiar functions:

The **Selection** tool ⊕ enables you to select graphs or reports (or portions of them) to be copied and pasted into Word, PowerPoint, or other applications.

Within the Toolbar, the following functions enable you to add comments or highlight areas of your graphs:

- **Annotate** [T] enables you to add a text box and comments in your graphs.

- **Lines** ⇒ enables you to draw lines on your graphs. For example, an arrow drawn from a text box to a point on a graph.

- **Polygon** and **Simple Shape** ◌ ◯ enable you to draw shapes in your graphs.

Note: The Toolbar features are also available from the Tools menu that appears at the top of the JMP window.

Figure 7.2 is an example of what these tools can do for you. You can summarize key aspects of your data with the Annotate tool and create pointers using the Lines tool to call out this information. Interpretation of this kind can provide valuable direction when you are not there to guide the report recipient. To generate these graphs in the **SATByYear** data table, select **Analyze ▶ Distribution ▶ Student/Faculty Ratio** and **ACT Score (1997), Y, Columns ▶ OK**.

Figure 7.2 JMP Results Window with Annotations

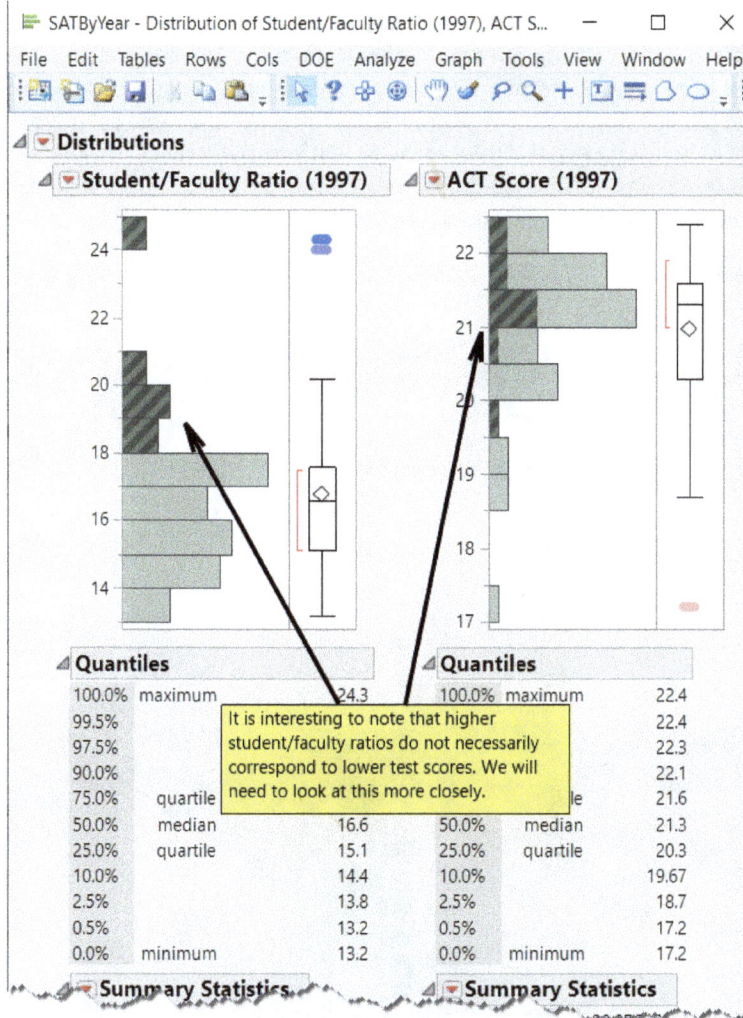

To create these features in JMP, do the following:

1. Select the **Annotate tool** ⊤ from the JMP Toolbar.

   Click on your report where you want to add an annotation. Type your comment and then left-click outside the annotation box. You can then move or resize the box as you would any text box or change the color of the background by right-clicking on the box.

2. To add lines or pointers, select the **Lines** icon ⇒ from the Toolbar and draw the line. Once the line has been drawn, right-click to access the options for pointers, thickness, style, and color.

Using Color

Changing the colors of your graphs is simple. Just right-click within the graph area. (See Figure 7.3.) In the **Distribution** platform, select **Histogram Color** and select a color. The method of accessing color options varies slightly by platform but generally can be found by right-clicking on the graph or through the red triangle. If you want your graphs to always use the same color, you can change the JMP defaults by selecting **File ▶ Preferences ▶ Platforms** and specifying the color, or **File ▶ Preferences ▶ Graphs**. (On the Mac, preferences are found under the JMP menu.) For more information about setting preferences, see Chapter 1.

Figure 7.3 Changing Histogram Color

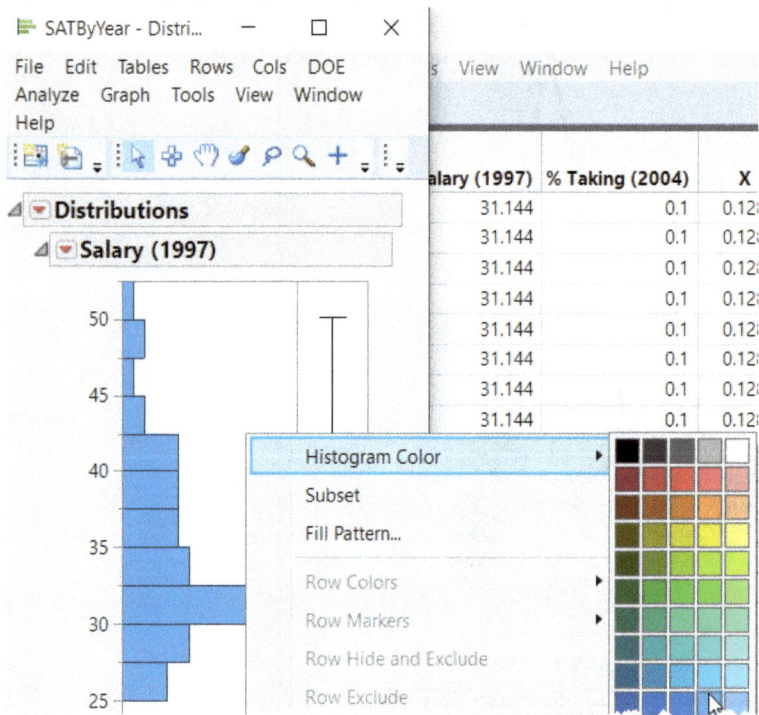

Coloring a graph in this manner does not affect graphs where individual rows are color-coded by a column value within the data table. (See Section 2.6 for more information about how to use color to denote a category or range within a graph.)

Background Color

You can change the background color of any graph by right-clicking in a graph area and selecting **Background Color**. (See Figure 7.4.) When using background colors, be mindful not to distort, distract, or obscure the information in the data.

Figure 7.4 Changing Background Color

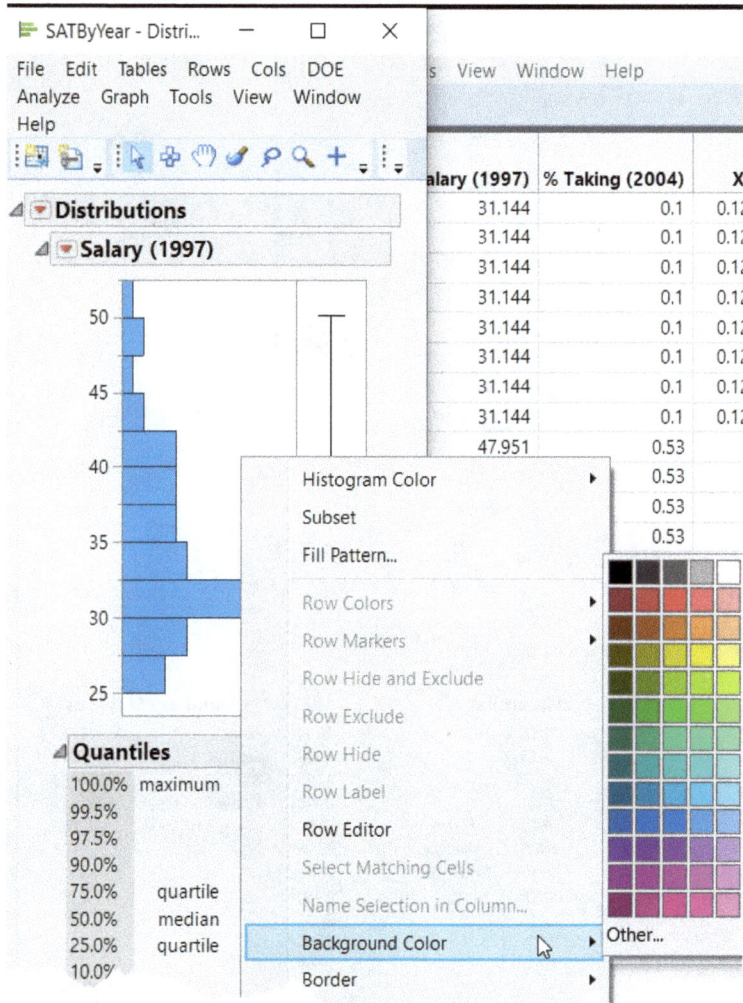

Horizontal Layout

Within the Distribution platform, the default display places the histograms in a vertical format (and side-by-side when there is more than one histogram generated). Clicking on any bin (or bar) within one histogram highlights those values in the remaining histograms. This is especially useful when you are exploring relationships among two or more columns and don't want to scroll down.

Sometimes, however, you would like to see the histogram in the more traditional horizontal format. Rotate the display by simply selecting **Display Options** under the red triangle after you have created a histogram and then selecting **Horizontal Layout**. (See Figures 7.5 and 7.6.)

Figure 7.5 Specifying Horizontal Layout

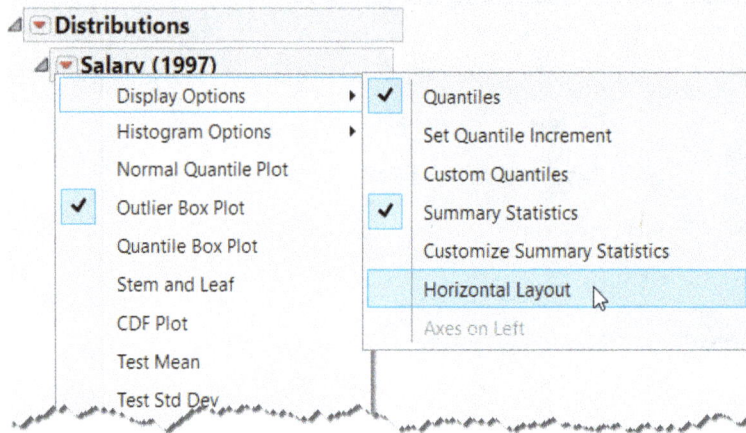

Figure 7.6 Horizontal Layout Result

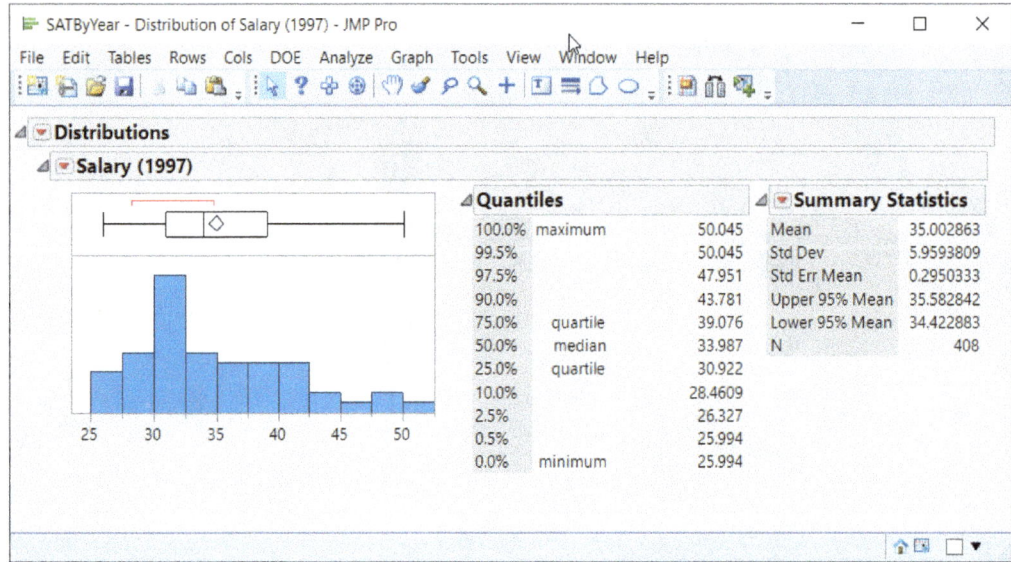

> **Note**
>
> If you want to stack multiple histograms horizontally, one on top of the other, you can use the Stack command under the Distributions red triangle (immediately above Salary's red triangle illustrated in this example).

Axis

JMP enables you to easily change the axis of any graph by simply grabbing the axis with your cursor and moving it. Alternatively, you can right-click or double-click on the axis to access the following menu. (See Figure 7.7.)

Figure 7.7 Axis Settings

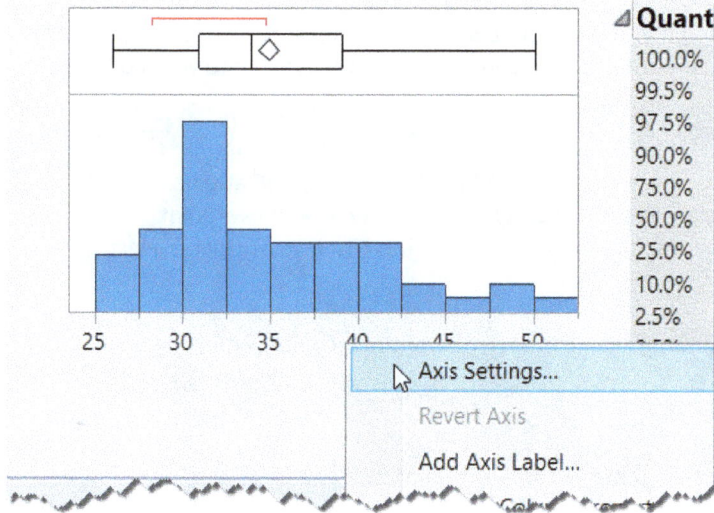

By selecting **Axis settings**, you access the Axis Specification window. (See Figure 7.8.) In it, you can customize the range, increment, and other axis options.

Figure 7.8 Axis Settings Window

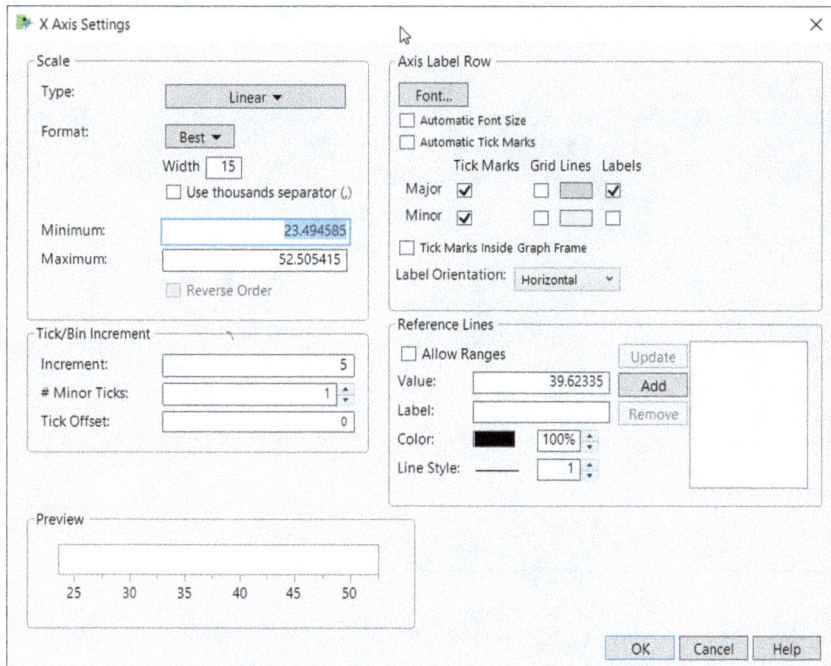

7.3 Placing Graphs into PowerPoint or Word

Copying a fixed or static JMP graph into Word or PowerPoint is simple. You just need to use the Selection tool to select what you want and then copy and paste the graph into your document.

> **Note**
>
> If you would like to move several graphs or output into PowerPoint, JMP also has a handy export option by simply selecting **File ▶ Save**, select **Save as type ▶ PowerPoint Presentation** from the desired results window. (On Mac, it is **File ▶ Export**, and select **PowerPoint Presentation**).

Let's illustrate this with the following example. Say you want to move the graph in Figure 7.9 from JMP into PowerPoint.

Figure 7.9 Moving a Graph

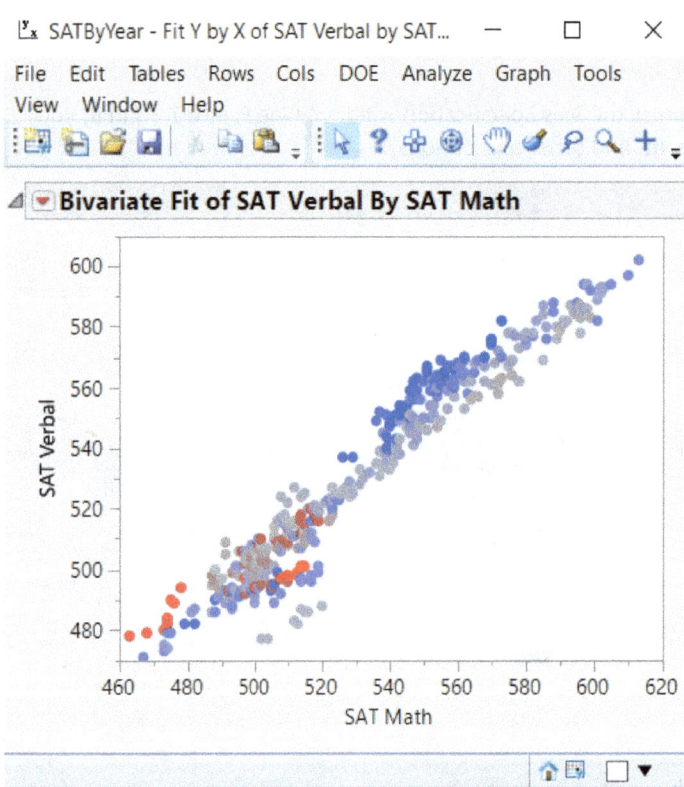

1. Select the **Selection** tool. (See Figure 7.10.)

   Figure 7.10 Use the Selection Tool

2. Click on the upper left corner of the graph, as shown, to capture the entire picture. (See Figure 7.11.)

   Figure 7.11 Selecting Graph Components

   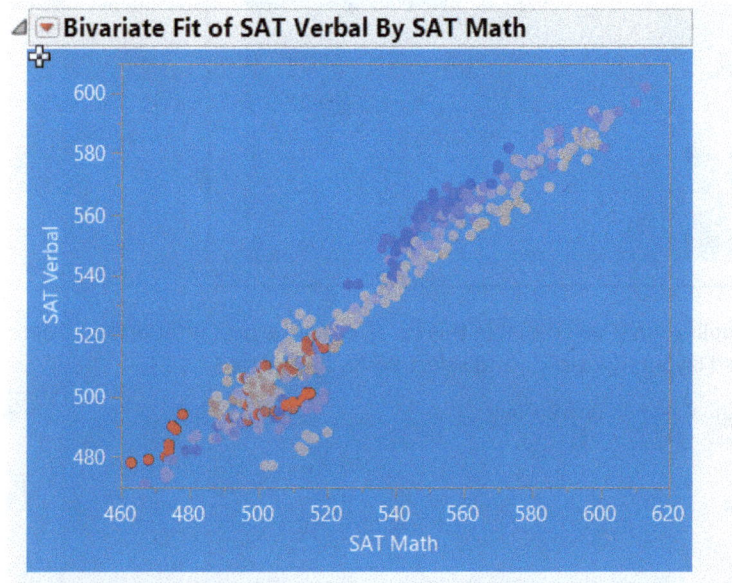

Alternatively, you might want to copy only selected elements of the graph, in which case you click and drag the cursor (or shift-click) to select the elements that you would like to copy.

3. Right-click and select **Copy**. (See Figure 7.12.)

Figure 7.12 Copy the Graph

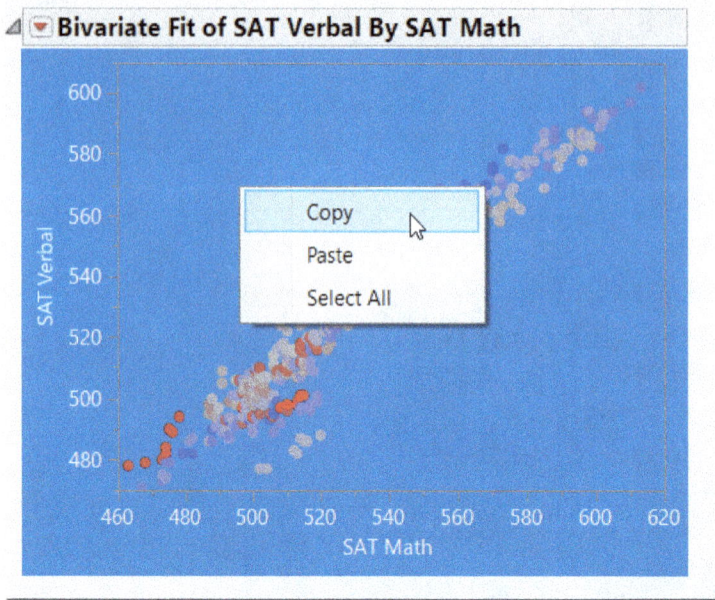

4. Open the desired application (PowerPoint in this case), open the presentation and create a slide with the desired layout, right-click, and select **Paste**. (See Figure 7.13.)

Figure 7.13 Pasting the Graph in PowerPoint

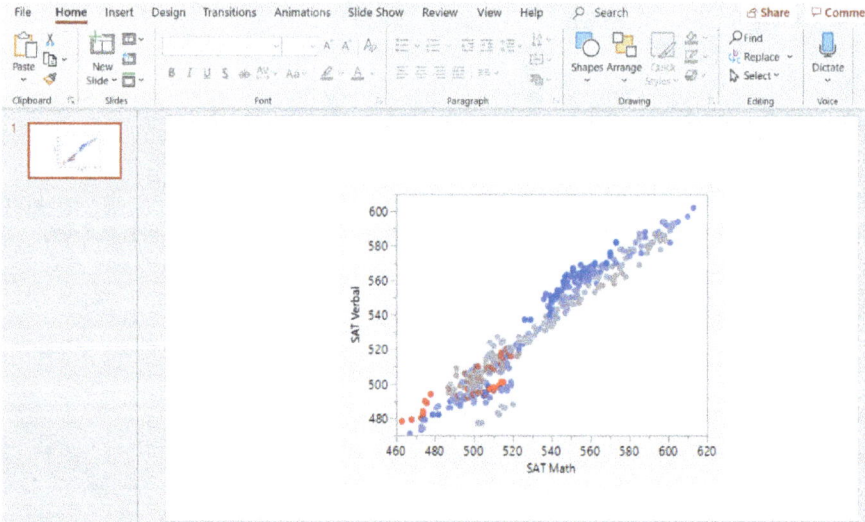

> **Note**
>
> When using JMP's default paste function, the object can be re-sized in the new application, but the contents are fixed.

Special Paste Functions

JMP also provides additional paste formats and functions. Sometimes you might want to edit the results or graph, or you need to provide the graph in a specific format. JMP has some options for you:

- Use **Paste Special**. This enables you to paste a .bmp or metafile. (See Figures 7.14 and 7.15.)
  - ○ **Metafile** is a vector art graphics format that can be edited in detail by Word and PowerPoint on Windows.

Figure 7.14 Paste Special

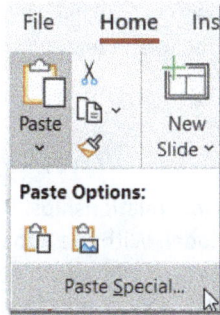

Figure 7.15 Paste Special Options

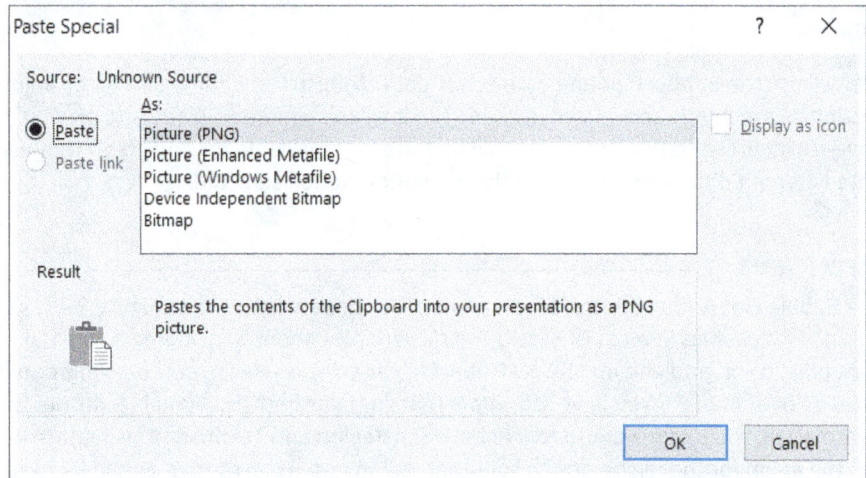

- If you want to change the formatting options of the graphs that you want to paste, change your preferences to reflect this:

1. Select **File ▶ Preferences**.
2. Select **Windows Specific** (or **MAC OS Settings**).
3. Under **Copy/Drag Graphic Formats**, check the formats that you want as options or remove options by unchecking them. (See Figure 7.16.)

Figure 7.16 Changing Formatting Options

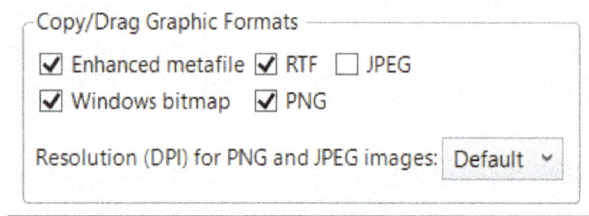

> **Note**
>
> If you simply want to save your results in a certain file type, such as a JPEG, EPS, PDF, or HTML, you can simply select **Save As ▶ type** from the File menu. (On Mac, **File ▶ Export**.) As an example, we will describe how to save an interactive HTML file in an upcoming section.

7.4 Creating and Sharing Animated Graphs in PowerPoint

Animated graphs are used to express trends among column/variable relationships. JMP provides the means to communicate a clear impression from a complex model. With the Bubble Plot and Profiler, you can export these to other environments such as HTML5 and PowerPoint or publish them to JMP Public, which is described later in this chapter. Unlike other graphs that are static or require direct interaction, this class of graphs brings an added dimension to your presentation by enabling you to view changes in data relationships over time, range, or category – very much like a movie.

We will use the Bubble Plot and time series data to illustrate what we mean by animation in this section. We will then show how you can save it as an animated GIF file and place it in a PowerPoint presentation. Animated GIF files are also available for any graph that uses a Local Data Filter or Column Switcher with the animation option selected.

Creating Exportable Animated Graphs

The Bubble Plot is found in the **Graph** menu. The window is shown in Figure 7.17. As you see, the Bubble Plot enables you to visualize up to six variables at once, including X and Y axes, time, size of bubble, color gradient, and By. JMP enables you to also select a second subgroup for your points. For example, if you have sales data that contains both individual territories and regional information, you might want to toggle between **Region** and **Territory**. This feature enables you to see the graph for the region and to split that region into its territories on the fly to see their individual performance.

1. Open your data file and select **Graph ▶ Bubble Plot**. Select columns to match desired roles (described in the following box). Only Y and X are required, but you will want to do more with this platform. (See Figure 7.17.)

Figure 7.17 Bubble Plot Launch Window

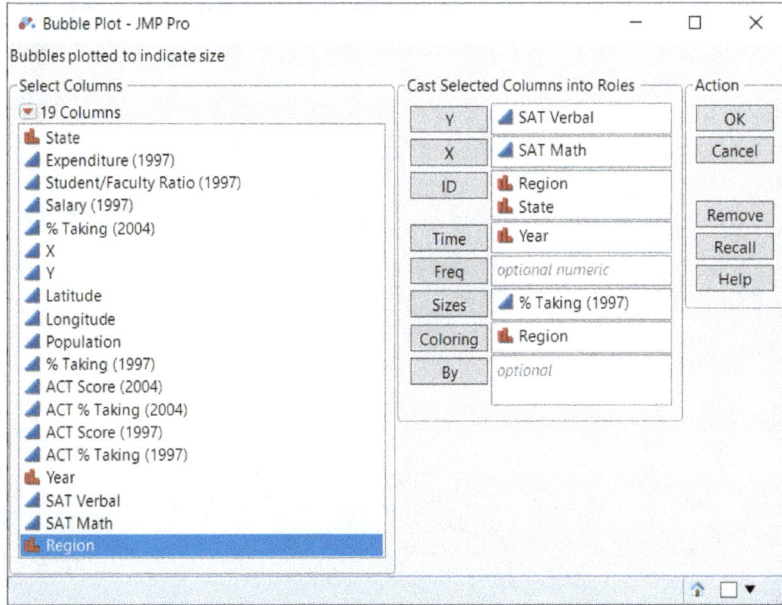

2. Using our chapter example SAT by Year, assign these columns to their corresponding roles:

Column		Role
SAT Verbal	>	Y
SAT Math	>	X
Region	>	ID
State	>	ID
Year	>	Time
% Taking (1997)	>	Sizes
Region	>	Coloring

3. After selecting **OK** (see Figure 7.18), locate and click on the blue play button in the lower left corner to see the animation over time.

Figure 7.18 Bubble Plot Results

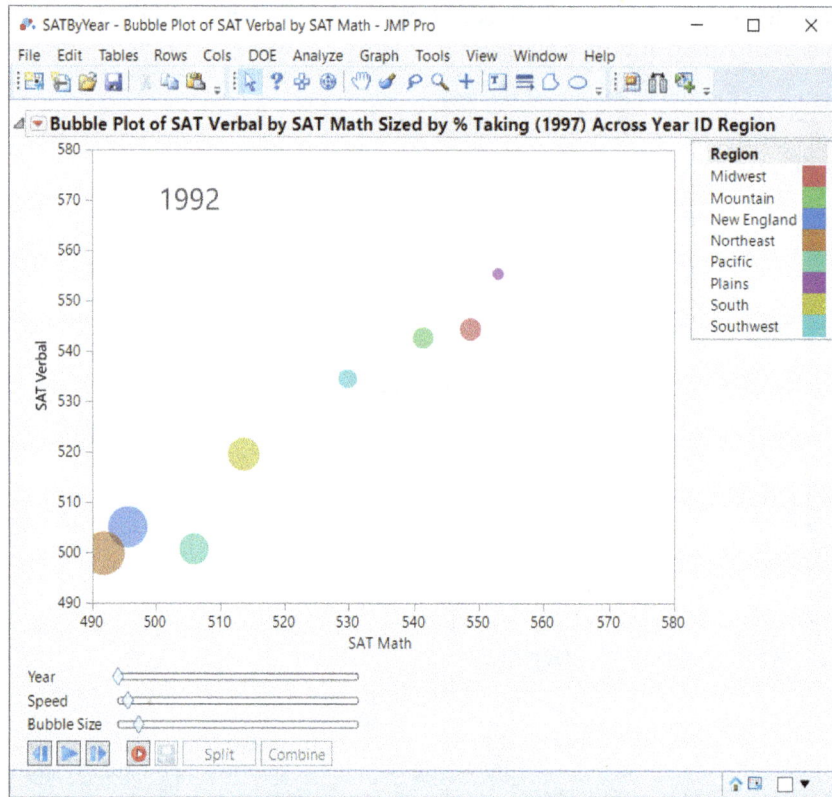

Now that you have created your animated graph, you will likely want to share that with others. There are a few ways to do this, including saving it as a GIF file and placing that in PowerPoint, which we will discuss in this section. There are other ways to share it such as with a JMP file with a built-in script (Section 7.9), as an HTML5 file (covered in the next section), or published to JMP Public (Section 7.6).

Note that the following instructions for creating a GIF format and placing that movie in PowerPoint is available only on the Windows version of JMP. Let's first record our animated bubble plot.

Picking up where we left off, you will also notice in the lower left corner of the Bubble Plot report window a red record button. Next to it, there is a save button that looks like an old floppy disk. (See Figure 7.19.) The record button will record what you play, and the save button will save that recording to a user-specified location.

Figure 7.19 Bubble Plot Recording Controls

4. **Click Play ▶ Record**, and then the **Record** button a second time to stop recording. Click the **Save** button (the floppy disk button). Name the file and save it in a memorable location. (See Figure 7.20.) It might take a few seconds to save.

Figure 7.20 Saving Bubble Plot Recording

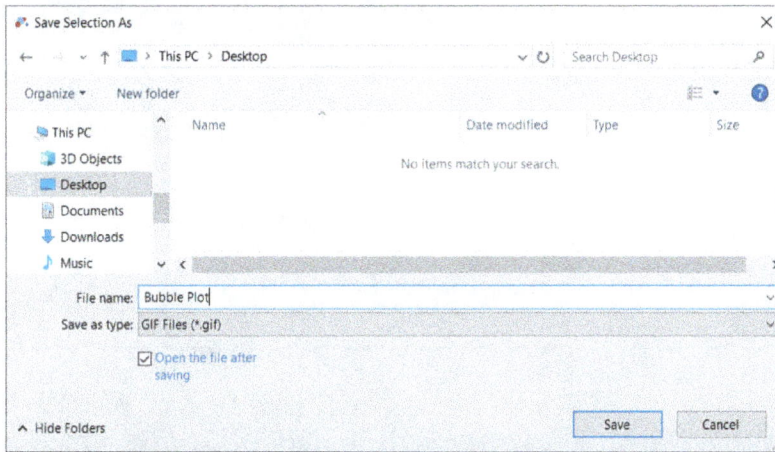

5. Open a PowerPoint presentation file and create a new slide with the desired layout. (Title and Content layout is recommended.)

6. With the empty slide, select **Insert ▶ Media ▶ Video ▶ Video on my PC**. Be sure to select **All Files** where it filters on file type. Select the file and **Insert.** It might take a few seconds to display. Once it appears, add a title. To play in PowerPoint, click the **Slide Show** button and it will automatically play. (See Figure 7.21.)

Figure 7.21 Bubble Plot Recording in PowerPoint

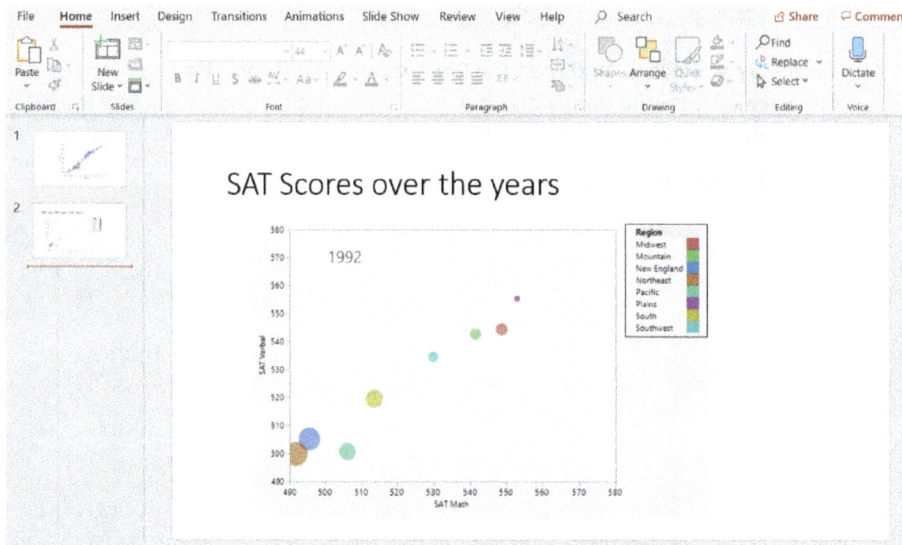

7.5 Sharing Dynamic Graphs with HTML5

Because JMP's graphs are dynamic and enable you to interact with them to see interesting things, it is reasonable that you would want to share these insights with others – including those who might not have a copy of JMP. JMP has an important way to share your discoveries with others by letting you save your reports as interactive HTML5. HTML5 is a standard format recognized by all browsers and enables you to save your results, which maintain much of the same interactivity that you get within JMP.

To save a report as HTML5, simply generate the results that you would like to share, go to **File ▶ Save**, select **Interactive HTML with Data** (under **Save as Type**), rename and locate your file as desired, and select **Save**.

You can simply send this file, which can then be opened with any browser (Figure 7.22). Note that at the time of publication, Interactive HTML has been implemented in most, but not all, platforms in JMP.

Figure 7.22 Interactive HTML5 Output in a Browser

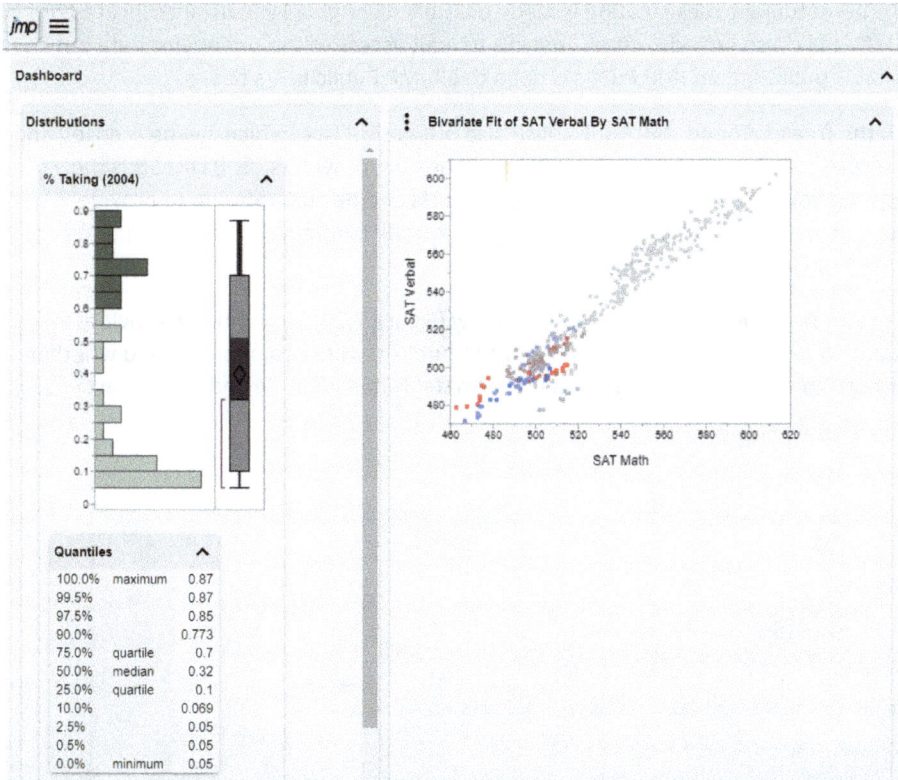

> **Note**
>
> When a file is saved in the HTML5 file type, the data accompanies the graph. Although it might not be obvious, the data cannot be encrypted in this format and can be extracted from the file. Thus, we caution users to be careful with sensitive or confidential data. Internal presentations might not be at issue, but sending Interactive HTML files externally can compromise confidentially.

7.6 Publishing Reports in JMP Public (and JMP Live)

A new way to share your discoveries with the world is through JMP Public (public.jmp.com) and JMP Live if your institution is so equipped. Unlike sharing results through a file, when we refer to publishing your results, we mean just that, to broadcast the results to the public or to select groups within an organization. JMP Public (and JMP Live) is accessed through your internet browser, so it uses HTML5 as we described in the previous section.

JMP Public is a free hosted site where you can publish results and their data if you so choose. JMP Public provides an added level of interactivity by allowing the viewer to not only interact with graphs but to also filter both graphical and numerical results and have them automatically recalculated. This automatic recalculation is made possible through a JMP instance located on the server. JMP Public also provides some controls to limit access to the underlying data but know that what is published on JMP Public is open to all JMP Public users to see.

In addition to the free and open JMP Public, JMP also offers JMP Live, which can be licensed and installed on secure company servers or through cloud hosting providers. JMP Live adds the ability for organizations to run regularly scheduled reports and dashboards and to provide security around group access. To illustrate JMP Public, we will continue with the completed Bubble Plot example from the previous section.

1. Go to **File ▶ Publish**. This launches a window (Figure 7.23). The window provides options to select the report that you want to publish, where to publish it and whether you want to also publish the accompanying data. Name your file and select **Next**.

Figure 7.23 Publish Window

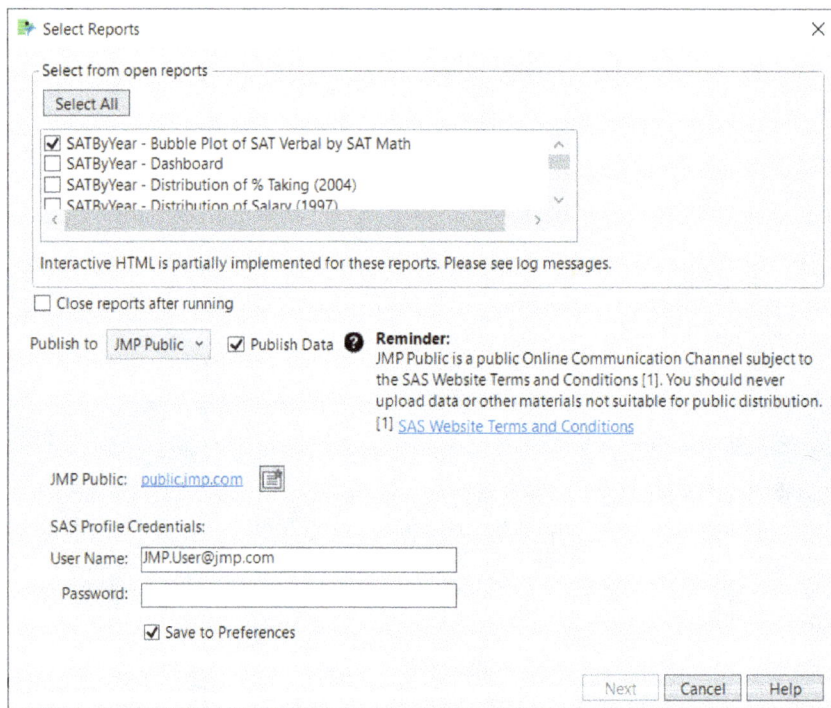

2. In the next screen you will see a thumbnail and have the option to edit the title and add a description to your report. (See Figure 7.24.) Once you are satisfied, select **Build Report**.

Figure 7.24 Preview of Published Report

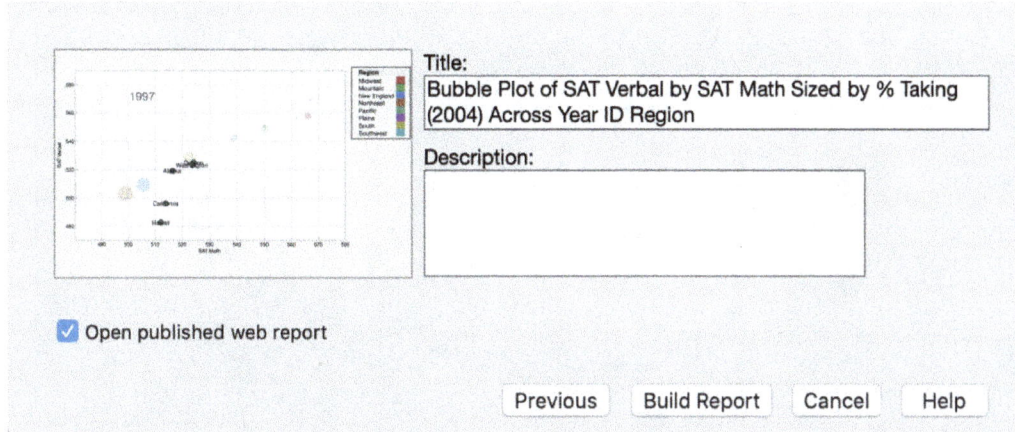

The report will now be published to the specified desired location. Note that with JMP Public, the posting will be reviewed prior to being posted.

7.7 Creating Interactive Dashboards

One of the fundamental features of JMP is its ability to dynamically link multiple graphs and view selected rows or observations in any number of graphs. Placing multiple graphs and filters in a single window, or *dashboard*, is the subject of this section. (See Figure 7.25.) JMP includes a variety of dashboard templates enabling you to simply drag and drop your graphs of choice into a template to build your dashboard.

Dashboards can be shared with others who also have JMP, shared in a variety of formats such as PowerPoint or as an interactive HTML5 file, or published to JMP Public or JMP Live as described in the previous section. Let's revisit the SATbyYear data table. In this data set, we would like to better understand SAT Math scores spatially in a map, but also in relationship with SAT Verbal scores.

Figure 7.25 Example of a JMP Dashboard

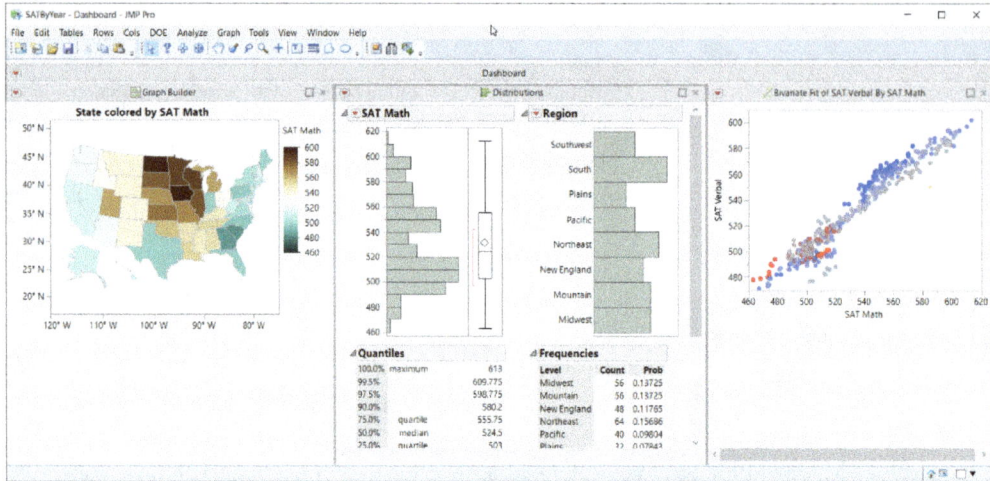

The first step in creating a dashboard is to create the individual graphs.

- In this example, we created a Map in Graph Builder (**Graph ▶ Graph Builder**) by dragging the **State** column into the lower left box titled "**Map/Shape**" and then dragging **SAT Math** into the resulting US map.

- We created distributions of SAT Math and Region by using the Distribution platform (**Analyze ▶ Distribution**) and placing **SAT Math** and **Region** into the **Y Columns** role and selecting **OK**.

- We created the scatterplot of SAT Math and SAT Verbal by using Fit Y by X (**Analyze ▶ Fit Y by X**) and casting **SAT Math** into the **Y** role and **SAT Verbal** into the **X** role and selecting **OK**.

Once we have created the three graphs (which will initially appear in separate windows), our next task is to combine them into a single Dashboard.

1. **File ▶ New ▶ Dashboard** launches the Dashboard builder that offers several template options (but you can also customize dashboards to your needs). For the purpose of illustration, select the 2x1 Dashboard template. (See Figure 7.26.)

Figure 7.26 Dashboard Builder

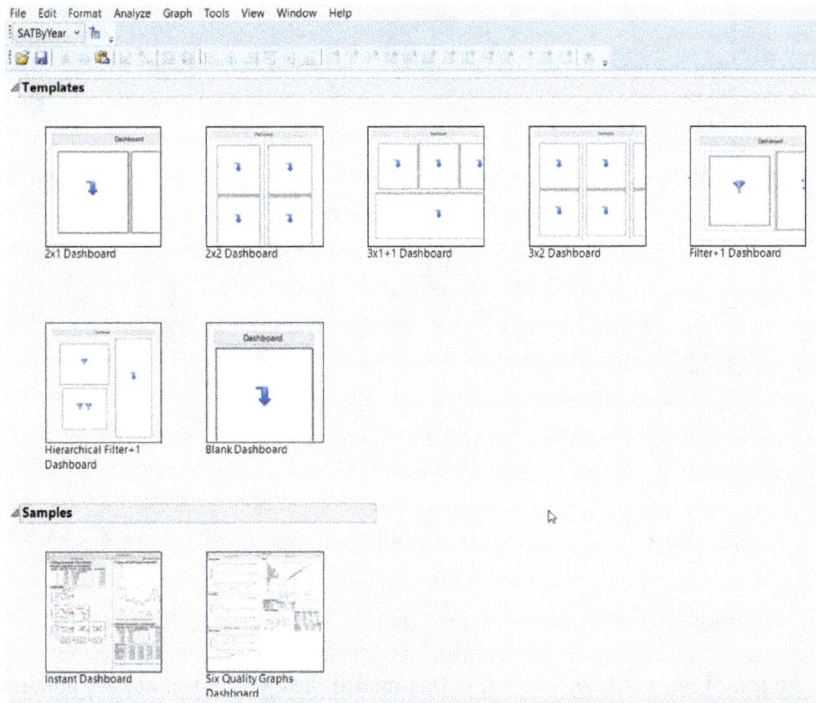

2. You will notice that the graphs you created earlier appear under Reports. From here you simply click and drag the reports to the preferred area of the dashboard template (Figure 7.27).

Figure 7.27 Selected Results in Dashboard Builder

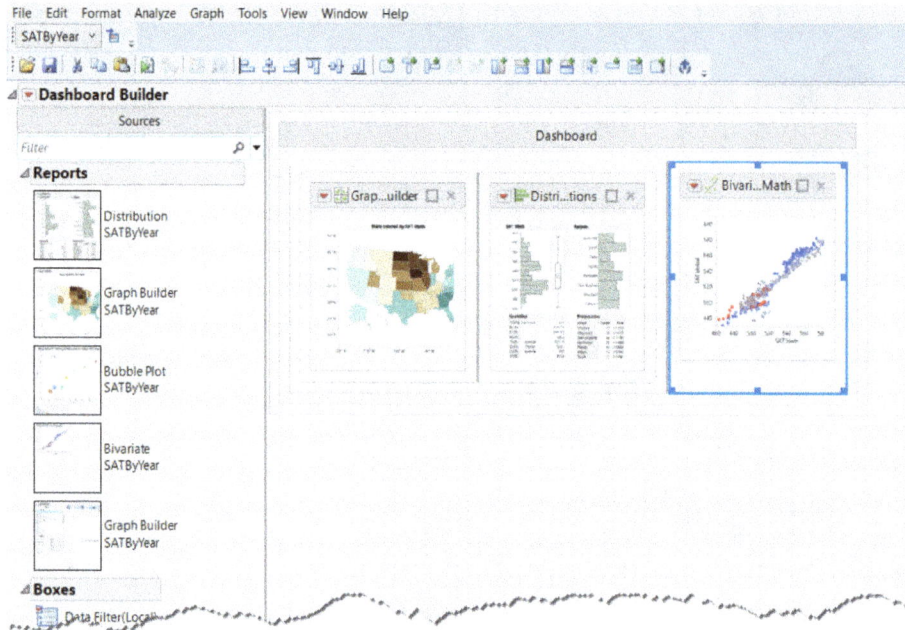

3. At the top of the window, select **Run Dashboard**. JMP will automatically generate the interactive dashboard. You will notice that if you select any point, range, or area of a graph within the dashboard, the corresponding observations in other graphs will also be selected (Figure 7.28).

Figure 7.28 Completed Dashboard

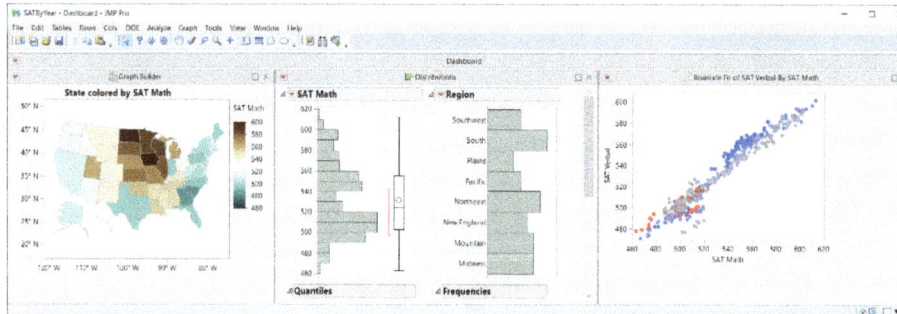

You can now save the Dashboard to run on new observations or share it as a JMP file, save it to PowerPoint or HTML5, or publish the results to JMP Public or JMP Live as described earlier in this chapter.

7.8 Using Scripts to Save or Share Work

In this chapter, we have covered a variety of ways of moving JMP graphs or output into other applications. However, on many occasions you will want to save or share your results in a manner that enables you to generate the same results again. JMP scripts capture steps, procedures, or actions which enable you to reproduce your past work, perhaps on new data. For example, you may want to generate a variety of graphs and output before deciding which ones are most useful. Or perhaps you would like to go back to an analysis that you have done and possibly add additional rows to your data table and re-run the analysis or even share your work with another JMP user who can pick up where you left off.

One of the things that even experienced JMP users might not realize is that JMP has a built-in scripting language (often referred to as JSL for JMP Scripting Language). JSL is always available and running in the background to capture the work that you have done so that you can easily reproduce it. It is available by simply clicking on the menu.

From the red triangle of any platform or graph is a **Script** option. When selecting the **Script** option, you see a submenu.

1. **Save Script to Data Table** translates this work into a script and places it in the Table panel of the Data Table. (See Figure 7.29.)

<table>
<tr><td>Note</td></tr>
<tr><td>If you ever find yourself needing to quickly exit and save all your work at once (including all open data sets and results in a session) and pick up where you left off later, select **File ▶ Save Session Script**. This saves all of your work to a specified location.</td></tr>
</table>

Figure 7.29 Saving a Script to a Data Table

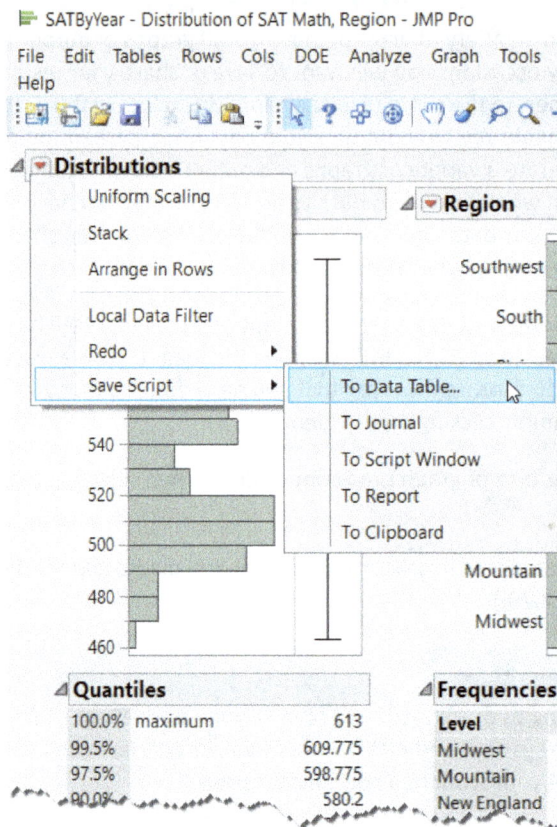

2. Once selected, a window gives you the option of re-naming the title of the script that might be more meaningful (Figure 7.30).

Figure 7.30 Window for Saving a Script to the Data Table

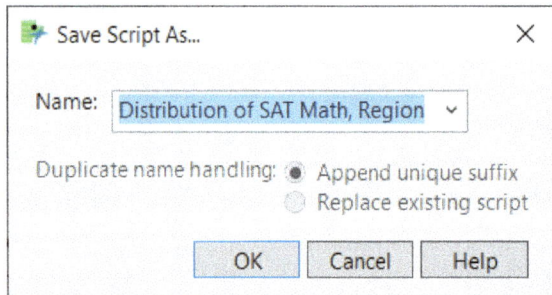

3. Go to the Data Table. You now see an item with the title specified in the Table panel with a green "play" button (Distribution of SAT Math, Region in this case). Click on the green **play** button, and you should now see exactly what you developed before. You can edit, save, and share these scripts (within the Data Table, in this case). (See Figure 7.31.)

Figure 7.31 Script on the Data Table

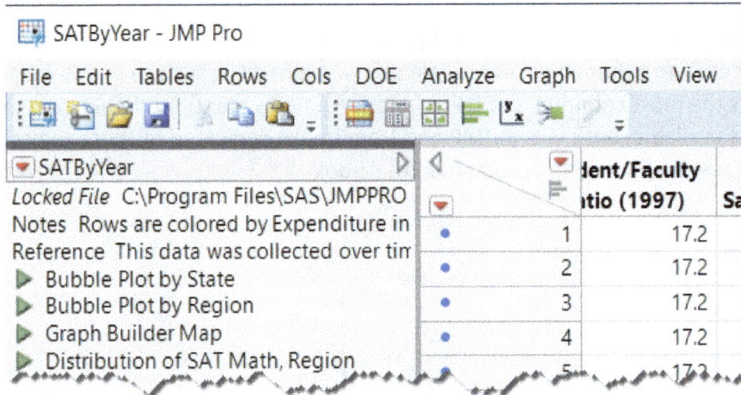

4. As the previous steps illustrate, you can use the JMP Scripting Language or JSL to capture and regenerate your work without looking at any code. If you are interested in seeing the code associated with the script or want to further customize the script, right-click on the **green arrow** of the script and select **Edit**. (See Figure 7.32.)

Figure 7.32 Simple Script

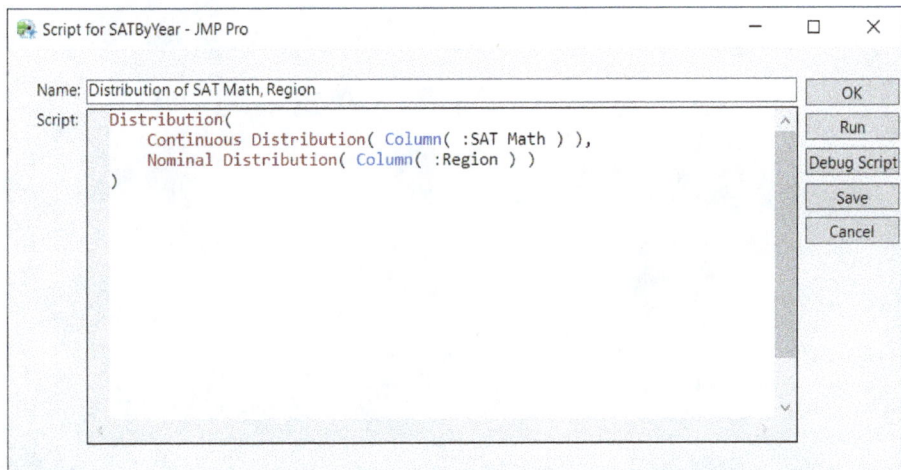

> **Note**
>
> If you would like to learn more about the JMP Scripting Language, go to **Help ▶ JMP Documentation Library ▶ JMP Scripting Guide** or the **JSL Syntax Reference**. For a comprehensive index of JSL functions, example code and linked help, go to **Help ▶ Scripting Index**.

7.9 Summary

In this chapter, we learned about communicating with graphs and about sharing JMP graphs and files and moving them into other applications. We only scratched the surface in this chapter, but we tried to address the most common and basic functionality. As mentioned, there are excellent reference materials available, both within JMP and from books dedicated to these topics. For more information, see the Bibliography section in this book.

Chapter 8: Getting Help

We designed this book to be a quick overview of the most commonly used features of JMP. As you begin to use JMP more frequently or dig deeper into its analytics, you will inevitably have questions or find places where you will want to learn more. This chapter is designed to help you navigate the extensive resources available to help you answer those questions. We begin with JMP's wide range of built-in resources.

A wide variety of documentation, tutorials, and other support materials live within JMP and are available on the web. We recommend you learn about these, because they were designed to assist you while you are using the software. These tools offer the most immediate answers.

Should you need additional assistance, other resources include outstanding technical support, teaching resources, live and on-demand Webcasts, user groups, blogs, and even social networking initiatives. JMP hosts dozens of events each year, including its annual Discovery Conference. JMP training courses are also available through SAS Education; many books, ranging from elementary to advanced, are referenced in the bibliography at the end of this book.

8.1 The Help Tool

When you have any question about what you are looking at in a JMP window, what a graphic or result is, or how it was generated, the Help tool (also known as the question mark tool or the **?** icon) in the Toolbar or the Tools menu is a great place to start (Figure 8.1).

Figure 8.1 The JMP Tools Menu and Toolbar

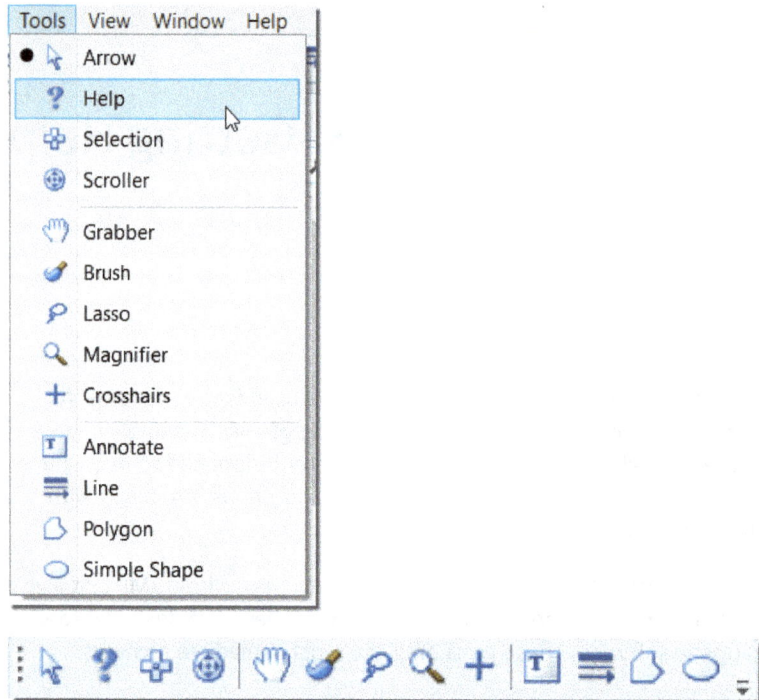

The **Help** tool provides a direct link to the documentation concerning the item in question.

1. Select the question mark from the toolbar and move the transformed "**?**" cursor to the item in question. (See Figure 8.2.)

Figure 8.2 The JMP Help Tool

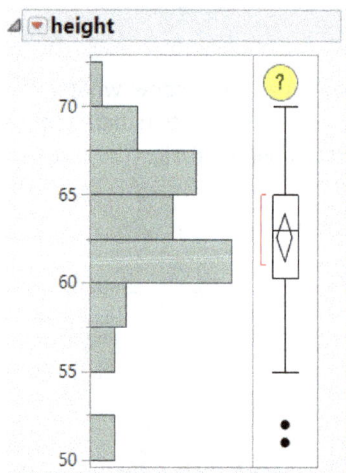

2. Click on the item in question, and JMP reveals the specific place within the documentation that addresses the concepts or output at hand. (See Figure 8.3.)

Figure 8.3 JMP Documentation accessed with the JMP Help Tool

Outlier Box Plot

Use the outlier box plot (also called a Tukey outlier box plot) to see the distribution and identify possible outliers. Generally, box plots show selected quantiles of continuous distributions.

Figure 3.8 Outlier Box Plot ▾

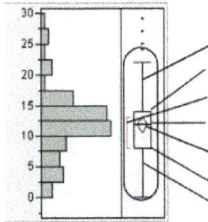

Note the following aspects about outlier box plots:

• The horizontal line within the box represents the median sample value.

• The confidence diamond contains the mean and the upper and lower 95% of the mean. If the mide would have

8.2 The Help Menu

From the **Help** menu (see Figure 8.4), you can access specific help topics along with all of JMP's documentation, tutorials, sample data, and shortcuts to statistical routines in JMP. Let's define each of these items in the menu.

Figure 8.4 The JMP Help Menu

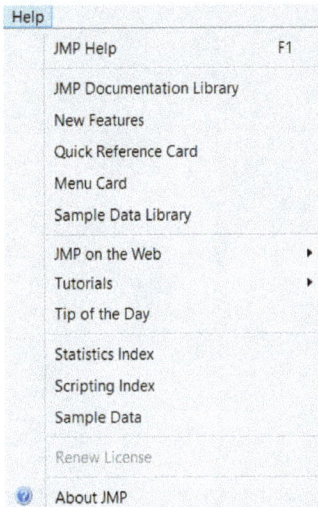

JMP Help

JMP Help takes you to the first page of the online JMP Documentation on the JMP website (Figure 8.5).

Figure 8.5 JMP Documentation Homepage

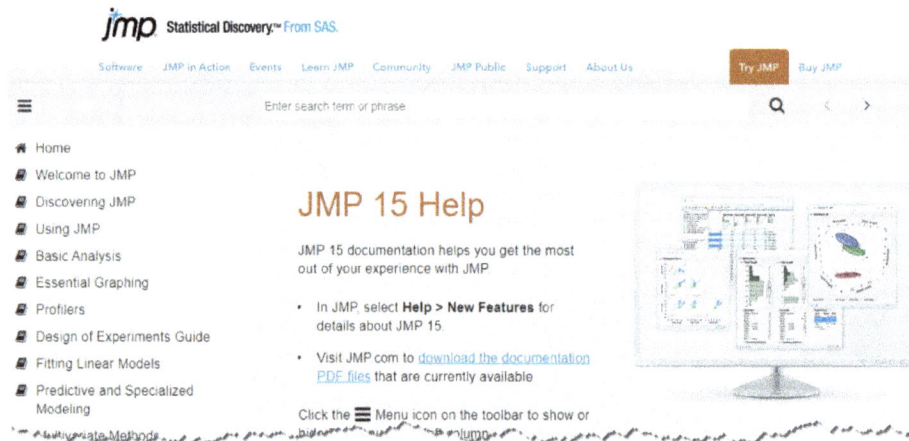

JMP Documentation Library

This menu item links to PDF file that contains all of the JMP user manuals. (See Figure 8.6.) Over 6,000 pages of JMP reference material are right at your fingertips. You can search and print any of these manuals. *Discovering JMP* is a logical next step for the reader of this book, as it covers the basics in slightly greater depth.

Figure 8.6 Documentation Included with JMP

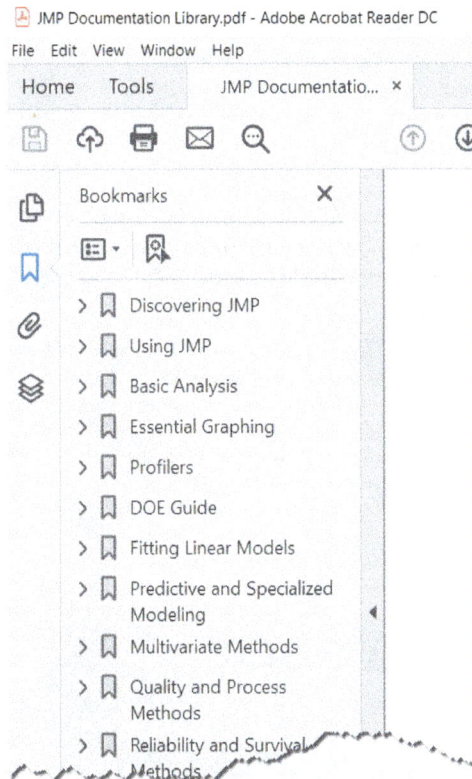

New Features

This menu item opens a PDF file that documents the features that are new to your version of JMP.

Quick Reference Card

The Quick Reference Card is a document listing common keyboard shortcuts in JMP for both Windows and Macintosh operating systems.

Menu Card

The Menu Card is a document providing a short explanation for every JMP menu item on the operating system.

Sample Data Library

This option opens the operating system folder that contains all of the JMP sample data.

JMP on the Web

This menu contains quick links to specific items on the JMP website: JMP Welcome Kit, JMP User Community, Statistical Thinking Course, and Statistics Knowledge Portal. See Section 8.4 for a description of each of these items.

Tutorials

The tutorials cover the more commonly used statistical tools in JMP. (See Figure 8.7.) Each of these can be completed quickly to help you not only master the JMP steps, but also review the concepts involved. Each tutorial walks through the concept at hand, illustrating how to perform the data management task or analysis with JMP while explaining what each step does.

Figure 8.7 JMP Tutorials

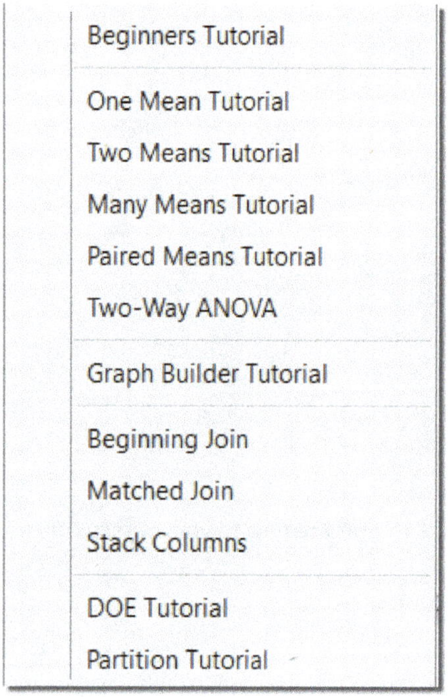

Beginners Tutorial
One Mean Tutorial
Two Means Tutorial
Many Means Tutorial
Paired Means Tutorial
Two-Way ANOVA
Graph Builder Tutorial
Beginning Join
Matched Join
Stack Columns
DOE Tutorial
Partition Tutorial

Tip of the Day

JMP has summarized the most common and important FAQs into a collection of more than 60 Tip of the Day boxes. Often these are somewhat hidden features, but worthy of special mention in this form.

After JMP is installed, Tip of the Day boxes will appear each time you launch JMP. (See Figure 8.8.) We encourage you to review these at least once, because many of them will save you valuable time and frustration in the long run. If you do not want these to appear at launch, you can remove this default by unchecking the box at the lower left corner of the window or by selecting **File ▶ Preferences ▶ General** and unchecking the **Show the Tip of the Day at startup**.

Figure 8.8 Tip of the Day with JMP Launch

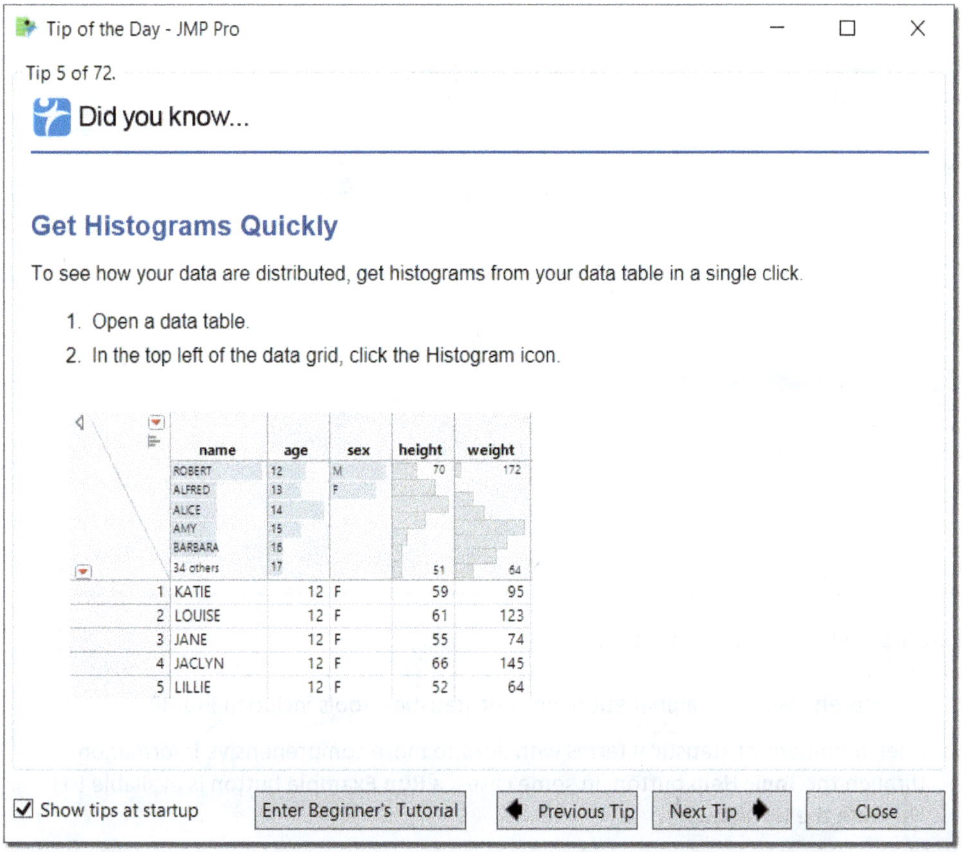

Statistics Index

The Statistics Index is the next item on the Help menu. Select **Statistics Index** when you know which statistical method you need but not where to find it in JMP. (See Figure 8.9.)

Figure 8.9 Statistics Index

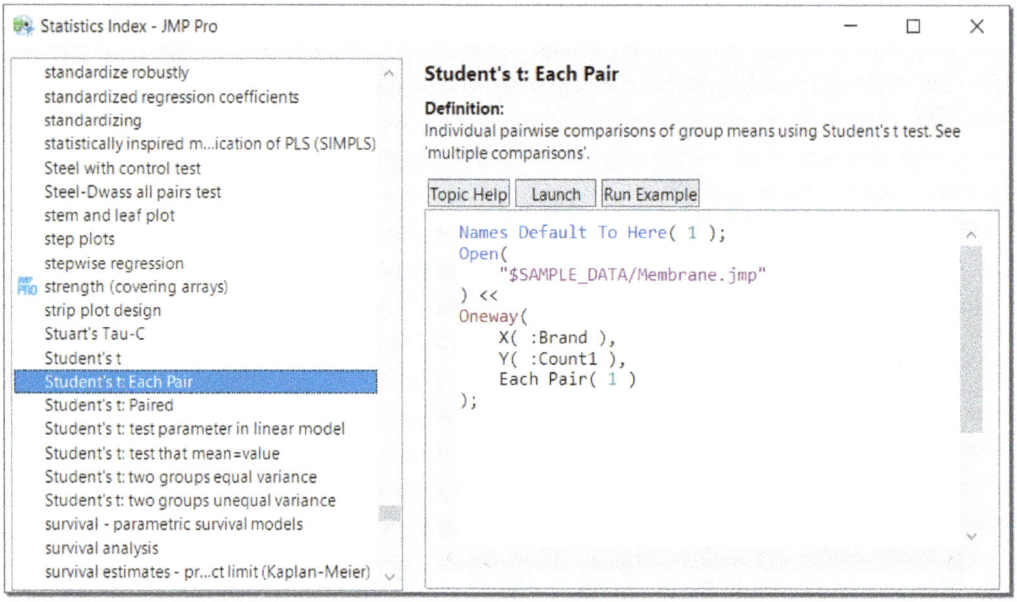

This feature provides the following information:

- A comprehensive and alphabetical index of statistical tools included in JMP.

- Brief definitions of statistical terms with links to more comprehensive information through the **Topic Help** button. In some cases, a **Run Example** button is available to illustrate the technique.

- A direct shortcut to the JMP window or platform that generates a desired statistic. Select **Help ▶ Statistics Index ▶** *[choose statistic]* **▶ Launch**. (See Figure 8.10.)

> **Note**
>
> When using the Launch button in the index as a shortcut, you need to first have your data open or follow the prompts to open it.

Figure 8.10 Launching a Platform from the Statistics Index

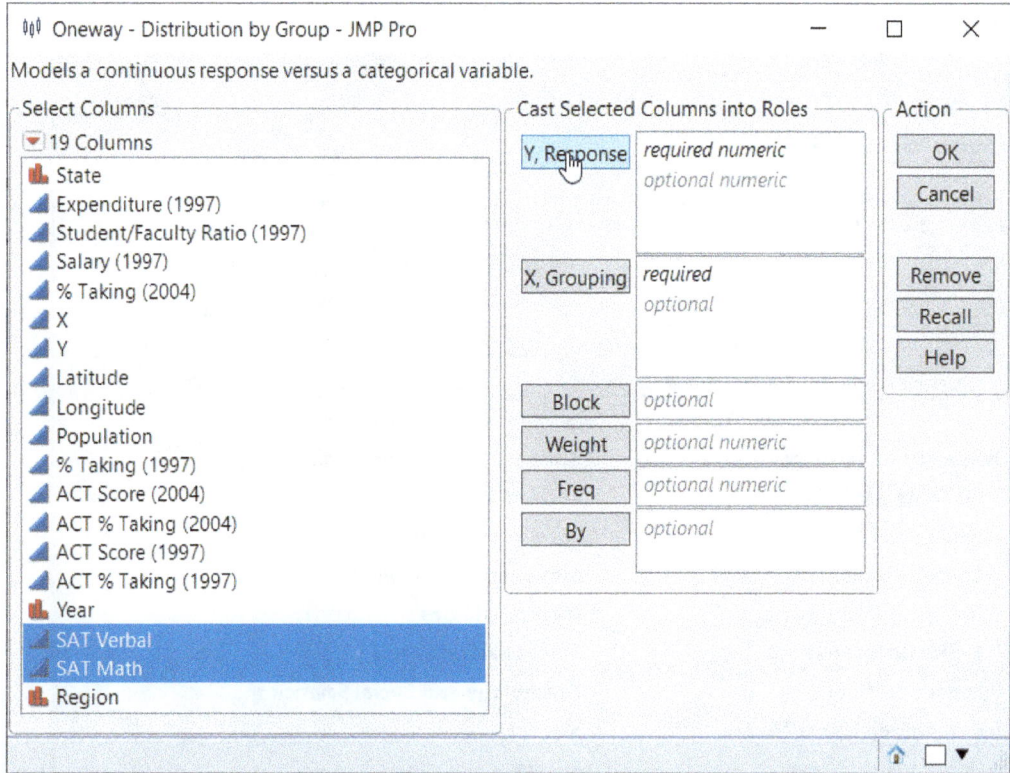

A Scripting Index follows the Statistics Index, which provides a comprehensive library of JMP scripting functions with links to examples and help. The Scripting Guide and JSL Syntax Reference (**Help ▶ JMP Documentation Library**) are also excellent resources for learning how to create custom applications using JMP Scripting Language (JSL).

Sample Data Index

The Sample Data Index (**Help ▶ Sample Data**) contains most of the data used in this book as well as in the JMP documentation. (See Figure 8.11.) It provides a convenient place to access data sets that you might want to use to see how JMP performs a specific method or type of problem, or to use for teaching purposes. It also contains other examples that might be of interest and provides a guide for how your data should be structured for a particular analysis.

Figure 8.11 Sample Data Index

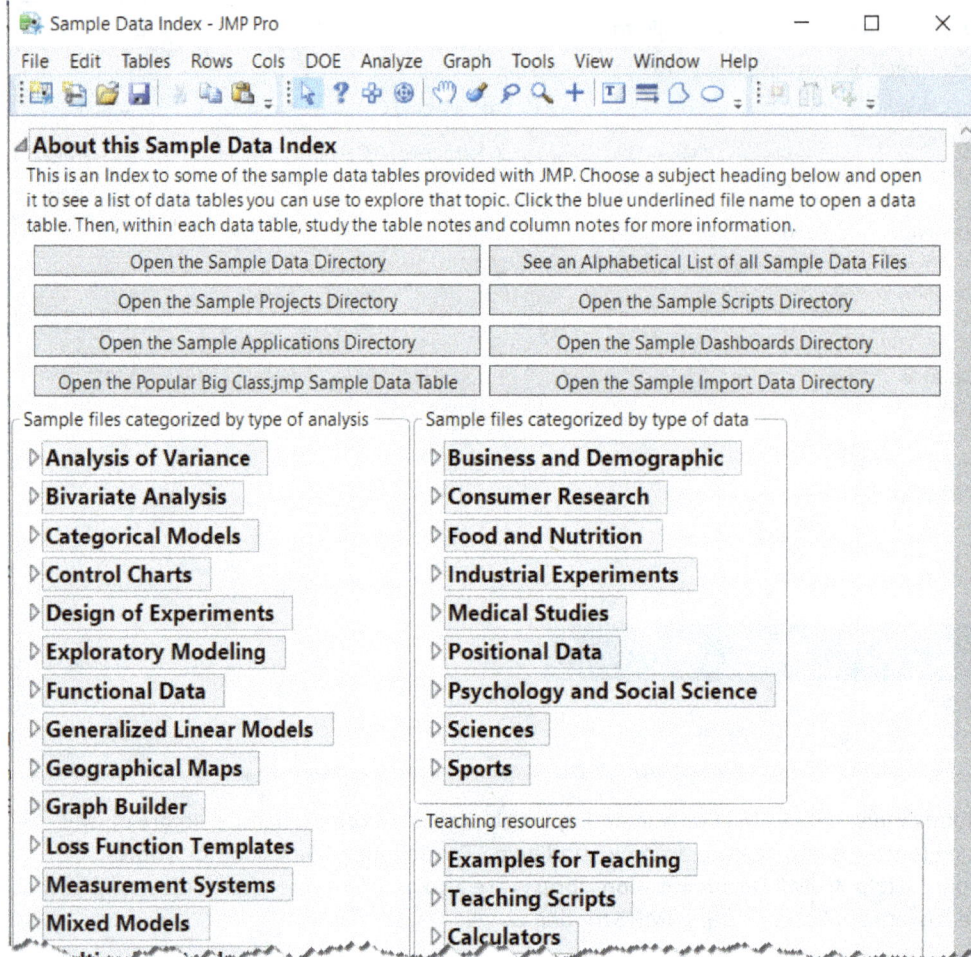

The Sample Data Index contains over 500 data sets. Many of these contain scripts that will generate an appropriate analysis, including those illustrated in the JMP documentation. Click on a gray triangle to disclose the data tables under a category of interest. To generate an analysis using a script, open a data table and locate a script in the Data Table's top left panel.

> **Note**
>
> The buttons at the top of this window enable you to quickly locate and access sample data tables, scripts, projects, dashboards, import data, and applications.

In addition to links to the sample data, the teaching resources panel in the Sample Data Index contains links to data tables that are particularly useful for teaching purposes as well as scripts that can be used to demonstrate introductory statistical concepts to students. Also found in this panel are calculators used to generate results from summary statistics common in a first statistics course that might not be found elsewhere in JMP.

8.3 The JMP Starter

The JMP Starter window provides an alternate means to navigate JMP along with a useful way to learn about the software. As discussed in Chapter 1, this book focuses on the menus that appear at the top of the JMP window. For some users who may be familiar with traditional statistical software, the JMP Starter can be a helpful and natural interface to JMP. For these users, we cover the JMP Starter essentials in this section. The JMP Starter consists of 19 categories with corresponding topics within each category.

- Select **View ▶ JMP Starter**. (See Figure 8.12.) Click through the items in the **Click Category** box on the left and see the corresponding topics within each category to the right. Notice that it organizes topics slightly differently and provides a description of the graphs and techniques within each option.

Figure 8.12 The JMP Starter

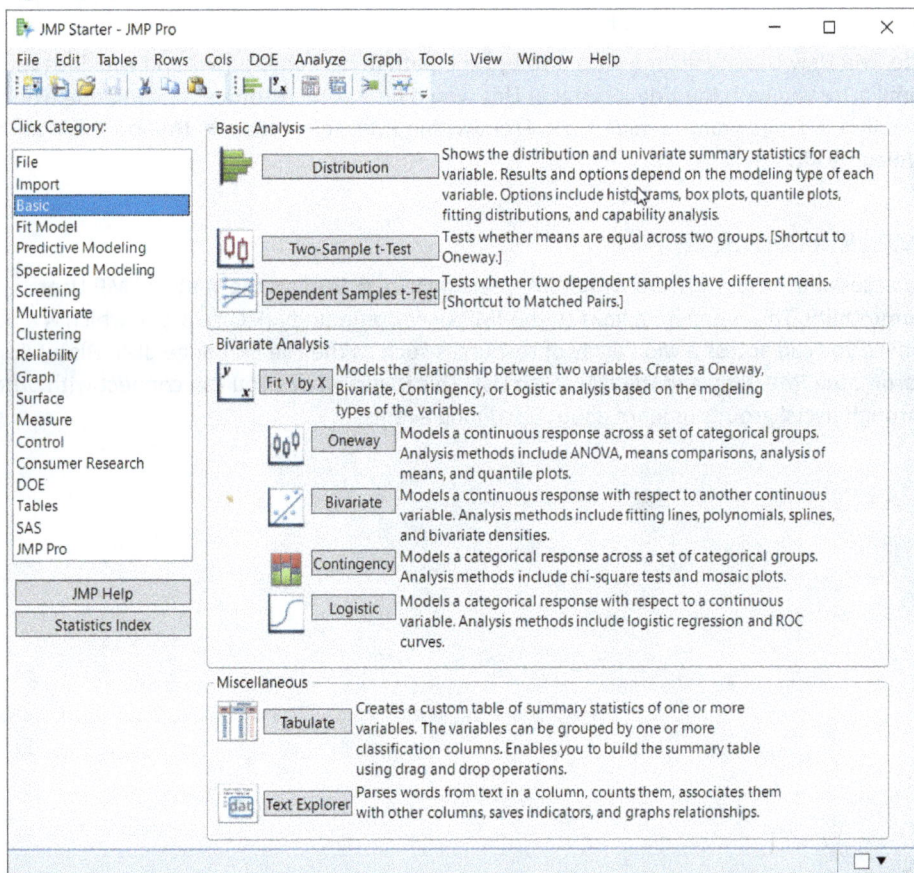

- Click on one of the topic buttons and a corresponding window launches to generate that output (similar to the Statistics Index in the Help menu).

There is extensive overlap with the menus that we have been discussing to this point, but the JMP Starter provides a little more description of and within each topic using statistical language to describe the button's functions. If you want to use the menus without reliance on statistical language to perform an analysis instead, see the Analyze menu description in Chapter 4.

8.4 The JMP Website

Think of the JMP website as the source for the latest information and a portal to the resources we discuss in this section. The website is a gateway to the **JMP User Community,** white papers, blogs, webcasts, and to downloads, such as maintenance upgrades. To learn about the latest JMP events and information, go to www.jmp.com or select **Help ▶ JMP on the Web ▶ JMP Website** from within the software.

JMP New User Welcome Kit

The JMP New User Welcome Kit is a self-paced path of demonstrations and activities to familiarize you with the JMP interface. This would be a good resource for someone who has just installed JMP and wants a quick start. From within JMP, select **Help ▶ JMP on the Web ▶ JMP Welcome Kit**.

JMP User Community

To access this resource from within JMP, select **Help ▶ JMP on the Web ▶ JMP User Community**. This menu item links to the JMP User Community (Figure 8.13), which is a portal where you can access a wide array of resources such as the File Exchange, JMP Discussion Forum, the JMP Blog, and the JMP Wish List. This is also where you can connect with other users through users' groups or learn about upcoming events.

Figure 8.13 The JMP User Community

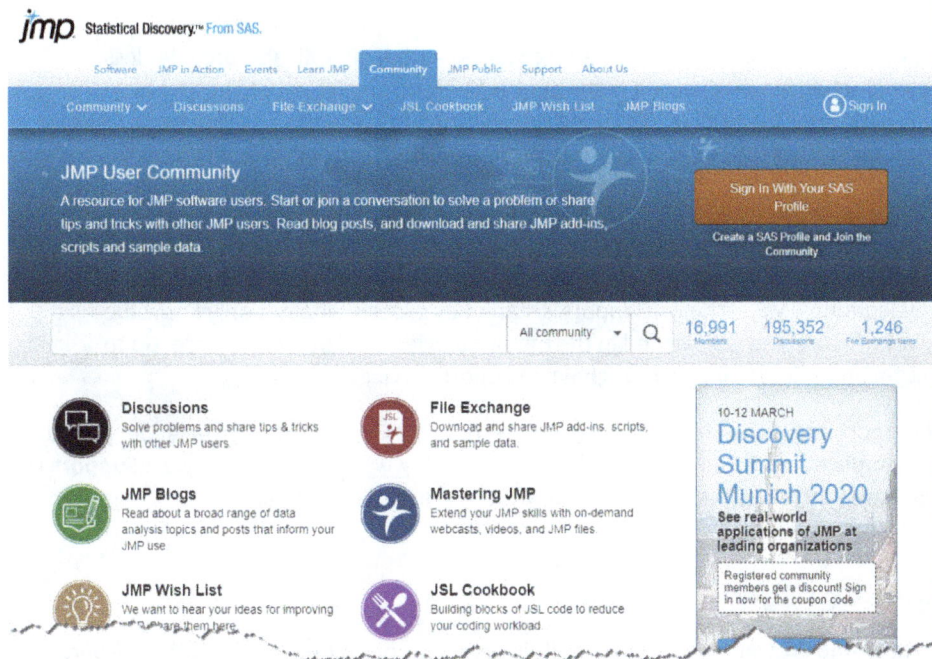

- The **File Exchange** is a database of user-developed scripts, helpful tools, and custom add-ins.

- **Discussions** is where you can ask other users usage questions and share tips and tricks.

- The **JMP Blogs** contain posts by experts about a wide range of analysis topics and JMP usage examples.

- The **JMP Wish List** is a place for users to share their requests for changes or additions to JMP.

Statistical Thinking for Industrial Problem Solving Course

This free online course is designed primarily to teach statistical concepts. At the same time, it introduces you to using JMP. The course consists of seven self-paced modules. Once you have completed the modules, you can demonstrate your mastery with the optional Certification exam. From within JMP, select **Help ▶ JMP on the Web ▶ Statistical Thinking Course**.

Statistics Knowledge Portal

Provides information about common statistical methods. This includes explanations, examples, and graphs. To access the portal from within JMP, select **Help ▶ JMP on the Web ▶ Statistics Knowledge Portal**.

JMP Webinars

JMP webcasts are offered each week for new users of the software at no cost. Go to www.jmp.com and select **Learn JMP ▶ Getting Started with JMP**. JMP also provides less frequent webcasts on more advanced features of JMP. These are live events hosted by JMP systems engineers and developers. Recorded or "on-demand" versions are also available for your convenience. Go to www.jmp.com and select **Learn JMP ▶ Mastering JMP**.

JMP Technical Support

JMP Technical Support provides support to registered users through the JMP website and with email and telephone. JMP Technical Support is staffed by experts who not only know the software, but who are also well-versed in the methodologies used in the software. Go to www.jmp.com and select **Support ▶ Technical Support Services**.

Technical Support maintains Frequently Asked Questions (FAQs) as well as installation, problem, usage, and sample notes. To access this information, go to www.jmp.com and select **Support ▶ Knowledge Base and FAQ**. To contact Tech Support by email, please email support@jmp.com.

8.5 Other Resources

When you can't find answers within JMP's built-in resources or on the website, other resources are available. JMP offers many events, including the annual Discovery Conference and regional user groups. Comprehensive JMP training is available through SAS Education at SAS Regional Training Centers, at company sites, and through the web. Additional books, written by JMP users, are also available.

JMP Events

JMP Discovery Summit Conferences are held in several locations throughout the world each year. These conferences bring together thought leaders, JMP users, and JMP development staff to share the latest in analytic methods and the use of JMP. JMP also offers Explorer events, which are typically day-long, knowledge-sharing events focused on analytics.

Users' Groups

There are several regional users' groups that have regular meetings. Users' groups are a convenient way to connect with other JMP users and to learn about some of the software's latest developments. For more information, go to www.jmp.com/about/events/usergroups/.

JMP Books

In addition to the JMP documentation and this book, there are many books written by JMP users. Topics include JMP Scripting and analytical methods such as quality control, predictive analytics, and design of experiments. Some of the books are focused on the use of JMP and others are focused on analytics using JMP. To discover more about the books available, go to www.jmp.com/en_us/software/books.html.

SAS Training for JMP

SAS Education offers many courses on JMP ranging from basic to advanced. These courses are offered at SAS training facilities throughout the world, at customer sites, and through the web. SAS trainers who teach the courses are well-known for their instructional skills and their subject matter expertise. Some of the courses currently offered include A Case Study Approach to Data Exploration, Statistical Decisions Using ANOVA and Regression, Custom Design of Experiments, Statistical Process Control, and Predictive Modeling. For a full list of offerings, see the website at www.jmp.com/training.

In addition to formal courses, SAS also offers a workforce transformation program and a mentoring service. The workforce transformation program allows you to work with the JMP team to develop and deliver a program that meets the needs of your team. The mentoring service allows a JMP instructor to spend time with one or more individuals, focusing on their analysis needs and application areas of interest.

For more information about current SAS Education offerings, see the JMP website at www.jmp.com/training or send a request for information to training@jmp.com.

Appendix A: Integrating with SAS

A.1 Working with SAS Data
A.2 Working with SAS Programs

When you use JMP, you are using one in a family of SAS products. JMP is designed as a nimble and fast desktop data discovery tool for use by a broad audience. At the beginning, SAS software was created to provide analytics on mainframe computers when few other options were available. Over the years, as problem sizes grew, so did SAS software. 92 of the top 100 companies on the 2018 FORTUNE Global 1000® list use SAS software on every size and type of problem.

For many years, JMP has been an integrated client to SAS, not just a stand-alone product as described in this book. This means that as your problem size grows or extends beyond that which is suitable for JMP alone, you can migrate smoothly from desktop JMP to other tools from SAS. When analyzing SAS data with JMP, you first need to import that data into JMP, which we will describe in the next section.

A.1 Working with SAS Data

SAS data files can be stored either on your local machine or on a network drive. You open them using the **File** menu.

Select **File ▶ Open**. Select **All JMP Files** to disclose all file types and select **SAS Data Sets** to highlight compatible SAS data formats. You see that JMP can read a variety of SAS data and file types. (See Figure A.1.) Locate the SAS file and select **Open**.

Figure A.1 Select SAS Data Sets

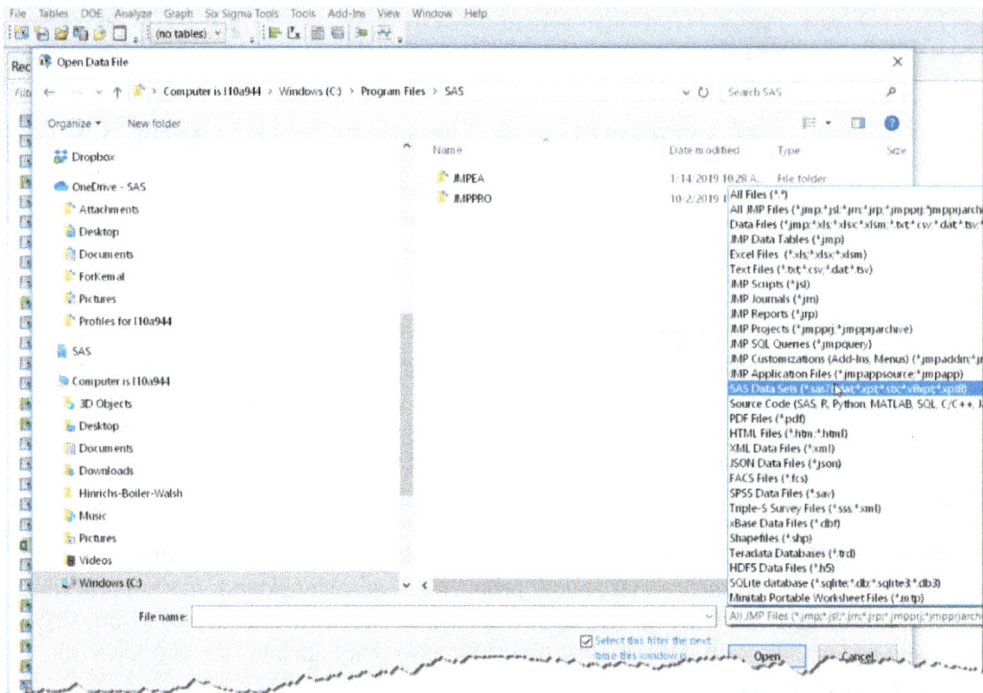

Many organizations connect different data sources using SAS to automate and streamline data integration. If you have an individual copy of SAS or if your organization has a SAS Metadata Server, then JMP can connect to data in these SAS frameworks.

If you have a metadata server or you have SAS installed locally, you can select **File ▶ SAS ▶ Browse Data.** (See Figure A.2.)

Figure A.2 Browse SAS Data

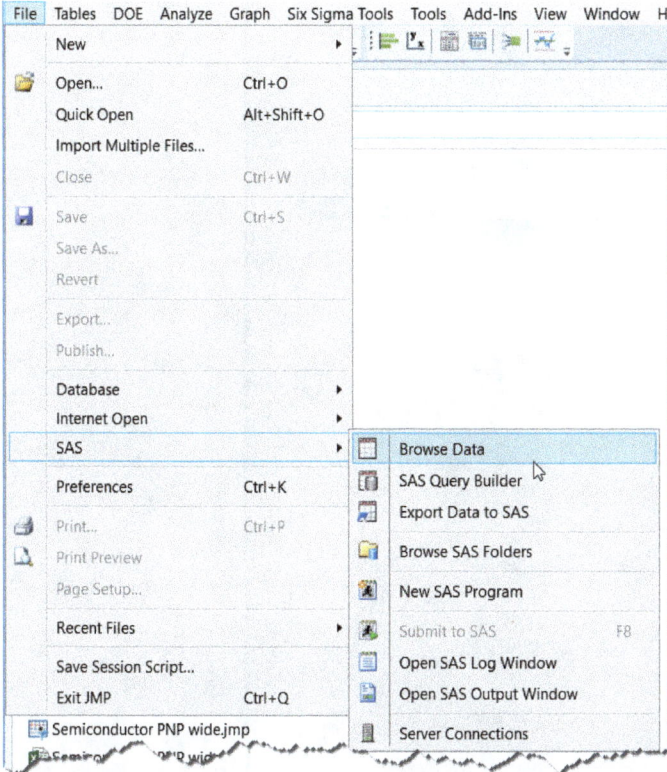

You are prompted to log in to the SAS Metadata Server. (See Figure A.3.) This is a powerful feature of SAS that provides a permission and security framework to keep data safe.

Figure A.3 Create SAS Profile

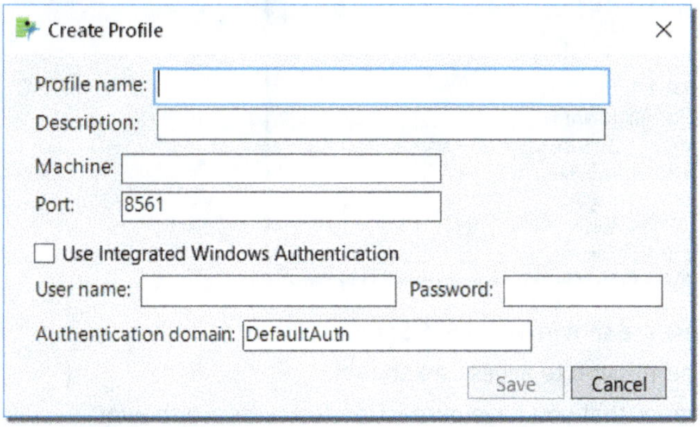

Enter your user name and password and click **OK**.

There is a convenient data import window that has many useful features. (See Figure A.4.)

Figure A.4 Data Import Window

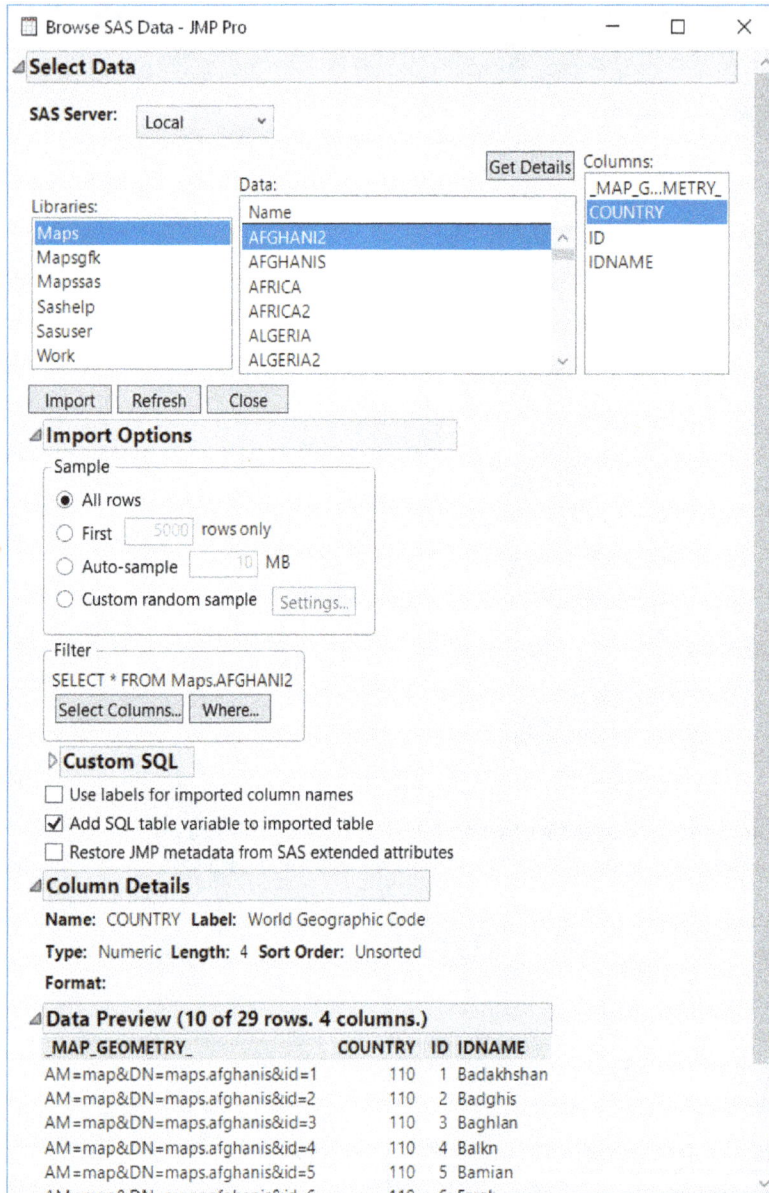

1. Local and network SAS **Servers** are shown in the SAS Server column.

2. SAS **Libraries** that you have permission to access are shown.

3. **Data** Tables within those libraries that you have permission to access are shown.

4. **Columns** within those data tables that you have permission to access are shown.

5. An interactive **Import Options** panel allows writing of SQL queries that can be run on the data. This is supported under the **Custom SQL** report tab.

6. A **Custom Random Sample** panel allows random sampling against the data. This is a critical feature when you are importing and managing large data sizes in JMP.

7. The **Column Details** panel shows details about the column(s).

8. A **Data Preview** panel details the file size in rows and columns before it is imported to JMP. This is also valuable when accessing especially large data sources that might be too large to fit on your desktop computer.

A SAS Metadata Server also provides tools to prepare and report data using server resources that JMP can leverage. We describe a few highlights in the next section.

A.2 Working with SAS Programs

SAS supports its own programming language that provides a rich set of tools, scaling to large and complex problems and bigger computer hardware including cloud-based environments.

When JMP has access to SAS, JMP can leverage these tools to extend its capabilities in several ways.

Opening a SAS Program

You may open an existing SAS program just as you would open any data file. Use **File ▶ Open** and then specify **Source Code (SAS, R, Python,...)** in the file of type submenu (Figure A.5). Locate the program and then select **Open**. This will open the program within a SAS Program Editor window where you may further edit the program or simply submit the program to SAS (Figure A.6).

Figure A.5 Open a SAS Program

JMP Application Files (*.jmpappsource
SAS Data Sets (*.sas7bdat;*.xpt;*.stx;*.v
Source Code (SAS, R, Python, MATLAB
PDF Files (*.pdf)
HTML Files (*.htm;*.html)
XML Data Files (*.xml)
JSON Data Files (*.json)
FACS Files (*.fcs)
SPSS Data Files (*.sav)
Triple-S Survey Files (*.sss;*.xml)
xBase Data Files (*.dbf)
Shapefiles (*.shp)
Teradata Databases (*.trd)
HDF5 Data Files (*.h5)
SQLite database (*.sqlite;*.db;*.sqlite3;
Minitab Portable Worksheet Files (*.m

Source Code (SAS, R, Python, M ⌄

Open   Cancel

Figure A.6 SAS Program Editor

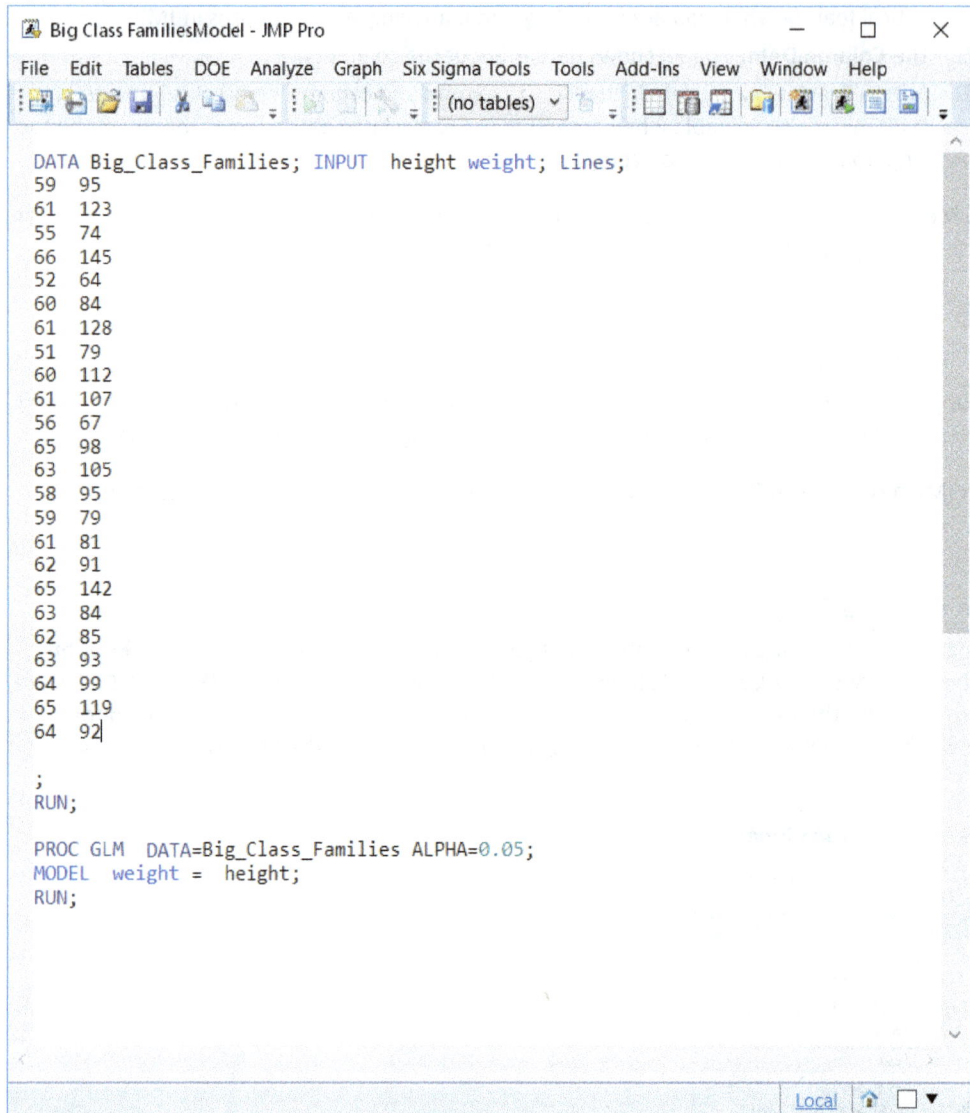

```
Big Class FamiliesModel - JMP Pro                           —    □    ×

File  Edit  Tables  DOE  Analyze  Graph  Six Sigma Tools  Tools  Add-Ins  View  Window  Help

    (no tables) 

DATA Big_Class_Families; INPUT  height weight; Lines;
59   95
61   123
55   74
66   145
52   64
60   84
61   128
51   79
60   112
61   107
56   67
65   98
63   105
58   95
59   79
61   81
62   91
65   142
63   84
62   85
63   93
64   99
65   119
64   92

;
RUN;

PROC GLM  DATA=Big_Class_Families ALPHA=0.05;
MODEL  weight =  height;
RUN;



                                                    Local
```

To submit to SAS, right-click within the window and select **Submit to SAS**. If you have not already established a connection to SAS either locally or to a server, JMP will prompt you at this point to specify the connection to SAS. On Windows this connection can either be to a local copy or a server instance of SAS whereas on Mac it must be to a server.

SAS Add-ins

SAS Add-ins are SAS Programs that are packaged in user-friendly JMP dialog windows. SAS Add-ins enable you to take advantage of SAS advanced analytics or custom applications conveniently and accessibly without needing to program in SAS. In this way, JMP can work as an easy-to-use client to SAS, assuming that SAS is installed and/or connected.

A variety of SAS Add-ins are available for download from the File Exchange located at the JMP User Community (Figure A.7). Go to **Help ▶ JMP User Community** and click the **File Exchange Tab**.

Figure A.7 JMP File Exchange

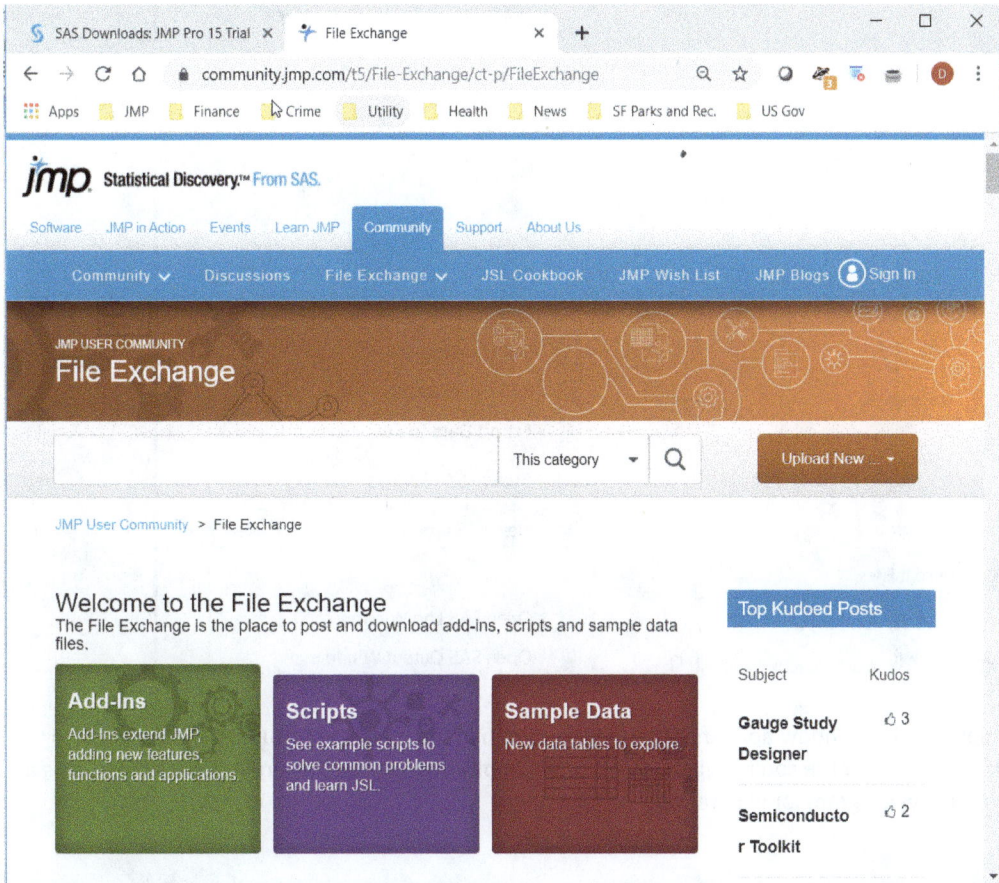

> **Note**
>
> You can create your own SAS Add-ins, which can be time savers, but we recommend that you consult with a SAS expert.

Writing a SAS Program

SAS programs can be written in JMP and submitted to SAS. In JMP, select **File ▶ SAS ▶ New SAS Program**. (See Figure A.8.)

Figure A.8 Write a SAS Program

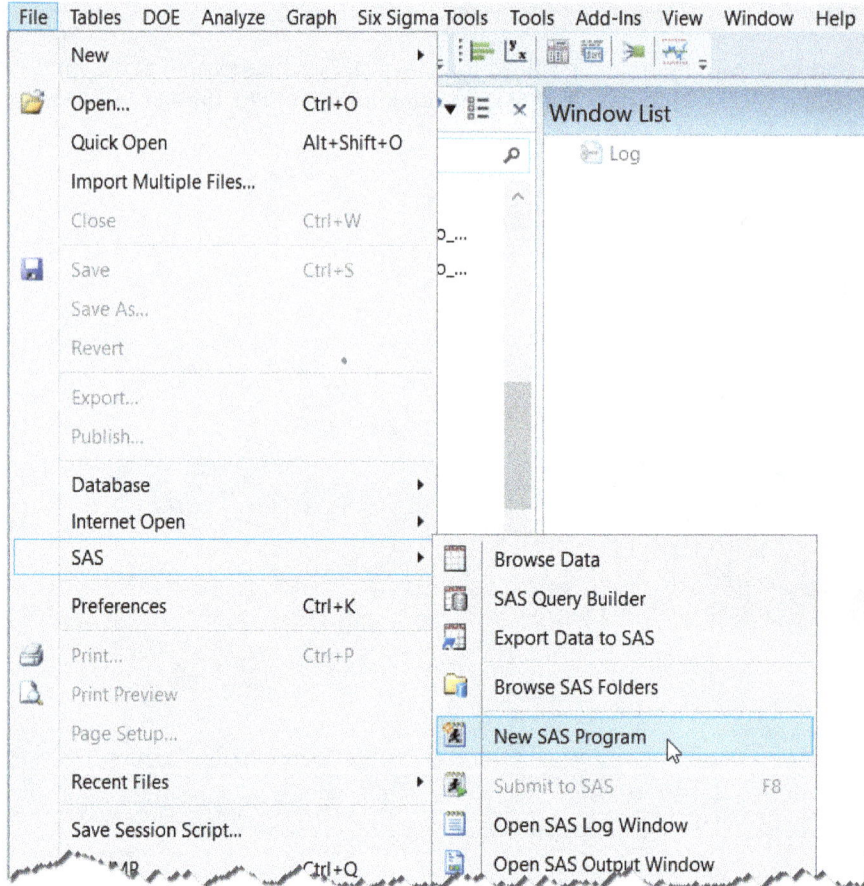

A blank coding window appears. Enter your program in the window. To submit the program to SAS, right-click in the coding window and select **Submit to SAS**. See Figure A.9, which contains a sample SAS program for illustration.

Figure A.9 Sample SAS Program

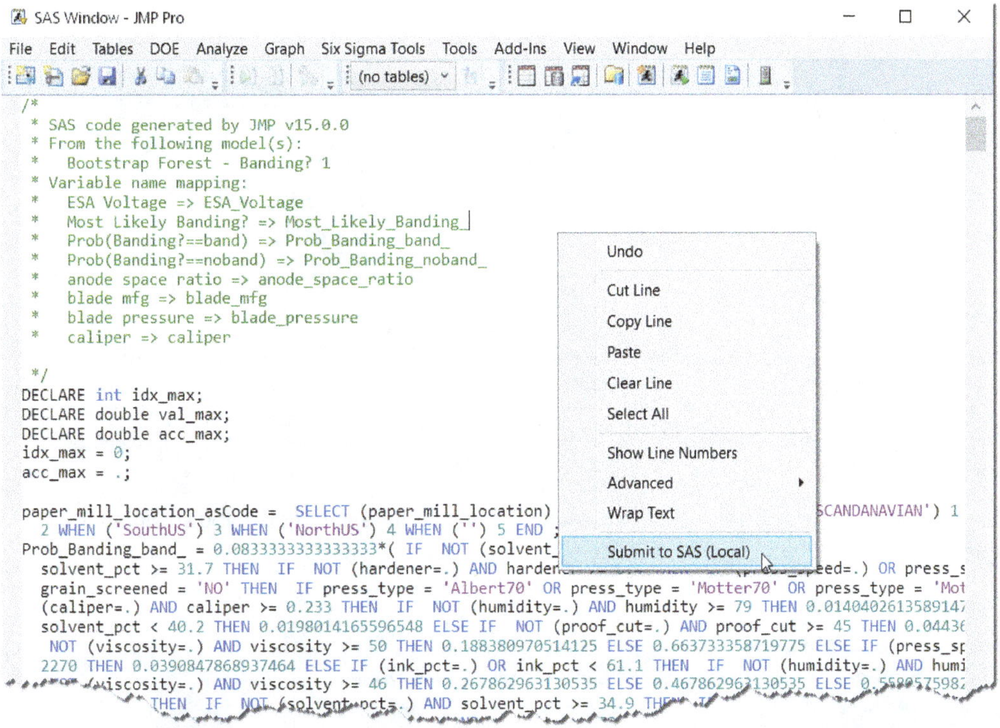

If you want to begin learning about SAS programming, we recommend *The Little SAS Book: A Primer* by Lora Delwiche and Susan Slaughter.

Automatically Generating a SAS Program

Some limited support for automatically generating a SAS program within JMP is available. For example, within the Fit Model platform (and ARIMA in Time series) in JMP, SAS program code can be automatically generated and then submitted by JMP. Let's try it out:

Select **Help ▶ Sample Data**, then select **Examples for Teaching ▶ Big Class**.

Within JMP, select **Analyze ▶ Fit Model**. Select **weight** for the Y role, select **height,** and click **Add**. Then, select the red triangle/hot spot in the upper left of the launch window and select **Create SAS Job**. (See Figure A.10.)

Figure A.10 Create a SAS Job from Fit Model

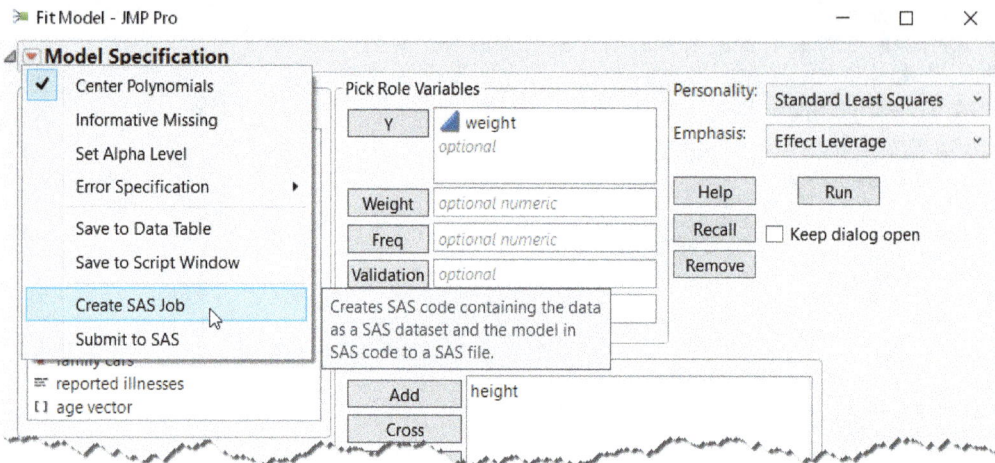

This creates the SAS program in a separate window for the data and model you specified. (See Figure A.11.)

Figure A.11 SAS Code

This SAS program can now be submitted to SAS by right-clicking in the window and selecting **Submit to SAS**.

Assuming you have a connection to SAS, the program runs in SAS and returns the results to JMP. (See Figure A.12.)

Figure A.12 SAS Report in JMP Report Window

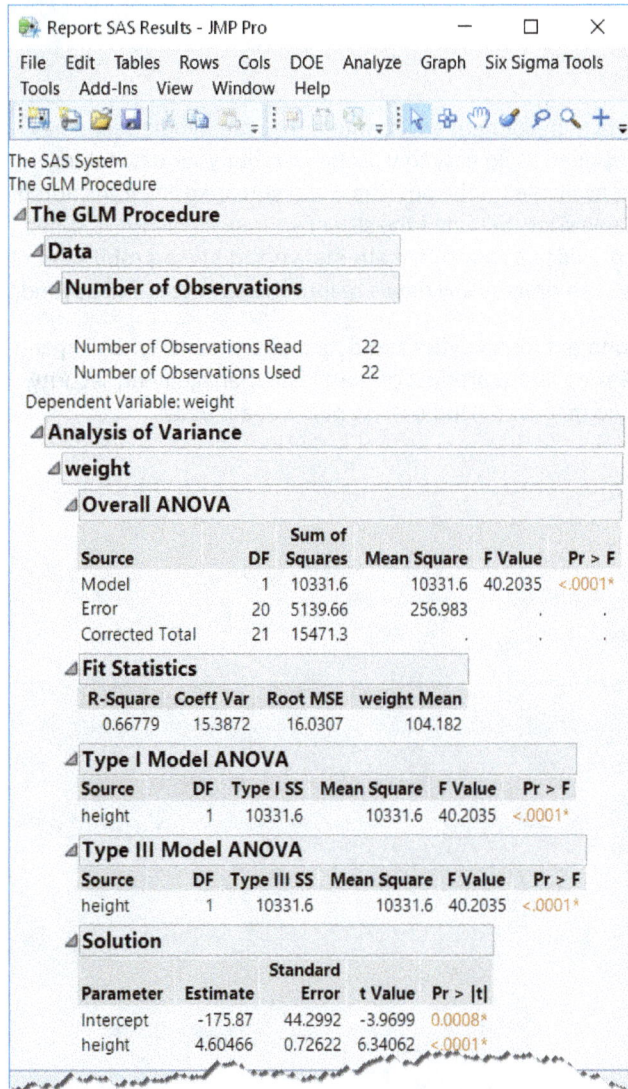

The SAS System
The GLM Procedure

⊿ **The GLM Procedure**

  ⊿ **Data**

   ⊿ **Number of Observations**

   Number of Observations Read   22
   Number of Observations Used   22

Dependent Variable: weight

⊿ **Analysis of Variance**

 ⊿ **weight**

  ⊿ **Overall ANOVA**

Source	DF	Sum of Squares	Mean Square	F Value	Pr > F
Model	1	10331.6	10331.6	40.2035	<.0001*
Error	20	5139.66	256.983	.	.
Corrected Total	21	15471.3	.	.	.

 ⊿ **Fit Statistics**

R-Square	Coeff Var	Root MSE	weight Mean
0.66779	15.3872	16.0307	104.182

⊿ **Type I Model ANOVA**

Source	DF	Type I SS	Mean Square	F Value	Pr > F
height	1	10331.6	10331.6	40.2035	<.0001*

⊿ **Type III Model ANOVA**

Source	DF	Type III SS	Mean Square	F Value	Pr > F
height	1	10331.6	10331.6	40.2035	<.0001*

⊿ **Solution**

Parameter	Estimate	Standard Error	t Value	Pr > |t|
Intercept	-175.87	44.2992	-3.9699	0.0008*
height	4.60466	0.72622	6.34062	<.0001*

Of course, in this simple model, JMP could easily generate the results, but when you want to run a very large or complex model, SAS can provide more options and will provide more detailed results.

Note

JMP Pro includes the Formula Depot that is a very useful platform for model comparison, creating ensemble models and deploying them into other environments such as SAS, Python,

> **Note**
>
> C, JavaScript, and SQL. This is useful when you want to explore your data and create and refine your models within JMP Pro and then deploy them into production environments.

Why SAS? Why JMP?

SAS is a highly scalable analytics and data management environment that can handle very small problems up to massive ones. JMP is designed to be easy to use and fast, but your data tables need to fit within the memory (or RAM) available on the desktop when you are using it. As noted in Chapter 2, JMP 15 can handle data table sizes up to half the size of your available RAM. When problem sizes exceed these limits, SAS provides a convenient alternative and hosts a robust set of capabilities to handle truly big data. SAS can process any data size into the terabytes and beyond.

SAS is also a very rich and robust environment for analytics and data management. When larger-sized problems appear, or if you are thinking about production-type data management, security, and reporting, SAS provides the tools and means to handle these needs seamlessly.

Appendix B: Understanding Results

In the process of analyzing data, you will encounter terms in the results that reference vital information. Throughout Chapters 5 and 6, we discussed many of the relevant terms and results in the context of the application and analysis that we were performing. This appendix is designed to provide a basic description of the common terms used in each of those platforms to help you understand your own results. We also point out where you can find complete descriptions in the JMP documentation. In most cases, you can simply use the "?" help tool and select the term in question to obtain more information. In other cases, by holding your mouse cursor over statistics, you can see Hover Help, which provides context-specific assistance in interpreting a statistic.

While a graph can be worth a thousand words and is more easily interpreted, the statistical results generated from the Analyze menu contain numerical results and terms that might be less clear. It is important that you understand the basic ideas behind the results. This section offers a basic and general reference rather than a comprehensive one. To increase your confidence in interpreting statistical results, we encourage you to seek advice from an experienced data analyst or reference the books that ship with JMP (see the Help menu) or those in the bibliography.

This appendix covers key terms and related concepts in the book, in alphabetical order. For the purpose of this appendix, the terms *column* and *variable* are interchangeable in these definitions.

Bivariate Plot: Also known as a *scatter diagram or scatterplot* where each point in the plot expresses both an *X* and *Y* value. These two-dimensional plots are used when comparing one continuous column with another continuous column. See **Help ▶ JMP Documentation Library ▶ Basic Analysis ▶ Chapter 5: Bivariate Analysis.**

Box and Whisker Plot: Also called a *box plot* or an *outlier box plot*. A graphical presentation of the important characteristics of a continuous variable. Box plots display the interquartile range of the data (the "box"), the spread (the whiskers), and potential outliers (disconnected points). Box plots are useful for describing variables that have a skewed distribution and for comparing two or more distributions. See **Help ▶ JMP Documentation Library ▶ Basic Analysis ▶ Chapter 3: Distributions.**

Confidence Interval: An interval within which we expect a value to fall with a given certainty. For a 95% confidence interval, we are 95% certain, or confident, that a new value will fall within this interval.

Contingency Table: A table showing the observed frequencies of two nominal variables, with the rows indicating one variable and the columns indicating another. The table reports a chi-square statistic. See **Help ▶ JMP Documentation Library ▶ Basic Analysis ▶ Chapter 7: Contingency Analysis.**

Count: The total number of members in a group.

Correlation: A measure of relationship between two continuous variables. It is a relationship where changes in the value of one variable are accompanied by changes in another variable or variables. For example, a correlation of 1 indicates that as one variable increases, the other also increases by some constant proportional amount. See **Help ▶ JMP Documentation Library ▶ Multivariate Methods ▶ Chapter 3: Correlations and Multivariate Techniques.**

Degrees of Freedom: Also abbreviated DF, degrees of freedom are associated with many statistical estimates. Intuitively, degrees of freedom are the number of freely varying observations in a data-based calculation. The larger the sample size, the larger the degrees of freedom and the stronger the inferences we can draw about population parameters. See **Help ▶ JMP Documentation Library ▶ Basic Analysis ▶ Oneway Analysis.**

Distribution: The values of a single column or variable in terms of frequency of occurrence, spread, and shape. A distribution can be observed (based on data) or theoretical. Some examples of theoretical distributions are normal (bell- shaped), binomial, and Poisson. Distribution is also the JMP univariate platform.

***F* Ratio:** In an ANOVA, the ratio of between-group variability to the within-group variability. (See One Way Analysis of Variance). The *F* ratio is used, in conjunction with Prob > *F*, to test the null hypothesis that the group means are equal (that there is no real difference between them). In general, larger *F* ratios indicate significant differences between at least two means. See **Help ▶ JMP Documentation Library ▶ Fitting Linear Models ▶ Stand Least Squares Report and Options.**

Frequency: The number of times a categorical value is observed, a type of event occurs, or the number of elements of a sample that belong to a specified group. It is also called *count*.

Interquartile Range: A measure of variability or dispersion for a continuous column calculated as the difference or distance between the 25th and 75th percentiles (the first and third quartiles, respectively).

Logistic Regression: A type of regression technique where the *Y*, dependent variable is nominal or ordinal and there is at least one *X*, independent variable. Logistic models (sometimes called *logit models*) are used to predict the probability of occurrence of an event based on the values of one or more variables. See **Help ▶ JMP Documentation Library ▶ Fitting Linear Models ▶ Chapter 11: Logistic Regression Models.**

Maximum: The largest value observed in the sample.

Mean: A measure of location or central tendency of a column of continuous data. It is the arithmetic average computed by summing all the values in a column and dividing by the number of non-missing rows.

Median: A measure of location or central tendency of a continuous column of data. It is the middle value in an ordered column, which divides a distribution exactly in half. Fifty percent of the values are higher than the median and 50% are lower.

Minimum: The smallest value observed in the sample.

Multiple Regression: An analysis involving two or more independent X variables as predictors to estimate the value of a single dependent variable. The dependent Y variable is usually continuous, but the independent X variables can be continuous or categorical. The model is usually estimated by the method of standard least squares. See **Help ▶ JMP Documentation Library ▶ Fitting Linear Models ▶ Chapter 2: Model Specification.**

One-Way Analysis of Variance (or One-Way ANOVA): A procedure involving a categorical X, independent variable and a continuous Y, dependent variable. One-way ANOVA is used to test for differences in means among two or more independent groups (though it is typically used to test for differences among at least three groups because the two- group case can be analyzed with a t-test). When there are only two means to compare, the t-test and the F-test (see F-test) are equivalent. See **Help ▶ JMP Documentation Library ▶ Basic Analysis ▶ Chapter 6: Oneway Analysis.**

Outlier: An observation that is so extreme that it stands apart from the rest of the observations; that is, it differs so greatly from the rest of the data that it gives rise to the question of whether it is from the same population or involves measurement error. One common rule of thumb for an outlier is any value that is 1.5 times the interquartile range if the distribution is approximately normally distributed.

Partition: The Partition platform iteratively separates data according to a predictive relationship between a Y and multiple X values, from strongest to weakest, forming a tree structure. Partition searches through the data table to find values within X columns that best predict the outcome of Y, your column of interest. Partition is a data mining or predictive modeling technique. See **Help ▶ JMP Documentation Library ▶ Predictive and Specialized Modeling ▶ Chapter 4: Partition Models.**

Prob > F: In ANOVA, the probability of obtaining (by chance alone) an F-value greater than the one calculated if, in reality, the null hypothesis is true. Prob (or "p" values) of 0.05 or less are often considered evidence that a model fits the data. See **Help ▶ JMP Documentation Library ▶ Basic Analysis ▶ Chapter 6: Oneway Analysis.**

Prob > t: A p-value or measure of significance for a t-test. For a one-sample test or for a test of differences between two means, it is the probability of obtaining a value more extreme than the hypothesized value, if the null hypothesis were true. Prob > t values of 0.05 or less are usually considered significant. See **Help ▶ JMP Documentation Library ▶ Basic Analysis ▶ Chapter 3: Distributions.**

Quantiles: Values that divide an ordered set of continuous data (from smallest to largest) into equal proportions. Related terms are *deciles* (dividing data into 10 parts) and *quartiles* (dividing data into four parts, or quarters). Values in the 97th percentile, or quantile, are equal to or larger than 97% of all values in the distribution.

Quartiles: Values in a continuous column of data that are first ordered (from smallest to largest) and then divided into four quarters, each of which contains 25% of the observed values. The 25th, 50th, and 75th percentiles are the same as the first, second, and third quartiles, respectively. See also **Quantiles**.

Regression: A statistical procedure that shows how two or more variables are related, which is represented in the simple case by a fitted line in a bivariate scatterplot. The fitted line, along with its regression equation, allows one to predict values of *Y* based on observations of *X*. The simple form of this equation is expressed as *y=mx+b*, which determines the extent to which one variable changes with another.

RSquare: A measure of the adequacy of a model defined as a proportion of variability that is accounted for by the statistical model. RSquare provides a measure of how well future outcomes are likely to be predicted by the model. See **Help ▶ JMP Documentation Library ▶ Basic Analysis ▶ Chapter 6: Bivariate Analysis.**

Standard Deviation: A measure of variability or dispersion of a data set calculated by taking the square root of the variance. It can be interpreted as the average distance of the individual observations from the mean. The standard deviation is expressed in the same units as the measurement in question. It is usually employed in conjunction with the mean to summarize a continuous column.

Standard Least Squares: A method of fitting a line to data in a bivariate plot or multiple regression model. Least squares is used where there is one continuous Y and at least one X column in your model. Least squares fits a model that minimizes the total sum of squares in the data, hence the name "least squares." The least squares method is used extensively for prediction and for calculating the relationship between two or more variables. See **Help ▶ JMP Documentation Library ▶ Fitting Linear Models ▶ Chapter 3: Standard Least Squares Report and Options.**

Sum of Squares: A measure of variation of the model to the observed data. This is calculated by squaring the errors (the vertical distance between each data point and the model fit) and summing those values. We square the values to obtain a positive sum of the errors regardless of whether the observed values are above or below the fit. The best model fit is one where the total sum of squares is minimized.

Appendix C: JMP Shortcuts

This appendix provides the essential menu steps necessary to generate a desired result with JMP. JMP launch windows require that you specify the right data and modeling types in order to generate a desired result. Should you have questions about modeling types or the options displayed in the windows, refer to Chapter 2.

Some launch windows contain options (for example, **Fit Model**) that should be selected in the Launch dialog window before you execute your results. More often however, you will find that desired results require a secondary step after you have generated some results. A semicolon (;) indicates that there is a secondary step in the process that occurs only after you have executed some base results. The ▼ > symbols indicate a selection is required from a hotspot (or red triangle) drop-down menu that appears in the base results.

There is often more than one way to complete a task (for example, Fit Y by X and Fit Model); we have used the most direct means to achieve them here.

Some of the tasks included in this section are not covered in the book, but you can reference these tools in the JMP built-in documentation at **Help ▶ JMP Documentation Library** and using the search within the PDF file.

Figure C.1 JMP Menus

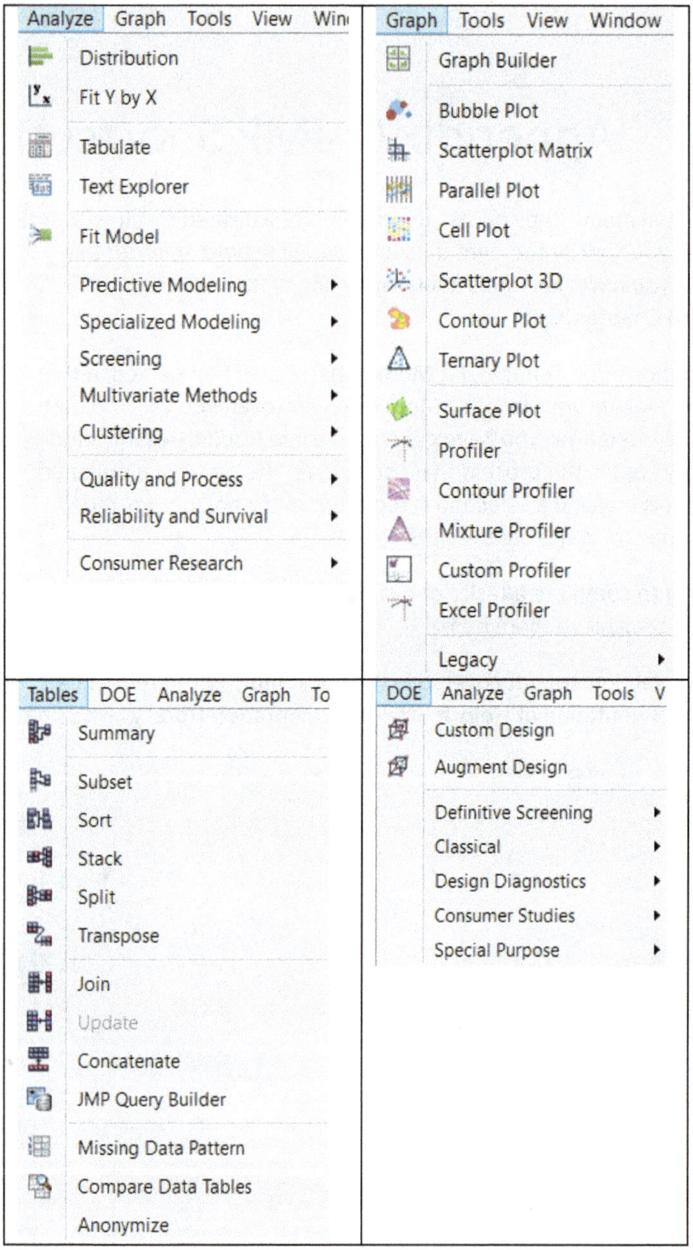

Analyze	Graph	Tools	View	Wind
	Distribution			
	Fit Y by X			
	Tabulate			
	Text Explorer			
	Fit Model			
	Predictive Modeling	▸		
	Specialized Modeling	▸		
	Screening	▸		
	Multivariate Methods	▸		
	Clustering	▸		
	Quality and Process	▸		
	Reliability and Survival	▸		
	Consumer Research	▸		

Graph	Tools	View	Window
Graph Builder			
Bubble Plot			
Scatterplot Matrix			
Parallel Plot			
Cell Plot			
Scatterplot 3D			
Contour Plot			
Ternary Plot			
Surface Plot			
Profiler			
Contour Profiler			
Mixture Profiler			
Custom Profiler			
Excel Profiler			
Legacy			▸

Tables	DOE	Analyze	Graph	To
Summary				
Subset				
Sort				
Stack				
Split				
Transpose				
Join				
Update				
Concatenate				
JMP Query Builder				
Missing Data Pattern				
Compare Data Tables				
Anonymize				

DOE	Analyze	Graph	Tools	V
Custom Design				
Augment Design				
Definitive Screening				▸
Classical				▸
Design Diagnostics				▸
Consumer Studies				▸
Special Purpose				▸

Task	Menu Selection
Adding Labels	*Click on column heading;* **Cols ▶ Label/Unlabel**
ANOVA, One Way	**Analyze ▶ Fit Y by X; ▼ ▶ Means/Anova**
ANOVA, two or more factors	**Analyze ▶ Fit Model**
Bar Chart	**Graph ▶ Graph Builder**
Basic Charts	**Graph ▶ Graph Builder**
Bivariate	**Analyze ▶ Fit Y by X**
Box Plots, one level	**Analyze ▶ Distribution; ▼ ▶ Outlier Box Plot**
Box Plot, two or more levels	**Analyze ▶ Fit Y by X; ▼ ▶ Display Options ▶ Box Plot**
Bubble Plot	**Graph ▶ Bubble Plot**
C-Chart	**Analyze ▶ Quality and Process ▶ Control Chart ▶ C Control Chart**
Chi-Square test	**Analyze ▶ Fit Y by X**
Color or Mark by	**Rows ▶ Color or Mark by Column**
Column Info Dialog	*Click on column heading;* **Cols ▶ Column Info**
Concatenate	**Tables ▶ Concatenate**
Contingency Platform	**Analyze ▶ Fit Y by X**
Control Charts	**Analyze ▶ Quality and Process ▶ Control Chart Builder**
Correlation	**Analyze ▶ Multivariate Methods ▶ Multivariate**

Task	Menu Selection
Covariance	**Analyze ▶ Multivariate Methods ▶ Multivariate ; ▼ ▶ Covariance Matrix**
CUSUM	**Analyze ▶ Quality and Process ▶ Control Chart ▶ CUSUM Control Chart**
Data Filter, Global	**Rows ▶ Data Filter**
Data Filter, Local	*In a results window;* ▼ ▶ **Local Data Filter**
Data Mining	*See* Partition, Neural Network, or Text Mining
Density Ellipses, Bivariate	**Analyze ▶ Fit Y by X;** ▼ ▶ **Density Ellipse ▶** *select confidence level*
Density Ellipses, Scatterplot Matrix	**Graph ▶ Scatterplot Matrix;** ▼ ▶ **Density Ellipses**
Descriptive Statistics	**Analyze ▶ Distribution;** ▼ ▶ *(under Summary Statistics)* ▶ **Customize Summary Statistics**
Design of Experiments	**DOE**
Distribution, Univariate	**Analyze ▶ Distribution**
Distribution Fitting	**Analyze ▶ Distribution;** ▼ ▶ **Continuous Fit >** *select distribution* **or** *All;* **OR** ▼ ▶ **Discrete Fit ▶** *select distribution*
Dot Plot	**Analyze ▶ Graph Builder;** ▼ ▶ *change the Jitter setting to Positive Grid*
Excel Files, to open	**File ▶ Open ▶** *Specify Excel file format*
Experimental Design	*See* Design of Experiments

Task	Menu Selection
Exponential Smoothing	**Analyze ▶ Specialized Modeling ▶ Time Series;** **▼ ▶ Smoothing Model**
Fit Line	**Analyze ▶ Fit Y by X; ▼ ▶ Fit Line**
Fit Polynomial	**Analyze ▶ Fit Y by X; ▼ ▶ Fit Polynomial ▶** *specify degree*
Fit Y by X Platform	**Analyze ▶ Fit Y by X**
Forecasting/Time Series	**Analyze ▶ Specialized Modeling ▶ Time Series**
Formula Editor	*Click on column heading;* **Cols ▶ Formula** (OR to create a new column) **Cols ▶ New; ▼ ▶ Column Properties ▶ Formula**
Frequency Distribution	**Analyze ▶ Distribution**
Full Factorial Design	**DOE ▶ Classical ▶ Full Factorial Design**
Gauge Chart	**Analyze ▶ Quality and Process ▶** **Variability/Attribute Gauge Chart**
Goodness of Fit - test for normality	**Analyze ▶ Distribution; ▼ ▶ Continuous Fit ▶ Normal;** **▼ ▶** *(Under Fitted Normal)* **▶ Goodness of Fit**
Graphs, Graph Builder	**Graph ▶ Graph Builder**
Graphs, Control Chart	**Analyze ▶ Quality and Process ▶ Control Chart Builder**
Graphs, Bubble Plot	**Graph ▶ Bubble Plot**
Graphs, Scatterplot	**Graph ▶ Graph Builder**
Graph Builder Platform	**Graph ▶ Graph Builder**
Histograms	**Analyze ▶ Distribution**

Task	Menu Selection
Histogram Color	*Right-click on histogram* ▶ **Histogram Color**
Holt-Winters	**Analyze ▶ Specialized Modeling ▶ Time Series;** ▼ ▶ **Smoothing Model ▶ Winters Method**
Horizontal Bar Chart	**Analyze ▶ Distribution;** ▼ ▶ **Display Options ▶ Horizontal**
IR Chart	**Analyze > Quality and Process > Control Chart > I/MR Control Chart**
Joining data tables	**Tables ▶ Join**
Kruskal-Wallis Test	**Analyze ▶ Fit Y by X;** ▼ ▶ **Nonparametric ▶ Wilcoxon Test**
Least Squares Regression	*See* Regression, Simple or Multiple
Line Chart	**Graph ▶ Graph Builder;** ▼ ▶ *choose the Line icon*
Logistic Regression	*See* Regression, Logistic
Mapping data	**Graph ▶ Graph Builder;** ▼ ▶ *drag shape column in the Map Shape drop zone*
Mosaic Plot	**Analyze ▶ Distribution;** ▼ ▶ **Mosaic Plot**
Mosaic Plot with two columns	**Analyze ▶ Fit Y by X**
Moving Averages	**Analyze ▶ Specialized Modeling ▶ Time Series;** ▼ ▶ **Smoothing Model ▶ Simple Moving Average**
Moving Average Control Chart	**Analyze ▶ Quality and Process ▶ Control Chart ▶ UWMA or EWMA**

Task	Menu Selection
Moving Range Chart	**Analyze ▶ Quality and Process ▶ Control Chart ▶ I/MR Control Chart**
Multiple Comparisons	**Analyze ▶ Fit Y by X; ▼ ▶ Compare Means**
Multiple Regression	*See* Regression, Multiple
Multivariate Platform	**Analyze ▶ Multivariate Methods**
Neural Network	**Analyze ▶ Predictive Modeling ▶ Neural**
Oneway ANOVA	**Analyze ▶ Fit Y by X ; ▼ ▶ Means/Anova/Pooled t**
Outlier Box Plot	*See* Box Plot
Overlay Plot	**Graph ▶ Graph Builder**
P Chart	**Analyze ▶ Quality and Process ▶ Control Chart ▶ P Control Chart**
Parallel Plot	**Graph ▶ Parallel Plot**
Pareto Diagram	**Analyze ▶ Quality and Process ▶ Pareto Plot**
Partition	**Analyze ▶ Predictive Modeling ▶ Partition**
Phase Chart	**Analyze ▶ Quality and Process ▶ Control Chart Builder; ▼ ▶** drag column to the Phase drop zone
Pie Chart	**Graph ▶ Graph Builder; ▼ ▶** choose the pie icon
Pivot-Table	*See* Tabulate
Point Chart	**Graph ▶ Graph Builder**

Task	Menu Selection
Power Calculations	**DOE ▶ Design Diagnostics ▶ Sample Size and Power**
Prediction Profiler	**Analyze ▶ Fit Model ; ▼ ▶ Factor Profiling > Profiler**
Predictive Modeling	*See* Partition, Profiler, Regression
Profiler	*See* Prediction Profiler
Process Control	*See* Control Charts
Query Builder	**File ▶ Database ▶ Query Builder**
R Chart	**Analyze ▶ Quality and Process ▶ Control Chart ▶ X Bar Control Chart**
Recursive Partitioning	*See* Partition
Regression, Logistic	**Analyze ▶ Fit Y by X or Analyze ▶ Fit Model**
Regression, Multiple	**Analyze ▶ Fit Model**
Regression, Simple with line fit	**Analyze ▶ Fit Y by X; ▼ ▶ Fit Line**
Regression, Simple with polynomial fit	**Analyze ▶ Fit Y by X; ▼ ▶ Fit Polynomial**
Regression Trees	*See* Partition
Residual Analysis	**Analyze ▶ Fit Model; ▼ ▶ Row Diagnostics**
Response Surface Design	**DOE ▶ Classical ▶ Response Surface Design**
Run Charts	**Analyze ▶ Quality and Process ▶ Control Chart ▶ Run Chart**

Task	Menu Selection
S Chart	**Analyze ▶ Quality and Process ▶ Control Chart Builder; ▼ ▶** *change the points on the second chart to Standard Deviation*
Sample Data	**Help ▶ Sample Data**
Scatterplot	**Analyze ▶ Fit Y by X**
Scatterplot, 3D	**Graph ▶ Scatterplot 3D**
Scatterplot Matrix	**Graph ▶ Scatterplot Matrix**
Scatterplot with Line	**Analyze ▶ Fit Y by X; ▼ ▶ Fit Line**
Scatterplot with Polynomial Fit	**Analyze ▶ Fit Y by X; ▼ ▶ Fit Polynomial**
Screening Design	**DOE ▶ Classical ▶ Screening Design**
Simple Regression	*See* Regression, Simple
Sort	**Tables ▶ Sort**
Spearman's Rho	**Analyze ▶ Distribution; ▼ ▶ Stem and Leaf**
Stepwise Regression	**Analyze ▶ Fit Model; ▼ ▶ Stepwise**
Subset	**Tables ▶ Subset**
Summary	**Tables ▶ Summary**
Summary Statistics	**Analyze ▶ Distribution**
t- or z-test, one sample	**Analyze ▶ Distribution; ▼ ▶ Test Mean**
t- or z-test, two sample	**Analyze ▶ Fit Y by X; ▼ ▶ t Test**

Task	Menu Selection
Tables, Concatenate	**Tables ▶ Concatenate**
Tables, Join	**Tables ▶ Join**
Tables, Sort	**Tables ▶ Sort**
Tables, Subset	**Tables ▶ Subset**
Tabulate Platform	**Analyze ▶ Tabulate**
Test for Equal/Unequal Variances	**Analyze ▶ Fit Y by X ; ▼ ▶ Unequal Variances**
Test for Proportions, one proportion	**Analyze ▶ Distribution; ▼ ▶ Test Probabilities**
Test for Proportions, two proportions	**Analyze ▶ Fit Y by X**
Test for Normality (Goodness of Fit)	**Analyze > Distribution; ▼ ▶ Continuous Fit ▶ Normal; ▼ ▶** *(Under Fitted Normal)* **▶ Goodness of Fit**
Text Mining	**Analyze ▶ Text Explorer**
Time Series Plot	**Analyze ▶ Modeling ▶ Time Series**
Tree Map	**Graph ▶ Graph Builder; ▼** *click on Treemap icon*
Two or more factor ANOVA	**Analyze ▶ Fit Model**
Tukey Box Plot	*See* Box Plot
U Chart	**Analyze ▶ Quality and Process ▶ Control Chart ▶ U Control Chart**

Task	Menu Selection
Univariate/Distribution Platform	**Analyze ▶ Distribution**
Variability Chart	**Analyze ▶ Quality and Process ▶ Variability/Gauge Chart**
Wilcoxon Rank Sum Test	**Analyze ▶ Fit Y by X; ▼ ▶ Nonparametric ▶ Wilcoxon Test**
Wilcoxon Signed Rank Test	**Analyze ▶ Distribution; ▼ ▶ Test Mean** *(check Wilcoxon Signed Rank)*
Word Cloud	**Analyze ▶ Text Explorer; ▼ ▶ Display Options ▶ Show Word Cloud**
XBar Chart	**Analyze ▶ Quality and Process ▶ Control Chart ▶ XBar Control Chart**

Bibliography

Carver, Robert. 2019. *Practical Data Analysis with JMP, Third Edition*. Cary, NC: SAS Institute, Inc.

Delwiche, Lora D., and Susan J. Slaughter. 2012. *The Little SAS Book: A Primer, Fifth Edition*. Cary, NC: SAS Institute, Inc.

De Veaux, Richard D., Paul F. Velleman, and David E. Bock. 2017. *Intro Stats, Fifth Edition*. Boston, MA: Pearson Education.

Few, Stephen. 2012. *Show Me the Numbers: Designing Tables and Graphs to Enlighten, Second Edition*. Oakland, CA: Analytics Press.

Few, Stephen. 2009. *Now You See It: Simple Visualization Techniques for Quantitative Analysis*. Oakland, CA: Analytics Press.

Goos, Peter and David Meintrup. 2015. *Statistics with JMP: Graphs, Descriptive Statistics and Probability*. Hoboken, NJ: John Wiley.

Moore, David S., George P. McCabe, and Bruce A. Craig. 2017. *Introduction to the Practice of Statistics, Ninth Edition*. New York: W. H. Freeman.

Olsen, Chris. 2011. *Teaching Elementary Statistics with JMP*. Cary, NC: SAS Institute, Inc.

Peck, Roxy and Tom Short. 2018. *Statistics: Learning from Data, Second Edition*. Stamford, CT: Cengage Learning.

Ramirez, Brenda S., and Jose Ramirez. 2018. *Douglas Montgomery's Introduction to Statistical Quality Control, A JMP Companion*. Cary, NC: SAS Institute, Inc.

Sahai, Hardeo and Anwer Khurshid. 2001. *Pocket Dictionary of Statistics*. Boston: McGraw-Hill.

Sall, John, Ann Lehman, Mia Stephens, and Sheila Loring. 2017. *JMP Start Statistics, Sixth Edition*. Cary, NC: SAS Institute Inc.

SAS Institute Inc. 2019. Using JMP 15. https://www.jmp.com/support/help/en/15.0/.

Shmueli, Galit, Peter C. Bruce, Mia L. Stephens, and Nitin R. Patel. 2016. *Data Mining for Business Analytics: Concepts, Techniques, and Applications with JMP Pro*. Hoboken, NJ: John Wiley and Sons.

Silvestrini, Rachel T. and Sarah E. Burke. 2018. *Linear Regression Analysis with JMP and R*. Milwaukee, WI: Quality Press.

Stine, Robert and Dean Foster. 2017. *Statistics for Business: Decision Making and Analysis, Third Edition*. Boston, MA: Pearson Addison-Wesley.

Tufte, Edward R. 1983. *The Visual Display of Quantitative Information*. Cheshire, CT: Graphics Press.

Utts, Jessica M. 2014. *Seeing Through Statistics, Fourth Edition*. Stamford, CT: Cengage Learning.

Index

V

value labels 54
value order 56–59
Variability chart 103–104, 339
variables
    See columns
versions (JMP) 2, 6
View menu 10
viewing preferences 19
visual dimension, adding to data 64–66
visualization 243–254

W

Web Resources 308–310
webcasts (JMP) 310
websites
    JMP 308
    users' groups 310
Wilcoxon Rank Sum Test, shortcuts for 339
Wilcoxon Signed Rank Test, shortcuts for 339
Window menu 10
Windows PCs, accessing JMP on 5
Word, placing graphs into 278–282
word cloud 135–139
writing SAS programs 317–324

X

X-Bar charts 98, 339
xBase data files, file extensions for 26
XML, importing 26

Y

Y, Columns 17

Z

z-Test, shortcuts for 337

www.ingramcontent.com/pod-product-compliance
Lightning Source LLC
Chambersburg PA
CBHW081046220326
41598CB00038B/7003